Principles of Digital Communication

This comprehensive and accessible text teaches the fundamentals of digital communication via a top-down-reversed approach, specifically formulated for a one-semester course. It offers students a smooth and exciting journey through the three sub-layers of the physical layer and is the result of many years of teaching digital communication.

The unique approach begins with the decision problem faced by the receiver of the first sub-layer (the decoder), hence cutting straight to the heart of digital communication, and enabling students to learn quickly, intuitively, and with minimal background knowledge.

A swift and elegant extension to the second sub-layer leaves students with an understanding of how a receiver works. Signal design is addressed in a seamless sequence of steps, and finally the third sub-layer emerges as a refinement that decreases implementation costs and increases flexibility.

The focus is on system-level design, with connections to physical reality, hardware constraints, engineering practice, and applications made throughout. Numerous worked examples, homework problems, and `MATLAB` simulation exercises make this text a solid basis for students to specialize in the field of digital communication and it is suitable for both traditional and flipped classroom teaching.

Bixio Rimoldi is a Professor at the Ecole Polytechnique Fédérale de Lausanne (EPFL), Switzerland, where he developed an introductory course on digital communication. Previously he was Associate Professor at Washington University and took visiting positions at Stanford, MIT, and Berkeley. He is an IEEE fellow, a past president of the IEEE Information Theory Society, and a past director of the communication system program at EPFL.

"This is an excellent introductory book on digital communications theory that is suitable for advanced undergraduate students and/or first-year graduate students, or alternatively for self-study. It achieves a nice degree of rigor in a clear, gentle and student-friendly manner. The exercises alone are worth the price of the book."

Dave Forney
MIT

"*Principles of Digital Communication: A Top-Down Approach*, 2015, Cambridge University Press, is a special and most attractive text to be used in an introductory (first) course on "Digital Communications". It is special in that it addresses the most basic features of digital communications in an attractive and simple way, thereby facilitating the teaching of these fundamental aspects within a single semester. This is done without compromising the required mathematical and statistical framework. This remarkable achievement is the outcome of many years of excellent teaching of undergraduate and graduate digital communication courses by the author.

The book is built as appears in the title in a top-down manner. It starts with only basic knowledge on decision theory and, through a natural progression, it develops the full receiver structure and the signal design principles. The final part addresses aspects of practical importance and implementation issues. The text also covers in a clear and simple way more advanced aspects of coding and the associated maximum likelihood (Viterbi) decoder. Hence it may be used also as an introductory text for a more advanced (graduate) digital communication course.

All in all, this extremely well-structured text is an excellent book for a first course on Digital Communications. It covers exactly what is needed and it does so in a simple and rigorous manner that the students and the tutor will appreciate. The achieved balance between theoretical and practical aspects makes this text well suited for students with inclinations to either an industrial or an academic career."

Shlomo Shamai
Technion, Israel Institute of Technology

"The Rimoldi text is perfect for a beginning mezzanine-level course in digital communications. The logical three layer – discrete-time, continuous-time, passband – approach to the problem of communication system design greatly enhances understanding. Numerous examples, problems, and `MATLAB` exercises make the book both student and instructor friendly. My discussions with the author about the book's development have convinced me that it's been a labor of love. The completed manuscript clearly bears this out."

Dan Costello
University of Notre Dame

Principles of Digital Communication
A Top-Down Approach

Bixio Rimoldi
School of Computer and Communication Sciences
Ecole Polytechnique Fédérale de Lausanne (EPFL)
Switzerland

CAMBRIDGE
UNIVERSITY PRESS

University Printing House, Cambridge CB2 8BS, United Kingdom

Cambridge University Press is part of the University of Cambridge.

It furthers the University's mission by disseminating knowledge in the pursuit of education, learning, and research at the highest international levels of excellence.

www.cambridge.org
Information on this title: www.cambridge.org/9781107116450

© Cambridge University Press 2016

This publication is in copyright. Subject to statutory exception and to the provisions of relevant collective licensing agreements, no reproduction of any part may take place without the written permission of Cambridge University Press.

First published 2016

Printed in the United Kingdom by TJ International Ltd. Padstow Cornwall

A catalog record for this publication is available from the British Library

Library of Congress Cataloging in Publication data
Rimoldi, Bixio.
Principles of digital communication : a top-down approach / Bixio Rimoldi, School of Computer and Communication Sciences, Ecole Polytechnique Fédérale de Lausanne (EPFL), Switzerland.
    pages   cm
Includes bibliographical references and index.
ISBN 978-1-107-11645-0 (Hardback : alk. paper)
1. Digital communications. 2. Computer networks.   I. Title.
TK5103.7.R56   2015
621.382–dc23   2015015425

ISBN 978-1-107-11645-0 Hardback

Additional resources for this publication at www.cambridge.org/rimoldi

Cambridge University Press has no responsibility for the persistence or accuracy of URLs for external or third-party internet websites referred to in this publication, and does not guarantee that any content on such websites is, or will remain, accurate or appropriate.

This book is dedicated to my parents,
for their boundless support and trust,
and to the late Professor James L. Massey,
whose knowledge, wisdom, and generosity
have deeply touched generations of students.

Contents

Preface		*page* xi
Acknowledgments		xviii
List of symbols		xx
List of abbreviations		xxii

1 Introduction and objectives   1

   1.1 The big picture through the OSI layering model   1
   1.2 The topic of this text and some historical perspective   5
   1.3 Problem formulation and preview   9
   1.4 Digital versus analog communication   13
   1.5 Notation   15
   1.6 A few anecdotes   16
   1.7 Supplementary reading   18
   1.8 Appendix: Sources and source coding   18
   1.9 Exercises   20

2 Receiver design for discrete-time observations: First layer   23

   2.1 Introduction   23
   2.2 Hypothesis testing   26
      2.2.1 Binary hypothesis testing   28
      2.2.2 m-ary hypothesis testing   30
   2.3 The Q function   31
   2.4 Receiver design for the discrete-time AWGN channel   32
      2.4.1 Binary decision for scalar observations   34
      2.4.2 Binary decision for n-tuple observations   35
      2.4.3 m-ary decision for n-tuple observations   39
   2.5 Irrelevance and sufficient statistic   41
   2.6 Error probability bounds   44
      2.6.1 Union bound   44
      2.6.2 Union Bhattacharyya bound   48
   2.7 Summary   51

	2.8	Appendix: Facts about matrices	53
	2.9	Appendix: Densities after one-to-one differentiable transformations	58
	2.10	Appendix: Gaussian random vectors	61
	2.11	Appendix: A fact about triangles	65
	2.12	Appendix: Inner product spaces	65
		2.12.1 Vector space	65
		2.12.2 Inner product space	66
	2.13	Exercises	74

3 Receiver design for the continuous-time AWGN channel: Second layer 95

	3.1	Introduction	95
	3.2	White Gaussian noise	97
	3.3	Observables and sufficient statistics	99
	3.4	Transmitter and receiver architecture	102
	3.5	Generalization and alternative receiver structures	107
	3.6	Continuous-time channels revisited	111
	3.7	Summary	114
	3.8	Appendix: A simple simulation	115
	3.9	Appendix: Dirac-delta-based definition of white Gaussian noise	116
	3.10	Appendix: Thermal noise	118
	3.11	Appendix: Channel modeling, a case study	119
	3.12	Exercises	123

4 Signal design trade-offs 132

	4.1	Introduction	132
	4.2	Isometric transformations applied to the codebook	132
	4.3	Isometric transformations applied to the waveform set	135
	4.4	Building intuition about scalability: n versus k	135
		4.4.1 Keeping n fixed as k grows	135
		4.4.2 Growing n linearly with k	137
		4.4.3 Growing n exponentially with k	139
	4.5	Duration, bandwidth, and dimensionality	142
	4.6	Bit-by-bit versus block-orthogonal	145
	4.7	Summary	146
	4.8	Appendix: Isometries and error probability	148
	4.9	Appendix: Bandwidth definitions	149
	4.10	Exercises	150

Contents

5 Symbol-by-symbol on a pulse train: Second layer revisited   159

- 5.1 Introduction   159
- 5.2 The ideal lowpass case   160
- 5.3 Power spectral density   163
- 5.4 Nyquist criterion for orthonormal bases   167
- 5.5 Root-raised-cosine family   170
- 5.6 Eye diagrams   172
- 5.7 Symbol synchronization   174
  - 5.7.1 Maximum likelihood approach   175
  - 5.7.2 Delay locked loop approach   176
- 5.8 Summary   179
- 5.9 Appendix: $\mathcal{L}_2$, and Lebesgue integral: A primer   180
- 5.10 Appendix: Fourier transform: A review   184
- 5.11 Appendix: Fourier series: A review   187
- 5.12 Appendix: Proof of the sampling theorem   189
- 5.13 Appendix: A review of stochastic processes   190
- 5.14 Appendix: Root-raised-cosine impulse response   192
- 5.15 Appendix: The picket fence "miracle"   193
- 5.16 Exercises   196

6 Convolutional coding and Viterbi decoding: First layer revisited   205

- 6.1 Introduction   205
- 6.2 The encoder   205
- 6.3 The decoder   208
- 6.4 Bit-error probability   211
  - 6.4.1 Counting detours   213
  - 6.4.2 Upper bound to P_b 216
- 6.5 Summary   219
- 6.6 Appendix: Formal definition of the Viterbi algorithm   222
- 6.7 Exercises   223

7 Passband communication via up/down conversion: Third layer   232

- 7.1 Introduction   232
- 7.2 The baseband-equivalent of a passband signal   235
  - 7.2.1 Analog amplitude modulations: DSB, AM, SSB, QAM   240
- 7.3 The third layer   243
- 7.4 Baseband-equivalent channel model   252
- 7.5 Parameter estimation   256

7.6	Non-coherent detection	260
7.7	Summary	264
7.8	Appendix: Relationship between real- and complex-valued operations	265
7.9	Appendix: Complex-valued random vectors	267
	7.9.1 General statements	267
	7.9.2 The Gaussian case	269
	7.9.3 The circularly symmetric Gaussian case	270
7.10	Exercises	275

Bibliography   284

Index   286

Preface

This text is intended for a one-semester course on the foundations of digital communication. It assumes that the reader has basic knowledge of linear algebra, probability theory, and signal processing, and has the mathematical maturity that is expected from a third-year engineering student.

The text has evolved out of lecture notes that I have written for EPFL students. The first version of my notes greatly profited from three excellent sources, namely the book *Principles of Communication Engineering* by Wozencraft and Jacobs [1], the lecture notes written by Professor Massey for his ETHZ course *Applied Digital Information Theory*, and the lecture notes written by Professors Gallager and Lapidoth for their MIT course *Introduction to Digital Communication*. Through the years the notes have evolved and although the influence of these sources is still recognizable, the text has now its own "personality" in terms of *content*, *style*, and *organization*.

The *content* is what I can cover in a one-semester course at EPFL.[1] The focus is the transmission problem. By staying focused on the transmission problem (rather than also covering the source digitization and compression problems), I have just the right content and amount of material for the goals that I deem most important, specifically: (1) cover to a reasonable depth the most central topic of digital communication; (2) have enough material to do justice to the beautiful and exciting area of digital communication; and (3) provide evidence that linear algebra, probability theory, calculus, and Fourier analysis are in the curriculum of our students for good reasons. Regarding this last point, the area of digital communication is an ideal showcase for the power of mathematics in solving engineering problems.

The digitization and compression problems, omitted in this text, are also important, but covering the former requires a digression into signal processing to acquire the necessary technical background, and the results are less surprising than those related to the transmission problem (which can be tackled right away, see Chapter 2). The latter is covered in all information theory courses and rightfully so. A more detailed account of the content is given below, where I discuss the text organization.

---

[1] We have six periods of 45 minutes per week, part of which we have devoted to exercises, for a total of 14 weeks.

In terms of *style*, I have paid due attention to proofs. The value of a rigorous proof goes beyond the scientific need of proving that a statement is indeed true. From a proof we can gain much insight. Once we see the proof of a theorem, we should be able to tell why the conditions (if any) imposed in the statement are necessary and what can happen if they are violated. Proofs are also important because the statements we find in theorems and the like are often not in the exact form needed for a particular application. Therefore, we might have to adapt the statement and the proof as needed.

An instructor should not miss the opportunity to share useful tricks. One of my favorites is the trick I learned from Professor Donald Snyder (Washington University) on how to label the Fourier transform of a rectangle. (Most students remember that the Fourier transform of a rectangle is a sinc but tend to forget how to determine its height and width. See Appendix 5.10.)

The remainder of this preface is about the text *organization*. We follow a *top-down* approach, but a more precise name for the approach is *top-down-reversed with successive refinements*. It is *top-down* in the sense of Figure 1.7 of Chapter 1, which gives a system-level view of the focus of this book. (It is also top-down in the sense of the OSI model depicted in Figure 1.1.) It is *reversed* in the sense that the receiver is treated before the transmitter. The logic behind this reversed order is that we can make sensible choices about the transmitter only once we are able to appreciate their impact on the receiver performance (error probability, implementation costs, algorithmic complexity). Once we have proved that the receiver and the transmitter decompose into blocks of well-defined tasks (Chapters 2 and 3), we *refine* our design, changing the focus from "what to do" to "how to do it effectively" (Chapters 5 and 6). In Chapter 7, we *refine* the design of the second layer to take into account the specificity of passband communication. As a result, the second layer splits into the second and the third layer of Figure 1.7.

In Chapter 2 we acquaint ourselves with the receiver-design problem for channels that have a discrete output alphabet. In doing so, we hide all but the most essential aspect of a channel, specifically that the input and the output are related stochastically. Starting this way takes us very quickly to the heart of digital communication, namely the decision rule implemented by a decoder that minimizes the error probability. The decision problem is an excellent place to begin as the problem is new to students, it has a clean-cut formulation in terms of minimizing an objective function (the error probability), the derivations rely only on basic probability theory, the solution is elegant and intuitive (the maximum a posteriori probability decision rule), and the topic is at the heart of digital communication. After a general start, the receiver design is specialized for the *discrete-time* AWGN (additive white Gaussian noise) channel that plays a key role in subsequent chapters. In Chapter 2, we also learn how to determine (or upper bound) the probability of error and we develop the notion of a sufficient statistic, needed in the following chapter. The appendices provide a review of relevant background material on matrices, on how to obtain the probability density function of a variable defined in terms of another, on Gaussian random vectors, and on inner product spaces. The chapter contains a large collection of exercises.

In Chapter 3 we make an important transition concerning the channel used to communicate, specifically from the rather abstract discrete-time channel to the

Preface

more realistic *continuous-time* AWGN channel. The objective remains the same, i.e. develop the receiver structure that minimizes the error probability. The theory of inner product spaces, as well as the notion of sufficient statistic developed in the previous chapter, give us the tools needed to make the transition elegantly and swiftly. We discover that the decomposition of the transmitter and the receiver, as done in the top two layers of Figure 1.7, is general and natural for the continuous-time AWGN channel. This constitutes the end of the first pass over the top two layers of Figure 1.7.

Up until Chapter 4, we assume that the transmitter has been given to us. In Chapter 4, we prepare the ground for the *signal-design*. We introduce the design parameters that we care about, namely transmission rate, delay, bandwidth, average transmitted energy, and error probability, and we discuss how they relate to one another. We introduce the notion of isometry in order to change the signal constellation without affecting the error probability. It can be applied to the encoder to minimize the average energy without affecting the other system parameters such as transmission rate, delay, bandwidth, error probability; alternatively, it can be applied to the waveform former to vary the signal's time/frequency features. The chapter ends with three case studies for developing intuition. In each case, we fix a signaling family, parameterized by the number of bits conveyed by a signal, and we determine the probability of error as the number of bits grows to infinity. For one family, the dimensionality of the signal space stays fixed, and the conclusion is that the error probability grows to 1 as the number of bits increases. For another family, we let the signal space dimensionality grow exponentially and, in so doing, we can make the error probability become exponentially small. Both of these cases are instructive but have drawbacks that make them unworkable solutions as the number of bits becomes large. The reasonable choice seems to be the "middle-ground" solution that consists in letting the dimensionality grow linearly with the number of bits. We demonstrate this approach by means of what is commonly called pulse amplitude modulation (PAM). We prefer, however, to call it symbol-by-symbol on a pulse train because PAM does not convey the idea that the pulse is used more than once and people tend to associate PAM to a certain family of symbol alphabets. We find symbol-by-symbol on a pulse train to be more descriptive and more general. It encompasses, for instance, phase-shift keying (PSK) and quadrature amplitude modulation (QAM).

Chapter 5 discusses how to choose the orthonormal basis that characterizes the waveform former (Figure 1.7). We discover the *Nyquist criterion* as a means to construct an orthonormal basis that consists of the T-spaced time translates of a single pulse, where T is the symbol interval. Hence we refine the n-tuple former that can be implemented with a single matched filter. In this chapter we also learn how to do symbol synchronization (to know when to sample the matched filter output) and introduce the eye diagram (to appreciate the importance of a correct symbol synchronization). Because of its connection to the Nyquist criterion, we also derive the expression for the *power spectral density* of the communication signal.

In Chapter 6, we design the encoder and refine the decoder. The goal is to expose the reader to a widely used way of encoding and decoding. Because there are several coding techniques – numerous enough to justify a graduate-level course – we approach the subject by means of a case study based on convolutional coding.

The minimum error probability decoder incorporates the Viterbi algorithm. The content of this chapter was selected as an introduction to coding and to introduce the reader to elegant and powerful tools, such as the previously mentioned Viterbi algorithm and the tools to assess the resulting bit-error probability, notably detour flow graphs and generating functions.

The material in Chapter 6 could be covered after Chapter 2, but there are some drawbacks in doing so. First, it unduly delays the transition from the discrete-time channel model of Chapter 2 to the more realistic continuous-time channel model of Chapter 3. Second, it makes more sense to organize the teaching into a *first pass* where we discover what to do (Chapters 2 and 3), and a *refinement* where we focus on how to do it effectively (Chapters 5, 6, and 7). Finally, at the end of Chapter 2, it is harder to motivate the students to invest time and energy into coding for the discrete-time AWGN channel, because there is no evidence yet that the channel plays a key role in practical systems. Such evidence is provided in Chapter 3. Chapters 5 and 6 could be done in the reverse order, but the chosen order is preferable for continuity reasons with respect to Chapter 4.

The final chapter, Chapter 7, is where the third layer emerges as a refinement of the second layer to facilitate *passband* communication.

The following diagram summarizes the main thread throughout the text.

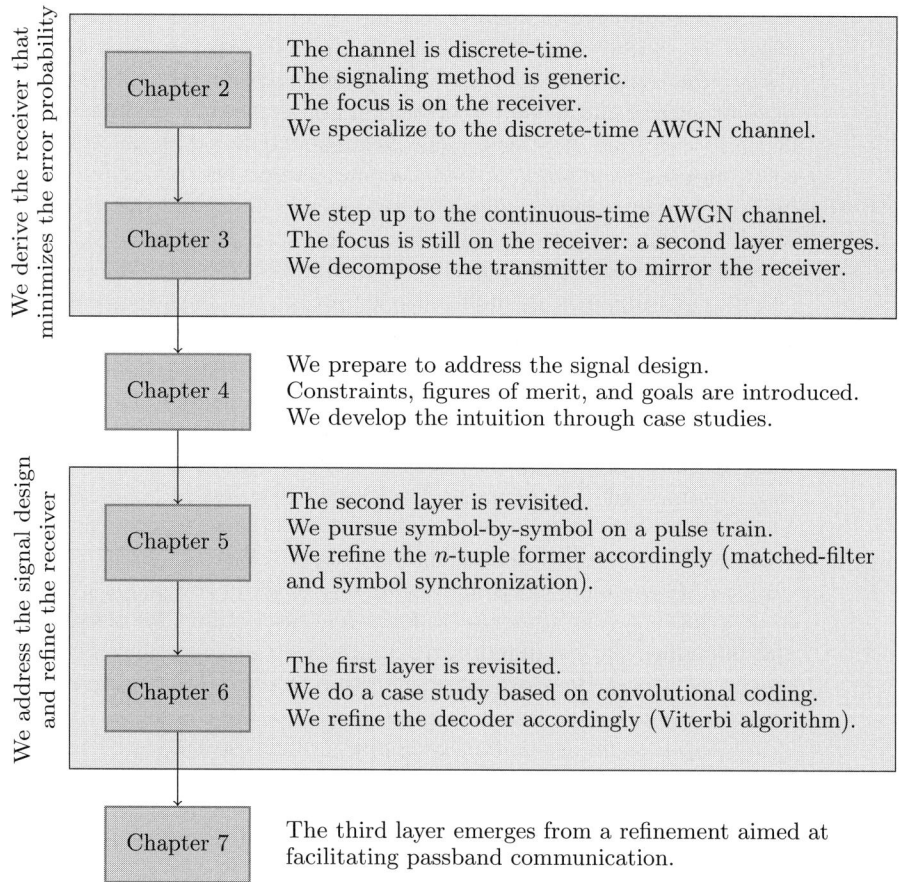

Preface

Each chapter contains one or more appendices, with either background or complementary material.

I should mention that I have made an important concession to mathematical rigor. This text is written for people with the mathematical background of an engineer. To be mathematically rigorous, the integrals that come up in dealing with Fourier analysis should be interpreted in the Lebesgue sense.[2] In most undergraduate curricula, engineers are not taught Lebesgue integration theory. Hence some compromise has to be made, and here is one that I find very satisfactory. In Appendix 5.9, I introduce the difference between the Riemann and the Lebesgue integrals in an informal way. I also introduce the space of $\mathcal{L}_2$ functions and the notion of $\mathcal{L}_2$ equivalence. The ideas are natural and can be understood without technical details. This gives us the language needed to rigorously state the sampling theorem and Nyquist criterion, and the insight to understand why the technical conditions that appear in those statements are necessary. The appendix also reminds us that two signals that have the same Fourier transform are $\mathcal{L}_2$ equivalent but not necessarily point-wise equal. Because we introduce the Lebesgue integral in an informal way, we are not in the position to prove, say, that we can swap an integral and an infinite sum. In some way, having a good reason for skipping such details is a blessing, because dealing with all technicalities can quickly become a major distraction. These technicalities are important at some level and unimportant at another level. They are important for ensuring that the theory is consistent and a serious graduate-level student should be exposed to them. However, I am not aware of a single case where they make a difference in dealing with finite-support functions that are continuous and have finite-energy, especially with the kind of signals we encounter in engineering. Details pertaining to integration theory that are skipped in this text can be found in Gallager's book [2], which contains an excellent summary of integration theory for communication engineers. Lapidoth [3] contains many details that are not found elsewhere. It is an invaluable text for scholars in the field of digital communication.

The last part of this preface is addressed to instructors. Instructors might consider taking a bottom-up approach with respect to Figure 1.7. Specifically, one could start with the passband AWGN channel model and, as the first step in the development, reduce it to the baseband model by means of the up/down converter. In this case the natural second step is to reduce the baseband channel to the discrete-time channel and only then address the communication problem across the discrete-time channel. I find such an approach to be pedagogically less appealing as it puts the communication problem last rather than first. As formulated by Claude Shannon, the father of modern digital communication, "The fundamental problem of communication is that of reproducing at one point either exactly or approximately a message selected at another point". This is indeed the problem that we address in Chapter 2. Furthermore, randomness is the most important aspect of a channel. Without randomness, there is no communication problem. The channels considered in Chapter 2 are good examples to start with, because they model randomness without additional distractions. However, the choice of

---

[2] The same can be said for the integrals involving the noise, but our approach avoids such integrals. See Section 3.2.

such abstract channels needs to be motivated. I motivate in two ways: (i) by asking the students to trust that the theory we develop for that abstract channel will turn out to be exactly what we need for more realistic channel models and (ii) by reminding them of the (too often overlooked) problem-solving technique that consists in addressing a difficult problem by first considering a simplified "toy version" of the same.

A couple of years ago, I flipped the classroom in the following sense. Rather than developing the theory in class via standard ex-cathedra lectures and letting the students work on problems at home, I have the students go over the theory at their own pace at home, and I devote the class time to exercises, to detecting difficulties, to filling the gaps, and to motivating the students. Almost the entire content of the book (appendices apart) is covered in the reading assignments.

In my case, flipping the classroom is the result of a process that began with the conviction that the time spent in class was not well spent for many students. There is a fair amount of math in the course *Principles of Digital Communication* and because it is mandatory at EPFL, there is quite a bit of disparity in terms of the rate at which a student can follow the math. Hence, no single pace of teaching is satisfactory, but the real issue has deeper roots. Learning is not about making one step after the other on a straight line at a constant pace, which is essentially what we do in a typical ex-cathedra lecture.

There are a number of things that can improve our effectiveness when we study and cannot be done in an ex-cathedra lecture. Ideally, we should be able to choose suitable periods of time and to decide when a break is needed. More importantly, we should be able to control the information flow, in the sense of being able to "pause" it, e.g. in order to think whether or not what we are learning makes sense to us, to make connections with what we know already, to work out examples, etc. We should also be able to "rewind", and to "fast forward". None of this can be done in an ex-cathedra lecture; however, all of this can be done when we study from a book.[3] Pausing to think, to make connections, and to work out examples is a particularly useful process that is not sufficiently ingrained in many undergraduate students, perhaps precisely because the ex-cathedra format does not permit it. The book has to be suitable (self-contained and sufficiently clear), the students should be sufficiently motivated to read, and they should be able to ask questions as needed.

Motivation is typically not a problem for the students when the reading is an essential ingredient for passing the class. In my course, the students quickly realize that they will not be able to solve the exercises if they do not read the theory, and there is little chance for them to pass the class without theory and exercises.

But what makes the flipped classroom today more interesting than in the past, is the availability of Web-based tools for posting and answering questions. For my class, I have been using Nota Bene.[4] Designed at MIT, this is a website on

---

[3] ...or when we watch a video. But a book can be more useful as a reference, because it is easier to find what you are looking for in a book than on a video, and a book can be annotated (personalized) more easily.

[4] http://nb.mit.edu.

Preface

which I post the reading assignments (essentially all sections). When students have a question, they go to the site, highlight the relevant part, and type a question in a pop-up window. The questions are summarized on a list that can be sorted according to various criteria. Students can "vote" on a question to increase its importance. Most questions are answered by students, and as an incentive to interact on Nota Bene, I give a small bonus for posting pertinent questions and/or for providing reasonable answers.[5] The teaching assistants (TAs) and myself monitor the site and we intervene as needed. Before I go to class, I take a look at the questions, ordered by importance; then in class I "fill the gaps" as I see fit.

Most of the class time is spent doing exercises. I encourage the students to help each other by working in groups. The TAs and myself are there to help. This way, I see who can do what and where the difficulties lie. Assessing the progress this way is more reliable than by grading exercises done at home. (We do not grade the exercises, but we do hand out solutions.) During an exercise session, I often go to the board to clarify, to help, or to complement, as necessary.

In terms of my own satisfaction, I find it more interesting to interact with the students in this way, rather than to give ex-cathedra lectures that change little from year to year. The vast majority of the students also prefer the flipped classroom: They say so and I can tell that it is the case from their involvement. The exercises are meant to be completed during the class time,[6] so that at home the students can focus on the reading. By the end of the semester[7] we have covered almost all sections of the book. (Appendices are left to the student's discretion.) Before a new reading assignment, I motivate the students to read by telling them why the topic is important and how it fits into the big picture. If there is something unusual, e.g. a particularly technical passage, I tell them what to expect and/or I give a few hints. Another advantage of the flipped classroom is never falling behind the schedule. At the beginning of the semester, I know which sections will be assigned which week, and prepare the exercises accordingly. After the midterm, I assign a MATLAB project to be completed in groups of two and to be presented during the last day of class. The students like this very much.[8]

---

[5] A pertinent intervention is worth half a percent of the total number of points that can be acquired over the semester and, for each student, I count at most one intervention per week. This limits the maximum amount of bonus points to 7% of the total.

[6] Six periods of 45 minutes at EPFL.

[7] Fourteen weeks at EPFL.

[8] The idea of a project was introduced with great success by my colleague, Rüdiger Urbanke, who taught the course during my sabbatical.

Acknowledgments

This book is the result of a slow process, which began around the year 2000, of seemingly endless revisions of my notes written for *Principles of Digital Communication* – a sixth-semester course that I have taught frequently at EPFL. I would like to acknowledge that the notes written by Professors Robert Gallager and Amos Lapidoth, for their MIT course *Introduction to Digital Communication*, as well as the notes by Professor James Massey, for his ETHZ course *Applied Digital Information Theory*, were of great help to me in writing the first set of notes that evolved into this text. Equally helpful were the notes written by EPFL Professor Emre Telatar, on matrices and on complex random variables; they became the core of some appendices on background and on complementary material.

A big thanks goes to the PhD students who helped me develop new exercises and write solutions. This includes Mani Bastani Parizi, Sibi Bhaskaran, László Czap, Prasenjit Dey, Vasudevan Dinkar, Jérémie Ezri, Vojislav Gajic, Michael Gastpar, Saeid Haghighatshoar, Hamed Hassani, Mahdi Jafari Siavoshani, Javad Ebrahimi, Satish Korada, Shrinivas Kudekar, Stéphane Musy, Christine Neuberg, Ayfer Özgür, Etienne Perron, Rajesh Manakkal, and Philippe Roud. Some exercises were created from scratch and some were inspired from other textbooks. Most of them evolved over the years and, at this point, it would be impossible to give proper credit to all those involved. The first round of teaching *Principles of Digital Communication* required creating a number of exercises from scratch. I was very fortunate to have Michael Gastpar (PhD student at the time and now an EPFL colleague) as my first teaching assistant. He did a fabulous job in creating many exercises and solutions.

I would like to thank my EPFL students for their valuable feedback. Pre-final drafts of this text were used at Stanford University and at UCLA, by Professors Ayfer Özgür and Suhas Diggavi, respectively. Professor Rüdiger Urbanke used them at EPFL during two of my sabbatical leaves. I am grateful to them for their feedback and for sharing with me their students' comments.

I am grateful to the following collaborators who have read part of the manuscript and whose feedback has been very valuable: Emmanuel Abbe, Albert Abilov, Nicolae Chiurtu, Michael Gastpar, Matthias Grossglauser, Paolo Ienne, Alberto Jimenez-Pacheco, Olivier Lévêque, Nicolas Macris, Stefano Rosati, Anja Skrivervik, and Adrian Tarniceriu.

Acknowledgments

I am particularly indebted to the following people for having read the whole manuscript and for giving me a long list of suggestions, while noting the typos and mistakes: Emre Telatar, Urs Niesen, Saeid Haghighatshoar, and Sepand Kashani-Akhavan.

Warm thanks go to Françoise Behn who learned LaTeX to type the first version of the notes, to Holly Cogliati-Bauereis for her infinite patience in correcting my English, to Emre Telatar for helping with LaTeX-related problems, and to Karol Kruzelecki and Damir Laurenzi for helping with computer issues.

Finally, I would like to acknowledge many interesting discussions with various colleagues, in particular those with Emmanuel Abbe, Michael Gastpar, Amos Lapidoth, Upamanyu Madhow, Emre Telatar, and Rüdiger Urbanke. I would also like to thank Rüdiger Urbanke for continuously encouraging me to publish my notes. Without his insistence and his jokes about my perpetual revisions, I might still be working on them.

List of symbols

$\mathcal{A}, \mathcal{B}, \ldots$	Sets.
$\mathbb{N}$	Set of natural numbers: $\{1, 2, 3, \ldots\}$.
$\mathbb{Z}$	Set of integers: $\{\ldots, -2, -1, 0, 1, 2, \ldots\}$.
$\mathbb{R}$	Set of real numbers.
$\mathbb{C}$	Set of complex numbers.
$\mathcal{H} := \{0, \ldots, m-1\}$	Message set.
$\mathcal{C} := \{c_0, \ldots, c_{m-1}\}$	Codebook (set of codewords).
$\mathcal{W} := \{w_0(t), \ldots, w_{m-1}(t)\}$	Set of waveform signals.
$\mathcal{V}$	Vector space or inner product space.
$u : \mathcal{A} \to \mathcal{B}$	Function u with domain $\mathcal{A}$ and range $\mathcal{B}$.
$H \in \mathcal{H}$	Random message (hypothesis) taking value in $\mathcal{H}$.
$N(t)$	Noise.
$N_E(t)$	Baseband-equivalent noise.
$R(t)$	Received (random) signal.
$Y = (Y_1, \ldots, Y_n)$	Random n-tuple observed by the decoder.
j	$\sqrt{-1}$.
$\{\,\}$	Set of objects.
A^T	Transpose of the matrix A. It may be applied to an n-tuple a.
$A^\dagger$	Hermitian transpose of the matrix A. It may be applied to an n-tuple a.
$\mathbb{E}[X]$	Expected value of X.
$\langle a, b \rangle$	Inner product between a and b (in that order).
$\|a\|$	Norm of the vector a.
$|a|$	Absolute value of a.
$a := b$	a is defined as b.
$\mathbb{1}\{S\}$	Indicator function. Its value is 1 if the statement S is true and 0 otherwise.
$\mathbb{1}_\mathcal{A}(x)$	Same as $\mathbb{1}\{x \in \mathcal{A}\}$.

List of symbols

$\mathcal{E}$	Average energy.
$K_N(t+\tau, t)$, $K_N(\tau)$	Autocovariance of $N(t)$.
$\Re$	Used to denote the end of theorems, definitions, examples, proofs, etc.
$\Re\{\cdot\}$	Real part of the enclosed quantity.
$\Im\{\cdot\}$	Imaginary part of the enclosed quantity.
$\angle$	Phase of the complex-valued number that follows.

List of abbreviations

AM	amplitude modulation.
bps	bits per second.
BSS	binary symmetric source.
DSB-SC	double-sideband modulation with suppressed carrier.
iid	independent and identically distributed.
l. i. m.	limit in $\mathcal{L}_2$ norm.
LNA	low-noise amplifier.
MAP	maximum a posteriori.
Mbps	megabits per second.
ML	maximum likelihood.
MMSE	minimum mean square error.
PAM	pulse amplitude modulation.
pdf	probability density function.
pmf	probability mass function.
PPM	pulse position modulation.
PSK	phase-shift keying.
QAM	quadrature amplitude modulation.
SSB	single-sideband modulation.
WSS	wide-sense stationary.

1 Introduction and objectives

This book focuses on the system-level engineering aspects of digital point-to-point communication. In a way, digital point-to-point communication is the building block we use to construct complex communication systems including the Internet, cellular networks, satellite communication systems, etc. The purpose of this chapter is to provide contextual information. Specifically, we do the following.

(i) Place digital point-to-point communication into the bigger picture. We do so in Section 1.1 where we discuss the Open System Interconnection (OSI) layering model.
(ii) Provide some historical context (Section 1.2).
(iii) Give a preview for the rest of the book (Section 1.3).
(iv) Clarify the difference between analog and digital communication (Section 1.4).
(v) Justify some of the choices we make about notation (Section 1.5).
(vi) Mention a few amusing and instructive anecdotes related to the history of communication (Section 1.6).
(vii) Suggest supplementary reading material (Section 1.7).

The reader eager to get started can skip this chapter without losing anything essential to understand the rest of the text.

1.1 The big picture through the OSI layering model

When we communicate using electronic devices, we produce streams of bits that typically go through various networks and are processed by devices from a variety of manufacturers. The system is very complex and there are a number of things that can go wrong. It is amazing that we can communicate as easily and reliably as we do. This could hardly be possible without layering and standardization. The Open System Interconnection (OSI) layering model of Figure 1.1 describes a framework for the definition of data-flow protocols. In this section we use the OSI model to convey the basic idea of how modern communication networks deal with the key challenges, notably routing, flow control, reliability, privacy, and authenticity. For the sake of concreteness, let us take e-mailing as a sample activity. Computers use bytes (8 bits) or multiples thereof to represent letters. So the content of an e-mail

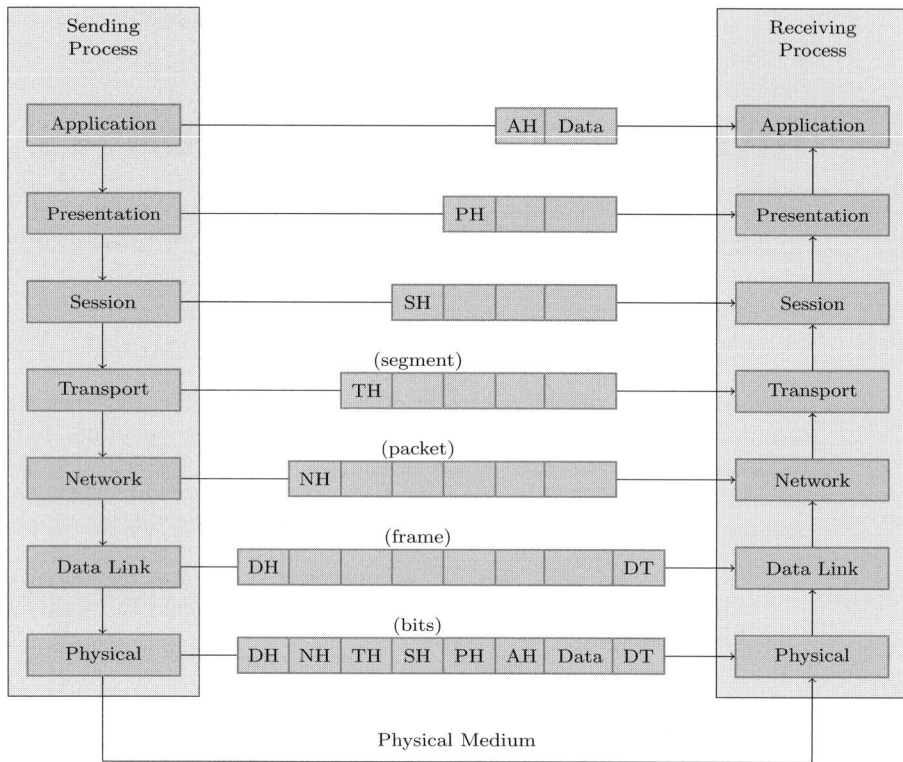

Figure 1.1. OSI layering model.

is represented by a stream of bytes. Received e-mails usually sit on a remote server. When we launch a program to read e-mail – hereafter referred to as the client – it checks into the server to see if there are new e-mails. It depends on the client's setting whether a new e-mail is automatically downloaded to the client or just a snippet is automatically downloaded until the rest is explicitly requested. The client tells the server what to do. For this to work, the server and the client not only need to be able to communicate the content of the mail message but they also need to talk to each other for the sake of coordination. This requires a protocol. If we use a dedicated program to do e-mail (as opposed to using a Web browser), the common protocols used for retrieving e-mail are the IMAP (Internet Message Access Protocol) and the POP (Post Office Protocol), whereas for sending e-mail it is common to use the SMTP (Simple Mail Transfer Protocol).

The idea of a protocol is not specific to e-mail. Every application that uses the Internet needs a protocol to interact with a peer application. The OSI model reserves the *application layer* for programs (also called processes) that implement application-related protocols. In terms of data traffic, the protocol places a so-called *application header* (AH) in front of the bits produced by the application. The top arrow in Figure 1.1 indicates that the two application layers talk to each other as if they had a direct link.

1.1. The big picture through the OSI layering model

Typically, there is no direct physical link between the two application layers. Instead, the communication between application layers goes through a shared network, which creates a number of challenges. To begin with, there is no guarantee of privacy for anything that goes through a shared network. Furthermore, networks carry data from many users and can get congested. Hence, if possible, the data should be compressed to reduce the traffic. Finally, there is no guarantee that the sending and the receiving computers represent letters the same way. Hence, the application header and the data need to be communicated by using a universal language. The *presentation layer* handles the encryption, the compression, and the translation to/from a universal language. The presentation layer also needs a protocol to talk to the peer presentation layer at the destination. The protocol is implemented by means of the *presentation header* (PH).

For the presentation layers to talk to each other, we need to make sure that the two hosting computers are connected. Establishing, maintaining, and ending communication between physical devices is the job of the *session layer*. The session layer also manages access rights. Like the other layers, the session layer uses a protocol to interact with the peer session layer. The protocol is implemented by means of the *session header* (SH).

The layers we have discussed so far would suffice if all the machines of interest were connected by a direct and reliable link. In reality, links are not always reliable. Making sure that from an end-to-end point of view the link appears reliable is one of the tasks of the *transport layer*. By means of parity check bits, the transport layer verifies that the communication is error-free and if not, it requests retransmission. The transport layer has a number of other functions, not all of which are necessarily required in any given network. The transport layer can break long sequences into shorter ones or it can multiplex several sessions between the same two machines into a single one. It also provides flow control by queueing up data if the network is congested or if the receiving end cannot absorb it sufficiently fast. The transport layer uses the *transport header* (TH) to communicate with the peer layer. The transport header followed by the data handed down by the session layer is called a segment.

Now assume that there are intermediate nodes between the peer processes of the transport layer. In this case, the *network layer* provides the routing service. Unlike the above layers, which operate on an end-to-end basis, the network layer and the layers below have a process also at intermediate nodes. The protocol of the network layer is implemented in the *network header* (NH). The network header contains, among other things, the source and the destination address. The network header followed by the segment (of the transport layer) is called a *packet*.

The next layer is the *data link* (DL) layer. Unlike the other layers, the DL puts a header at the beginning and a trailer at the end of each packet handed down by the network layer. The result is called a *frame*. Some of the overhead bits are parity-check bits meant to determine if errors have occurred in the link between nodes. If the DL detects errors, it might ask to retransmit or drop the frame altogether. If it drops the frame, it is up to the transport layer, which operates on an end-to-end basis, to request retransmission.

The *physical layer* – the subject of this text – is the bottom layer of the OSI stack. The physical layer creates a more-or-less reliable "bit pipe" out of the physical channel between two nodes. It does so by means of a transmitter/receiver pair, called *modem*,[1] on each side of the physical channel. We will learn that the physical-layer designer can trade reliability for complexity and delay.

In summary, the OSI model has the following characteristics. Although the actual data transmission is vertical, each layer is programmed as if the transmission were horizontal. For a process, whatever is not part of its own header is considered as being actual data. In particular, a process makes no distinction between the headers of the higher layers and the actual data segment. For instance, the presentation layer translates, compresses, and encrypts whatever it receives from the application layer, attaches the PH, and sends the result to its peer presentation layer. The peer in turn reads and removes the PH and decrypts, decompresses, and translates the data which is then passed to the application layer. What the application layer receives is identical to what the peer application layer has sent up to a possible language translation. The DL inserts a trailer in addition to a header. All layers, except the transport and the DL layer, assume that the communication to the peer layer is error-free. If it can, the DL layer provides reliability between successive nodes. Even if the reliability between successive nodes is guaranteed, nodes might drop packets due to queueing overflow. The transport layer, which operates at the end-to-end level, detects missing segments and requests retransmission.

It should be clear that a layering approach drastically simplifies the tasks of designing and deploying communication infrastructures. For instance, a programmer can test the application layer protocol with both applications running on the same computer – thus bypassing all networking problems. Likewise, a physical-layer specialist can test a modem on point-to-point links, also disregarding networking issues. Each of the tasks of compressing, providing reliability, privacy, authenticity, routing, flow control, and physical-layer communication requires specific knowledge. Thanks to the layering approach, each task can be accomplished by people specialized in their respective domain. Similarly, equipment from different manufacturers work together, as long as they respect the protocols.

The OSI architecture is a generic model that does not prescribe a specific protocol. The Internet uses the TCP/IP protocol stack, which is essentially compatible with the OSI architecture but uses five instead of seven layers [4]. The reduction is mainly obtained by combining the OSI application, presentation, and session layers into a single layer called the application layer. The transport layer

---

[1] *Modem* is the result of contracting the terms *mod*ulator and *dem*odulator. In analog modulation, such as frequency modulation (FM) and amplitude modulation (AM), the source signal modulates a parameter of a high-frequency oscillation, called the carrier signal. In AM it modulates the carrier's amplitude and in FM it modulates the carrier's frequency. The modulated signal can be transmitted over the air and in the absence of noise (which is never the case) the demodulator at the receiver reconstructs an exact copy of the source signal. In practice, due to noise, the reconstruction only approximates the source signal. Although modulation and demodulation are misnomers in digital communication, the term modem has remained in use.

is implemented either by the *Transmission Control Protocol* (TCP) or by the *User Datagram Protocol* (UDP). The network layer implements addressing and routing via the *Internet Protocol* (IP). The DL and the physical layers complete the stack.

1.2 The topic of this text and some historical perspective

This text is about the theory that governs the physical-layer design (bottom layer in Figure 1.1), referred to as communication theory. Of course, other layers are about communication as well, and the reader might wonder why communication theory is not about all the layers. The terminology became established in the early days, prior to the OSI model, when communication was mainly point-to-point. Rather than including the other layers as they became part of the big picture, communication theory remained "faithful" to its original domain. The reason is most likely due to the dissimilarity between the body of knowledge needed to reason about the objectives of different layers. To gain some historical perspective, we summarize the key developments that have led to communication theory.

Electromagnetism was discovered in the 1820s by Ørsted and Ampère. The wireline telegraph was demonstrated by Henry and Morse in the 1830s. Maxwell's theory of electromagnetic fields, published in 1865, predicted that electromagnetic fields travel through space as waves, moving at the speed of light. In 1876, Bell invented the telephone. Around 1887, Hertz demonstrated the transmission and reception of the electromagnetic waves predicted by Maxwell. In 1895, Marconi built a wireless system capable of transmitting signals over more than 2 km. The invention of the vacuum-tube diode by Fleming in 1904 and of the vacuum-tube triode amplifier by Forest in 1906 enabled long-distance communication by wire and wireless. The push for sending many phone conversations over the same line led, in 1917, to the invention of the wave filter by Campbell.

The beginning of digital communication theory is associated with the work of Nyquist (1924) and Hartley (1928), both at Bell Laboratories. Quite generally, we communicate over a channel by choosing the value of one or more parameters of a carrier signal. Intuitively, the more parameters we can choose independently, the more information we can send with one signal, provided that the receiver is capable of determining the value of those parameters.

A good analogy to understand the relationship between a signal and its parameters is obtained by comparing a signal with a point in a three-dimensional (3D) space. A point in 3D space is completely specified by the three coordinates of the point with respect to a Cartesian coordinate system. Similarly, a signal can be described by a number n of parameters with respect to an appropriately chosen reference system called the orthonormal basis. If we choose each coordinate as a function of a certain number of bits, the more coordinates n we have the more bits we can convey with one signal.

Nyquist realized that if the signal has to be confined to a specified time interval of duration T seconds (e.g. the duration of the communication) and frequency interval of width B Hz (e.g. the channel bandwidth), then the integer n can be

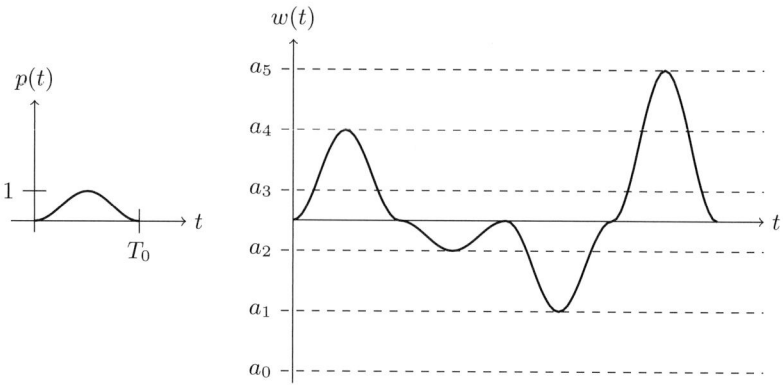

Figure 1.2. Information-carrying pulse train. It is the scaling factor of each pulse, called symbol, that carries information. In this example, the symbols take value in $\{a_0, a_1, a_2, a_3, a_4, a_5\}$.

chosen to be close to ηBT, where η is some positive number that depends on the definition of duration and bandwidth. A good value is $\eta = 2$.

As an example, consider Figure 1.2. On the left of the figure is a pulse $p(t)$ that we use as a building block for the communication signal.[2] On the right is an example of a pulse train of the form $w(t) = \sum_{i=0}^{3} s_i p(t - iT_0)$, obtained from shifted and scaled replicas of $p(t)$. We communicate by scaling the pulse replica $p(t - iT_0)$ by the information-carrying *symbol* s_i. If we could substitute $p(t)$ with a narrower pulse, we could fit more such pulses in a given time interval and therefore we could send more information-carrying symbols. But a narrower pulse uses more bandwidth. Hence there is a limit to the pulse width. For a given pulse width, there is a limit to the number of pulses that we can pack in a given time interval if we want the receiver to be able to retrieve the symbol sequence from the received pulse train. Nyquist's result implies that we can fit essentially $2BT$ non-interfering pulses in a time interval of T seconds if the bandwidth is not to exceed B Hz.

In trying to determine the maximum number of bits that can be conveyed with one signal, Hartley introduced two constraints that make good engineering sense. First, in a practical realization, the symbols cannot take arbitrarily large values in $\mathbb{R}$ (the set of real numbers). Second, the receiver cannot estimate a symbol with infinite precision. This suggests that, to avoid errors, symbols should take values in a discrete subset of some interval $[-A, A]$. If $\pm \Delta$ is the receiver's precision in determining the amplitude of a pulse, then symbols should take a value in some alphabet $\{a_0, a_1, \ldots, a_{m-1}\} \subset [-A, A]$ such that $|a_i - a_j| \geq 2\Delta$ when $i \neq j$. This implies that the alphabet size can be at most $m = 1 + \frac{A}{\Delta}$ (see Figure 1.3).

There are m^n distinct n-length sequences that can be formed with symbols taken from an alphabet of size m. Now suppose that we want to communicate a sequence

---

[2] A pulse is not necessarily rectangular. In fact, we do not communicate via rectangular pulses because they use too much bandwidth.

1.2. The topic of this text and some historical perspective

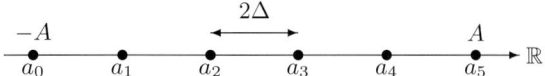

Figure 1.3. Symbol alphabet of size $m = 6$.

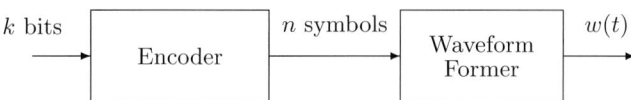

Figure 1.4. Transmitter.

of k bits. There are 2^k distinct such sequences and each such sequence should be mapped into a distinct symbol sequence (see Figure 1.4). This implies

$$2^k \leq m^n. \tag{1.1}$$

EXAMPLE 1.1 *There are $2^4 = 16$ distinct binary sequences of length $k = 4$ and there are $4^2 = 16$ distinct symbol sequences of length $n = 2$ with symbols taking value in an alphabet of size $m = 4$. Hence we can associate a distinct length-2 symbol sequence to each length-4 bit sequence. The following is an example with symbols taken from the alphabet $\{a_0, a_1, a_2, a_3\}$.*

bit sequence	symbol sequence
0000	$a_0 a_0$
0001	$a_0 a_1$
0010	$a_0 a_2$
0011	$a_0 a_3$
0100	$a_1 a_0$
⋮	⋮
1111	$a_3 a_3$

□

Inserting $m = 1 + \frac{A}{\Delta}$ and $n = 2BT$ in (1.1) and solving for $\frac{k}{T}$ yields

$$\frac{k}{T} \leq 2B \log_2 \left(1 + \frac{A}{\Delta}\right) \tag{1.2}$$

as the highest possible rate in bits per second that can be achieved reliably with bandwidth B, symbol amplitudes within $\pm A$, and receiver accuracy $\pm \Delta$.

Unfortunately, (1.2) does not provide a fundamental limit to the bit rate, because there is no fundamental limit to how small Δ can be made.

The missing ingredient in Hartley's calculation was the noise. In 1926 Johnson, also at Bell Labs, realized that every conductor is affected by thermal noise. The idea that the received signal should be modeled as the sum of the transmitted signal plus noise became prevalent through the work of Wiener (1942). Clearly the noise prevents the receiver from retrieving the symbols' values with infinite precision, which is the effect that Hartley wanted to capture with the introduction of Δ, but unfortunately there is no way to choose Δ as a function of the noise. In fact, in the presence of thermal noise, error-free communication becomes impossible. (But we can make the error probability as small as desired.)

Prior to the publication of Shannon's revolutionary 1948 paper, the common belief was that the error probability induced by the noise could be reduced only by increasing the signal's power (e.g. by increasing A in the example of Figure 1.3) or by reducing the bit rate (e.g. by transmitting the same bit multiple times). Shannon proved that the noise can set a limit to the number of bits per second that can be transmitted reliably, but as long as we communicate below that limit, the error probability can be made as small as desired without modifying the signal's power and bandwidth. The limit to the bit rate is called *channel capacity*. For the setup of interest to us it is the right-hand side of

$$\frac{k}{T} \leq B \log_2 \left(1 + \frac{P}{N_0 B}\right), \tag{1.3}$$

where P is the transmitted signal's power and $N_0/2$ is the power spectral density of the noise (assumed to be white and Gaussian). If the bit rate of a system is above channel capacity then, no matter how clever the design, the error probability is above a certain value. The theory that leads to (1.3) is far more subtle and far more beautiful than the arguments leading to (1.2); yet, the two expressions are strikingly similar.

What we mentioned here is only a special case of a general formula derived by Shannon to compute the capacity of a broad class of channels. As he did for channels, Shannon also posed and answered fundamental questions about sources. For the purpose of this text, there are two lessons that we should retain about sources. (1) The essence of a source is its randomness. If a listener knew exactly what a speaker is about to say, there would be no need to listen. Hence a source should be modeled by a random variable (or a sequence thereof). In line with the topic of this text, we assume that the source is digital, meaning that the random variable takes values in a discrete set. (See Appendix 1.8 for a brief summary of various kind of sources.) (2) For every such source, there exists a source encoder that converts the source output into the shortest (in average) binary string and a source decoder that reconstructs the source output from the encoder output. The encoder output, for which no further compression is possible, has the same statistic as a sequence of unbiased coin flips, i.e. it is a sequence of independent and uniformly distributed bits. Clearly, we can minimize the storage and/or communicate more efficiently if we compress the source into the shortest binary string. In this text, we are not concerned with source coding but, for the above-mentioned reasons, we model the source as a generator of independent and uniformly distributed bits.

Like many of the inventors mentioned above, Shannon worked at Bell Labs. His work appeared one year after the invention of the solid-state transistor, by Bardeen, Brattain, and Shockley, also at Bell Labs. Figure 1.5 summarizes the various milestones.

1.3. Problem formulation and preview

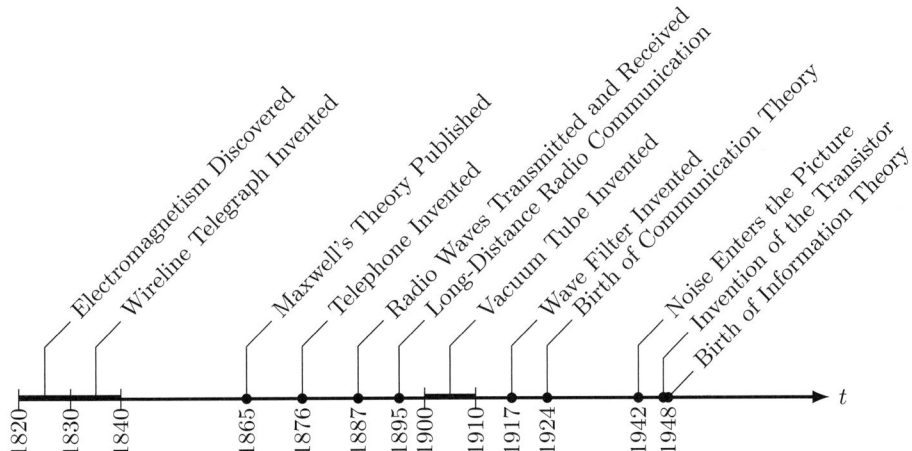

Figure 1.5. Technical milestones leading up to information theory.

Information theory, which is mainly concerned with discerning between what can be done and what cannot, regardless of complexity, led to coding theory – a field mainly concerned with implementable ways to achieve what information theory proves as achievable. In particular, Shannon's 1948 paper triggered a race for finding implementable ways to communicate reliably (i.e. with low error probability) at rates close to the channel capacity. Coding theory has produced many beautiful results: too many to summarize here; yet, approaching channel capacity in wireline and wireless communication was out of reach until the discovery of turbo codes by Berrou, Glavieux, and Thitimajshima in 1993. They were the first to demonstrate a method that could be used to communicate reliably over wireline and wireless channels, with less than 1 dB of power above the theoretical limit. As of today, the most powerful codes for wireless and wireline communication are the low-density parity-check (LDPC) codes that were invented in 1960 by Gallager in his doctoral dissertation at MIT and reinvented in 1996 by MacKay and Neal. When first invented, the codes did not receive much attention because at that time their extraordinary performance could not be demonstrated for lack of appropriate analytical and simulation tools. What also played a role is that the mapping of LDPC codes into hardware requires many connections, which was a problem before the advent of VLSI (very-large-scale integration) technology in the early 1980s.

1.3 Problem formulation and preview

Our focus is on the system aspects of digital point-to-point communication. By the term *system aspect* we mean that we remain at the level of building blocks rather than going into electronic details; *digital* means that the message is taken from a finite set of possibilities; and we restrict ourselves to *point-to-point* communication as it constitutes the building block of all communication systems.

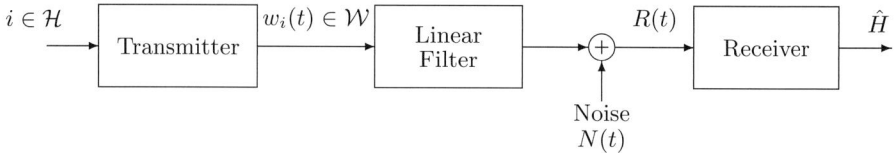

Figure 1.6. Basic point-to-point communication system over a bandlimited Gaussian channel.

Digital communication is a rather unique field in engineering in which theoretical ideas from probability theory, stochastic processes, linear algebra, and Fourier analysis have had an extraordinary impact on actual system design. The mathematically inclined will appreciate how these theories seem to be so appropriate to solve the problems we will encounter.

Our main target is to acquire a solid understanding about how to communicate through a channel, modeled as depicted in Figure 1.6. The source chooses a message represented in the figure by the index i, which is the realization of a random variable H that takes values in some finite alphabet $\mathcal{H} = \{0, 1, \ldots, m-1\}$. As already mentioned, in reality, the message is represented by a sequence of bits, but for notational convenience it is often easier to label each sequence with an index and use the index to represent the message. The transmitter maps a message i into a signal $w_i(t) \in \mathcal{W}$, where $\mathcal{W}$ and $\mathcal{H}$ have the same cardinality. The channel filters the signal and adds Gaussian noise $N(t)$. The receiver's task is to guess the message based on the channel output $R(t)$. So $\hat{H}$ is the receiver's guess of H. (Owing to the random behavior of the noise, $\hat{H}$ is a random variable even under the condition that $H = i$.)

In a typical scenario, the channel is given and the communication engineer designs the transmitter/receiver pair, taking into account objectives and constraints. The objective could be to maximize the number of bits per second being communicated, while keeping the error probability below some threshold. The constraints could be expressed in terms of the signal's power and bandwidth.

The noise added by the channel is Gaussian because it represents the contribution of various noise sources.[3] The filter has both a physical and a conceptual justification. The conceptual justification stems from the fact that most wireless communication systems are subject to a license that dictates, among other things, the frequency band that the signal is allowed to occupy. A convenient way for the system designer to deal with this constraint is to assume that the channel contains an ideal filter that blocks everything outside the intended band. The physical reason has to do with the observation that the signal emitted from the transmitting antenna typically encounters obstacles that create reflections and scattering. Hence the receiving antenna might capture the superposition of a number of delayed and attenuated replicas of the transmitted signal (plus noise). It is a straightforward

---

[3] Individual noise sources do not necessarily have Gaussian statistics. However, due to the central limit theorem, their aggregate contribution is often quite well approximated by a Gaussian random process.

1.3. Problem formulation and preview

exercise to check that this physical channel is linear and time-invariant. Thus it can be modeled by a linear filter as shown in Figure 1.6.[4] Additional filtering may occur due to the limitations of some of the components at the sender and/or at the receiver. For instance, this is the case for a linear amplifier and/or an antenna for which the amplitude response over the frequency range of interest is not flat and/or the phase response is not linear. The filter in Figure 1.6 accounts for all linear time-invariant transformations that act upon the communication signals as they travel from the sender to the receiver. The channel model of Figure 1.6 is meaningful for both wireline and wireless communication channels. It is referred to as the *bandlimited Gaussian channel*.

Mathematically, a transmitter implements a one-to-one mapping between the message set and a set of signals. Without loss of essential generality, we may let the message set be $\mathcal{H} = \{0, 1, \ldots, m-1\}$ for some integer $m \geq 2$. For the channel model of Figure 1.6, the signal set $\mathcal{W} = \{w_0(t), w_1(t), \ldots, w_{m-1}(t)\}$ consists of continuous and finite-energy signals. We think of the signals as stimuli used by the transmitter to excite the channel input. They are chosen in such a way that the receiver can tell, with high probability, which channel input produced an observed channel output.

Even if we model the source as producing an index from $\mathcal{H} = \{0, 1, \ldots, m-1\}$ rather than a sequence of bits, we can still measure the communication rate in terms of bits per second (bps). In fact the elements of the message set can be labeled with distinct binary sequences of length $\log_2 m$. Every time that we communicate a message, we equivalently communicate $\log_2 m$ bits. If we can send a signal from the set $\mathcal{W}$ every T seconds, then the message rate is $1/T$ [messages per second] and the bit rate is $(\log_2 m)/T$ [bits per second].

Digital communication is a field that has seen many exciting developments and is still in vigorous expansion. Our goal is to introduce the reader to the field, with emphasis on fundamental ideas and techniques. We hope that the reader will develop an appreciation for the trade-offs that are possible at the transmitter, will understand how to design (at the building-block level) a receiver that minimizes the error probability, and will be able to analyze the performance of a point-to-point communication system.

We will discover that a natural way to design, analyze, and implement a transmitter/receiver pair for the channel of Figure 1.6 is to think in terms of the modules shown in Figure 1.7. As in the OSI layering model, peer modules are designed as if they were connected by their own channel. The bottom layer reduces the passband channel to the more basic baseband-equivalent channel. The middle layer further reduces the channel to a discrete-time channel that can be handled by the encoder/decoder pair.

We conclude this section with a very brief overview of the chapters.

Chapter 2 addresses the *receiver-design problem for discrete-time observations*, in particular in relationship to the channel seen by the top layer of Figure 1.7, which is the discrete-time additive white Gaussian noise (AWGN) channel. Throughout

---

[4] If the scattering and reflecting objects move with respect to the transmitting/receiving antenna, then the filter is time-varying. We do not consider this case.

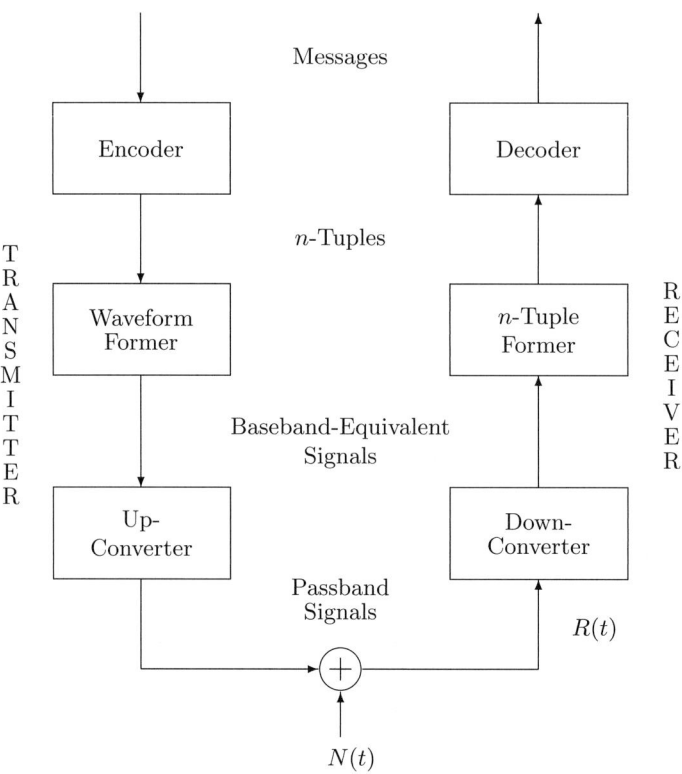

Figure 1.7. Decomposed transmitter and receiver and corresponding (sub-) layers of the physical layer.

the text the receiver's objective is to minimize the probability of an incorrect decision.

In Chapter 3 we "upgrade" the channel model to a *continuous-time* AWGN channel. We will discover that all we have learned in the previous chapter has a direct application for the new channel. In fact, we will discover that, without loss of optimality, we can insert what we call a waveform former at the channel input and the corresponding n-tuple former at the output and, in so doing, we turn the new channel model into the one already considered.

Chapter 4 develops intuition about the high-level implications of the *signal set* used to communicate. It is in this chapter that we start shifting attention from the problem of designing the receiver for a given set of signals to the problem of designing the signal set itself.

The next two chapters are devoted to practical signaling. In Chapter 5, we focus on the waveform former for what we call *symbol-by-symbol on a pulse train*. Chapter 6 is a case study on *coding*. The encoder is of *convolutional* type and the decoder is based on the *Viterbi algorithm*.

1.4. Digital versus analog communication

Chapter 7 is about passband communication. A typical passband channel is the radio channel. What we have learned in the previous chapters can, in principle, be applied directly to passband channels; but there are several reasons in favor of a design that consists of a baseband transmitter followed by an up-converter that shifts the spectrum to the desired frequency interval. The receiver reflects the transmitter's structure, with the down-converter that shifts the spectrum back to baseband. An obvious advantage of this approach is that we decouple most of the transmitter/receiver design from the carrier frequency (also called center frequency) of the transmitted signal. If we decide to shift the carrier frequency, like when we change the channel in a walkie-talkie, we just act on the up/down-converter, and this can be done very easily. Furthermore, having the last stage of the transmitter operate in its own frequency band prevents the output signal from feeding back "over the air" into the earlier stages and producing the equivalent of the annoying "audio feedback" that occurs when we put a microphone next to the corresponding speaker.

1.4 Digital versus analog communication

The meaning of *digital* versus *analog* communication needs to be clarified, in particular because it should not be confused with their meaning in the context of electronic circuits. We can communicate digitally by means of analog or digital electronics and the same is true for analog communication.

We speak of *digital* communication when the transmitter sends one of a finite set of possible signals. For instance, if we communicate 1000 bits, we are communicating one out of 2^{1000} possible binary sequences of length 1000. To communicate our choice, we use signals that are appropriate for the channel at hand. No matter which signals we use, the result will be digital communication. One of the simplest ways to do this is that each bit determines the amplitude of a carrier over a certain duration of time. So the first bit could determine the amplitude from time 0 to T, the second from T to $2T$, etc. This is the simplest form of pulse amplitude modulation. There are many sensible ways to map bits to waveforms that are suitable to a channel, and regardless of the choice, it will be a form of digital communication.

We speak of *analog* communication when the transmitter sends one of a *continuum* of possible signals. The transmitted signal could be the output of a microphone. Any tiny variation of the signal can constitute another valid signal. More likely, in analog communication we use the source signal to vary a parameter of a *carrier signal*. Two popular ways to do analog communication are amplitude modulation (AM) and frequency modulation (FM). In AM we let the carrier's amplitude depend on the source signal. In FM it is the carrier's frequency that varies as a function of the source signal.

The difference between analog and digital communication might seem to be minimal at this point, but actually it is not. It all boils down to the fact that in digital communication the receiver has a chance to exactly reconstruct the transmitted signal because there is a finite number of possibilities to choose from.

The signals used by the transmitter are chosen to facilitate the receiver's decision. One of the performance criteria is the error probability, and we can design systems that have such a small error probability that for all practical purposes it is zero. The situation is quite different in analog communication. As there is a continuum of signals that the transmitter could possibly send, there is no chance for the receiver to reconstruct an *exact* replica of the transmitted signal from the noisy received signal. It no longer makes sense to talk about error probability. If we say that an error occurs every time that there is a difference between the transmitted signal and the reconstruction provided by the receiver, then the error probability is always 1.

EXAMPLE 1.2 *Consider a very basic transmitter that maps a sequence b_0, b_1, b_2, b_3 of numbers into a sequence $w(t)$ of rectangular pulses of a fixed duration as shown in Figure 1.8. (The ith pulse has amplitude b_i.)*

Is this analog or digital communication? It depends on the alphabet of b_i, $i = 0, \ldots, 3$.

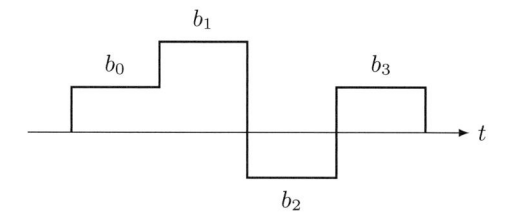

Figure 1.8. Transmitted signal $w(t)$.

If it is a discrete alphabet, like $\{0.9, 2, -1.3\}$, then we speak of digital communication. In this case there are only m^4 valid sequences b_0, b_1, b_2, b_3, where m is the alphabet size, and equally many possibilities for $w(t)$. In principle, the receiver can compare the noisy channel output waveform against all these possibilities and choose the most likely sequence. If the alphabet is $\mathbb{R}$, then the communication is analog. In this case the noise will make it virtually impossible for the receiver to guess the correct sequence. □

The difference, which may still seem insignificant at this point, is made significant by the notion of *channel capacity*. For every channel, there is a rate, called channel capacity, with the following meaning. Digital communication across the channel can be made as reliable as desired at any rate below channel capacity. At rates above channel capacity, it is impossible to reduce the error probability below a certain value. Now we can see where the difference between analog and digital communication becomes fundamental. For instance, if we want to communicate at 1 gigabit per second (Gbps) from Zurich to Los Angeles by using a certain type of cable, we can cut the cable into pieces of length L, chosen in such a way that the channel capacity of each piece is greater than 1 Gbps. We can then design a transmitter and a receiver that allow us to communicate virtually error-free at 1 Gbps over distance L. By concatenating many such links, we can cover any desired distance at the same rate. By making the error probability over each link sufficiently small, we can meet the desired end-to-end probability of error. The situation is very different in analog communication, where every piece of cable contributes to a degradation of the reconstruction.

Need another example? Compare faxing a text to sending an e-mail over the same telephone line. The fax uses analog technology. It treats the document as a continuum of gray levels (in two dimensions). It does not differentiate between text or images. The receiver prints a degraded version of the original. And if we repeat the operation multiple times by re-faxing the latest reproduction it will not take long until the result is dismal. E-mail on the other hand is a form of digital communication. It is almost certain that the receiver reconstructs an identical replica of the transmitted text.

Because we can turn a continuous-time source into a discrete one, as described in Appendix 1.8, we always have the option of doing digital rather than analog communication. In the conversion from continuous to discrete, there is a deterioration that we control and can make as small as desired. The result can, in principle, be communicated over unlimited distance and over arbitrarily poor channels with no further degradation.

1.5 Notation

In Chapter 2 and Chapter 3 we use a discrete-time and a continuous-time channel model, respectively. Accordingly, the signals we use to communicate are n-tuples in Chapter 2 and functions of time in Chapter 3. The transition from one set of signals to the other is made smoothly via the elegant theory of inner product spaces. This requires seeing both n-tuples and functions as vectors of an appropriate inner product space, which is the reason we have opted to use the same fonts for both kind of signals. (Many authors use bold-faced fonts for n-tuples.)

Some functions of time are referred to as *waveforms*. These are functions that typically represent voltages or currents within electrical circuits. An example of a waveform is the signal we use to communicate across a continuous-time channel. *Pulses* are waveforms that serve as building blocks for more complex waveforms. An example of a pulse is the impulse response of a linear time-invariant filter (LTI). From a mathematical point of view it is by no means essential to make a distinction between a function, a waveform, and a pulse. We use these terms because they are part of the language used by engineers and because it helps us associate a physical meaning with the specific function being discussed.

In this text, a generic function such as $g : \mathcal{I} \to \mathcal{B}$, where $\mathcal{I} \subseteq \mathbb{R}$ is the domain and $\mathcal{B}$ is the range, is typically a function of time or a function of frequency. Engineering texts underline the distinction by writing $g(t)$ and $g(f)$, respectively. This is an abuse of notation, which can be very helpful. We will make use of this abuse of notation as we see fit. By writing $g(t)$ instead of $g : \mathcal{I} \to \mathcal{B}$, we are effectively seeing t as representing $\mathcal{I}$, rather than representing a particular value of $\mathcal{I}$. To refer to a particular moment in time, we use a subscripts such as in t_0. So, $g(t_0)$ refers to the value that the function g takes at $t = t_0$. Similarly, $g(f)$ refers to a function of frequency and $g(f_0)$ is the value that g takes at $f = f_0$.

The Fourier transform of a function $g : \mathbb{R} \to \mathbb{C}$ is denoted by $g_\mathcal{F}$. Hence

$$g_\mathcal{F}(f) = \int_{-\infty}^{\infty} g(t) e^{-\mathsf{j} 2\pi f t} dt,$$

where $\mathsf{j} = \sqrt{-1}$.

By $[a,b]$ we denote the set of real numbers between a and b, including a and b. We write $(a,b]$, $[a,b)$, and (a,b) to exclude a, b, or both a and b from the set, respectively. A set that consists of the elements $a_1, a_2, \ldots, a_m$ is denoted by $\{a_1, a_2, \ldots, a_m\}$.

If $\mathcal{A}$ is a set and S is a statement that can be true or false, then $\{x \in \mathcal{A} : S\}$ denotes the subset of $\mathcal{A}$ for which the statement S is true. For instance, $\{x \in \mathbb{Z} : 3 \text{ divides } x\}$ is the set of all integers that are an integer multiple of 3.

We write $\mathbb{1}\{S\}$ for the indicator of the statement S. The indicator of S is 1 if S is true and 0 if S is false. For instance $\mathbb{1}\{t \in [a,b]\}$ takes value 1 when $t \in [a,b]$ and 0 elsewhere. When the indicator is a function like in this case, we call it *indicator function*. As another example, $\mathbb{1}\{k = l\}$ is 1 when $k = l$ and 0 when $k \neq l$.

As is customary in mathematics, we use the letters i, j, k, l, m, n for integers. (The integer j should not be confused with the complex number j.)

The convolution between $u(t)$ and $v(t)$ is denoted by $(u \star v)(t)$. Here $(u \star v)$ should be seen as the name of a new function obtained by convolving the functions $u(t)$ and $v(t)$. Sometimes it is useful to write $u(t) \star v(t)$ for $(u \star v)(t)$.

1.6 A few anecdotes

This text is targeted mainly at engineering students. Throughout their careers some will make inventions that may or may not be successful. After reading *The Information: A History, a Theory, a Flood* by James Gleick[5] [5], I felt that I should pass on some anecdotes that nicely illustrate one point, specifically that no matter how great an idea or an invention is, there will be people that will criticize it.

The printing press was invented by Johannes Gutenberg around 1440. It is now recognized that it played an essential role in the transition from medieval to modern times. Yet in the sixteenth century, the German priest Martin Luther decried that the "multitude of books [were] a great evil ..."; in the seventeenth century, referring to the "horrible mass of books" Leibniz feared a return to barbarism for "in the end the disorder will become nearly insurmountable"; in 1970 the American historian Lewis Mumford predicted that "the overproduction of books will bring about a state of intellectual enervation and depletion hardly to be distinguished from massive ignorance".

The telegraph was invented by Claude Chappe during the French Revolution. A telegraph was a tower for sending optical signals to other towers in line of sight. In 1840 measurements were made to determine the transmission speed. Over a stretch of 760 km, from Toulon to Paris, comprising 120 stations, it was determined

---
[5] A copy of the book was generously offered by our dean, Martin Vetterli, to each professor as a 2011 Christmas gift.

1.6. A few anecdotes

that two out of three messages arrived within a day during the warm months and that only one in three arrived in winter. This was the situation when F. B. Morse proposed to the French government a telegraph that used electrical wires. Morse's proposal was rejected because "No one could interfere with telegraph signals in the sky, but wire could be cut by saboteurs" [5, Chapter 5].

In 1833 the lawyer and philologist John Pickering, referring to the American version of the French telegraph on Central Wharf (a Chappe-like tower communicating shipping news with three other stations in a 12-mile line across Boston Harbor) asserted that "It must be evident to the most common observer, that no means of conveying intelligence can ever be devised, that shall exceed or even equal the rapidity of the Telegraph, for, with the exception of the scarcely perceptible relay at each station, its rapidity may be compared with that of light itself". In today's technology we can communicate over optical fiber at more than 10^{12} bits per second, which may be 12 orders of magnitude faster than the telegraph referred to by Pickering. Yet Pickering's flawed reasoning may have seemed correct to most of his contemporaries.

The electrical telegraph eventually came and was immediately a great success, yet some feared that it would put newspapers out of business. In 1852 it was declared that "All ideas of connecting Europe with America, by lines extending directly across the Atlantic, is utterly impracticable and absurd". Six years later Queen Victoria and President Buchanan were communicating via such a line.

Then came the telephone. The first experimental applications of the "electrical speaking telephone" were made in the US in the 1870s. It quickly became a great success in the USA, but not in England. In 1876 the chief engineer of the General Post Office, William Preece, reported to the British Parliament: "I fancy the descriptions we get of its use in America are a little exaggerated, though there are conditions in America which necessitate the use of such instruments more than here. Here we have a superabundance of messengers, errand boys and things of that kind ... I have one in my office, but more for show. If I want to send a message – I use a sounder or employ a boy to take it".

Compared to the telegraph, the telephone looked like a toy because any child could use it. In comparison, the telegraph required literacy. Business people first thought that the telephone was not serious. Where the telegraph dealt in facts and numbers, the telephone appealed to emotions. Seeing information technology as a threat to privacy is not new. Already at the time one commentator said, "No matter to what extent a man may close his doors and windows, and hermetically seal his key-holes and furnace-registers, with towels and blankets, whatever he may say, either to himself or a companion, will be overheard".

In summary, the printing press has been criticized for promoting barbarism; the electrical telegraphy for being vulnerable to vandalism, a threat to newspapers, and not superior to the French telegraph; the telephone for being childish, of no business value, and a threat to privacy. We could of course extend the list with comments about typewriters, cell phones, computers, the Internet, or about applications such as e-mail, SMS, Wikipedia, Street View by Google, etc. It would be good to keep some of these examples in mind when attempts to promote new ideas are met with resistance.

1.7 Supplementary reading

Here is a small selection of recommended textbooks for background material, to complement this text, or to venture into more specialized topics related to communication theory.

There are many excellent books on background material. A recommended selection is: [6] by Vetterli, Kovačević and Goyal for signal processing; [7] by Ross for probability theory; [8] by Rudin and [9] by Apostol for real analysis; [10] by Axler and [11] by Hoffman and Kunze for linear algebra; [12] by Horn and Johnson for matrices.

A very accessible undergraduate textbook on communication, with background material on signals and systems, as well as on random processes, is [13] by Proakis and Salehi. For the graduate-level student, [2] by Gallager is a very lucid exposition on the principles of digital communication with integration theory done at the "right level". The more mathematical reader is referred to [3] by Lapidoth. For breadth, more of an engineering perspective, and synchronization issues see [14] by Barry, Lee, and Messerschmitt. Other recommended textbooks that have a broad coverage are [15] by Madhow and [16] by Wilson. Note that [1] by Wozencraft and Jacobs is a somewhat dated classic textbook but still a recommended read.

To specialize in wireless communication, the recommended textbooks are [17] by Tse and Viswanath and [18] by Goldsmith. The standard textbooks for a first course in information theory are [19] by Cover and Thomas and [20] by Gallager. Reference [21] by MacKay is an original and refreshing textbook for information theory, coding, and statistical inference; [22] by Lin and Costello is recommended for a broad introduction to coding theory, whereas [23] by Richardson and Urbanke is the reference for low-density parity-check coding, and [4] by Kurose and Ross is recommended for computer networking.

1.8 Appendix: Sources and source coding

We often assume that the message to be communicated is the realization of a sequence of *independent* and *identically distributed* binary symbols. The purpose of this section is to justify this assumption. The results summarized here are given without proof.

In communication, it is always a bad idea to assume that a source produces a deterministic output. As mentioned in Section 1.2, if a listener knew exactly what a speaker is about to say, there would be no need to listen. Therefore, a source output should be considered as the realization of a stochastic process. Source coding is about the representation of the source output by a string of symbols from a finite (often binary) alphabet. Whether this is done in one, two, or three steps, depends on the kind of source.

Discrete sources   A discrete source is modeled by a discrete-time random process that takes values in some finite alphabet. For instance, a computer file is represented as a sequence of bytes, each of which can take on one of 256 possible values.

1.8. Appendix: Sources and source coding

So when we consider a file as being the source signal, the source can be modeled as a discrete-time random process taking values in the finite alphabet $\{0, 1, \ldots, 255\}$. Alternatively, we can consider the file as a sequence of bits, in which case the stochastic process takes values in $\{0, 1\}$.

For another example, consider the sequence of pixel values produced by a digital camera. The color of a pixel is obtained by mixing various intensities of red, green, and blue. Each of the three intensities is represented by a certain number of bits. One way to exchange images is to exchange one pixel at a time, according to some predetermined way of serializing the pixel's intensities. Also in this case we can model the source as a discrete-time process.

A discrete-time sequence taking values in a finite alphabet can always be converted into a *binary* sequence. The resulting average length depends on the source statistic and on how we do the conversion. In principle we could find the minimum average length by analyzing all possible ways of making the conversion. Surprisingly, we can bypass this tedious process and find the result by means of a simple formula that determines the so-called *entropy* (of the source). This was a major result in Shannon's 1948 paper.

EXAMPLE 1.3 *A discrete memoryless source is a discrete source with the additional property that the output symbols are independent and identically distributed. For a discrete memoryless source that produces symbols taking values in an m-letter alphabet the entropy is*

$$-\sum_{i=1}^{m} p_i \log_2 p_i,$$

where p_i, $i = 1, \ldots, m$ is the probability associated to the ith letter. For instance, if $m = 3$ and the probabilities are $p_1 = 0.5$, $p_2 = p_3 = 0.25$ then the source entropy is 1.5 bits. Shannon's result implies that it is possible to encode the source in such a way that, on average, it requires 1.5 bits to describe a source symbol. To see how this can be done, we encode two ternary source symbols into three binary symbols by mapping the most likely source letter into 1 and the other two letters into 01 and 00, respectively. Then the average length of the binary representation is $1 \times 0.5 + 2 \times 0.25 + 2 \times 0.25 = 1.5$ bits per source symbol, as predicted by the source entropy. There is no way to compress the source to fewer than 1.5 bits per source symbol and be able to recover the original from the compressed description. □

Any book on information theory will prove the stated relationship between the entropy of a memoryless source and the minimum average number of bits needed to represent a source symbol. A standard reference is [19].

If the output of the encoder that produces the shortest binary sequence can no longer be compressed, it means that it has entropy 1. One can show that to have entropy 1, a binary source must produce independent and uniformly distributed symbols. Such a source is called a *binary symmetric source (BSS)*. We conclude that the binary output of a source encoder can either be further compressed or it has the same statistic as the output of a BSS. This is the main reason a communication-link designer typically assumes that the source is a BSS.

Discrete-time continuous-alphabet sources These are modeled by a discrete-time random process that takes values in some continuous alphabet. For instance, if we measure the temperature of a room at regular time-intervals, we obtain a sequence of real-valued numbers, modeled as the realization of a discrete-time continuous alphabet random process. (This is assuming that we measure with infinite precision. If we use a digital thermometer, then we are back to the discrete case.) To store or to transmit the realization of such a source, we first round up or down the number to the nearest element of some fixed discrete set of numbers (as the digital thermometer does). This is called *quantization* and the result is the quantized process with the discrete set as its alphabet. Quantization is irreversible, but by choosing a sufficiently dense alphabet, we can make the difference between the original and the quantized process as small as desired. As described in the previous paragraph, the quantized sequence can be mapped in a reversible way into a binary sequence. If the resulting binary sequence is as short as possible (in average) then, once again, it is indistinguishable from the output of a BSS.

Continuous-time sources These are modeled by a continuous-time random process. The electrical signal produced by a microphone can be seen as a sample path of a continuous-time random process. In all practical applications, such signals are either bandlimited or can be lowpass-filtered to make them bandlimited. For instance, even the most sensitive human ear cannot detect frequencies above some value (say 25 kHz). Hence any signal meant for the human ear can be made bandlimited through a lowpass filter. The sampling theorem (Theorem 5.2) asserts that a bandlimited signal can be represented by a discrete-time sequence, which in turn can be made into a binary sequence as described.

In this text we will not need results from information theory, but we will often assume that the message to be communicated is the output of a BSS. Because we can always map a block of k bits into an index taking value in $\{0, 1, \ldots, 2^k - 1\}$, an essentially equivalent assumption is that the source produces independent and identically distributed (iid) random variables that are uniformly distributed over some finite alphabet.

1.9 Exercises

Note: The exercises in this first chapter are meant to test if the reader has the expected knowledge in probability theory.

EXERCISE 1.1 (Probabilities of basic events) *Assume that X_1 and X_2 are independent random variables that are uniformly distributed in the interval $[0, 1]$. Compute the probability of the following events. Hint: For each event, identify the corresponding region inside the unit square.*

(a) $0 \leq X_1 - X_2 \leq \frac{1}{3}$.
(b) $X_1^3 \leq X_2 \leq X_1^2$.
(c) $X_2 - X_1 = \frac{1}{2}$.

1.9. Exercises

(d) $(X_1 - \frac{1}{2})^2 + (X_2 - \frac{1}{2})^2 \leq (\frac{1}{2})^2$.
(e) Given that $X_1 \geq \frac{1}{4}$, compute the probability that $(X_1 - \frac{1}{2})^2 + (X_2 - \frac{1}{2})^2 \leq (\frac{1}{2})^2$.

EXERCISE 1.2 (Basic probabilities) Find the following probabilities.

(a) A box contains m white and n black balls. Suppose k balls are drawn. Find the probability of drawing at least one white ball.
(b) We have two coins; the first is fair and the second is two-headed. We pick one of the coins at random, toss it twice and obtain heads both times. Find the probability that the coin is fair.

EXERCISE 1.3 (Conditional distribution) Assume that X and Y are random variables with joint density function

$$f_{X,Y}(x,y) = \begin{cases} A, & 0 \leq x < y \leq 1 \\ 0, & \text{otherwise.} \end{cases}$$

(a) Are X and Y independent?
(b) Find the value of A.
(c) Find the density function of Y. Do this first by arguing informally using a sketch of $f_{X,Y}(x,y)$, then compute the density formally.
(d) Find $\mathbb{E}[X|Y=y]$. Hint: Try to find it from a sketch of $f_{X,Y}(x,y)$.
(e) The $\mathbb{E}[X|Y=y]$ found in part (d) is a function of y, call it $f(y)$. Find $\mathbb{E}[f(Y)]$.
(f) Find $\mathbb{E}[X]$ (from the definition) and compare it to the $\mathbb{E}[\mathbb{E}[X|Y]] = \mathbb{E}[f(Y)]$ that you have found in part (e).

EXERCISE 1.4 (Playing darts) Assume that you are throwing darts at a target. We assume that the target is one-dimensional, i.e. that the darts all end up on a line. The "bull's eye" is in the center of the line, and we give it the coordinate 0. The position of a dart on the target can then be measured with respect to 0. We assume that the position X_1 of a dart that lands on the target is a random variable that has a Gaussian distribution with variance σ_1^2 and mean 0. Assume now that there is a second target, which is further away. If you throw a dart to that target, the position X_2 has a Gaussian distribution with variance σ_2^2 (where $\sigma_2^2 > \sigma_1^2$) and mean 0. You play the following game: You toss a "coin" which gives you $Z = 1$ with probability p and $Z = 0$ with probability $1 - p$ for some fixed $p \in [0,1]$. If $Z = 1$, you throw a dart onto the first target. If $Z = 0$, you aim for the second target instead. Let X be the relative position of the dart with respect to the center of the target that you have chosen.

(a) Write down X in terms of X_1, X_2 and Z.
(b) Compute the variance of X. Is X Gaussian?
(c) Let $S = |X|$ be the score, which is given by the distance of the dart to the center of the target (that you picked using the coin). Compute the average score $\mathbb{E}[S]$.

EXERCISE 1.5 (Uncorrelated vs. independent random variables)

(a) Let X and Y be two continuous real-valued random variables with joint probability density function f_{XY}. Show that if X and Y are independent, they are also uncorrelated.

(b) Consider two independent and uniformly distributed random variables $U \in \{0,1\}$ and $V \in \{0,1\}$. Assume that X and Y are defined as follows: $X = U+V$ and $Y = |U - V|$. Are X and Y independent? Compute the covariance of X and Y. What do you conclude?

EXERCISE 1.6 (Monty Hall) Assume you are participating in a quiz show. You are shown three boxes that look identical from the outside, except they have labels 0, 1, and 2, respectively. Only one of them contains one million Swiss francs, the other two contain nothing. You choose one box at random with a uniform probability. Let A be the random variable that denotes your choice, $A \in \{0,1,2\}$.

(a) What is the probability that the box A contains the money?
(b) The quizmaster of course knows which box contains the money. He opens one of the two boxes that you did not choose, being careful not to open the one that contains the money. Now, you know that the money is either in A (your first choice) or in B (the only other box that could contain the money). What is the probability that B contains the money?
(c) If you are now allowed to change your mind, i.e. choose B instead of sticking with A, would you do it?

2 Receiver design for discrete-time observations: First layer

2.1 Introduction

The focus of this and the next chapter is the receiver design. The task of the receiver can be appreciated by considering a very noisy channel. The "GPS channel" is a good example. Let the channel input be the electrical signal applied to the transmitting antenna of a GPS satellite orbiting at an altitude of 20 200 km, and let the channel output be the signal at the antenna output of a GPS receiver at sea level. The signal of interest at the output of the receiving antenna is very weak. If we were to observe the receiving antenna output with a general-purpose instrument, such as an oscilloscope or a spectrum analyzer, we would not be able to distinguish the signal from the noise. Yet, most of the time the receiver manages to reproduce the bit sequence transmitted by the satellite. This is the result of sophisticated operations performed by the receiver.

To understand how to deal with the randomness introduced by the channel, it is instructive to start with channels that output n-tuples (possibly $n = 1$) rather than waveforms. In this chapter, we learn all we need to know about decisions based on the output of such channels. As a prominent special case, we will consider the discrete-time additive white Gaussian noise (AWGN) channel. In so doing, by the end of the chapter we will have derived the receiver for the first layer of Figure 1.7.

Figure 2.1 depicts the communication systems considered in this chapter. Its components are as follows.

- *A source:* The source (not represented in the figure) produces the message to be transmitted. In a typical application, the message consists of a sequence of bits but this detail is not fundamental for the theory developed in this chapter. It is fundamental that the source chooses one "message" from a set of possible messages. We are free to choose the "label" we assign to the various messages and our choice is based on mathematical convenience. For now the mathematical model of a source is as follows. If there are m possible choices, we model the source as a random variable H that takes values in the message set $\mathcal{H} = \{0, 1, \ldots, (m-1)\}$. More often than not, all messages are assumed to have the same probability but for generality we allow message i to occur with probability $P_H(i)$. The message set $\mathcal{H}$ and the probability distribution P_H are assumed to be known to the system designer.

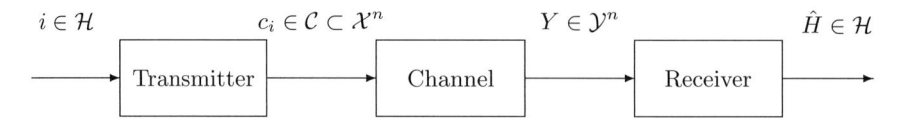

Figure 2.1. General setup for Chapter 2.

- *A channel:* The system designer needs to be able to cope with a broad class of channel models. A quite general way to describe a channel is by specifying its input alphabet $\mathcal{X}$ (the set of signals that are physically compatible with the channel input), the channel output alphabet $\mathcal{Y}$, and a statistical description of the output given the input. Unless otherwise specified, in this chapter the output alphabet $\mathcal{Y}$ is a subset of $\mathbb{R}$. A convenient way to think about the channel is to imagine that for each letter $x \in \mathcal{X}$ that we apply to the channel input, the channel outputs the realization of a random variable $Y \in \mathcal{Y}$ of statistic that depends on x. If Y is a discrete random variable, we describe the probability distribution (also called probability mass function, abbreviated to pmf) of Y given x, denoted by $P_{Y|X}(\cdot|x)$. If Y is a continuous random variable, we describe the probability density function (pdf) of Y given x, denoted by $f_{Y|X}(\cdot|x)$. In a typical application, we need to know the statistic of a sequence $Y_1, \ldots, Y_n$ of channel outputs, $Y_k \in \mathcal{Y}$, given a sequence $X_1, \ldots, X_n$ of channel inputs, $X_k \in \mathcal{X}$, but our typical channel is memoryless, meaning that

$$P_{Y_1,\ldots,Y_n|X_1,\ldots,X_n}(y_1,\ldots,y_n|x_1,\ldots,x_n) = \prod_{i=1}^{n} P_{Y_i|X_i}(y_i|x_i)$$

for discrete-output alphabets and

$$f_{Y_1,\ldots,Y_n|X_1,\ldots,X_n}(y_1,\ldots,y_n|x_1,\ldots,x_n) = \prod_{i=1}^{n} f_{Y_i|X_i}(y_i|x_i)$$

for continuous-output alphabets. A discrete-time channel of this generality might seem to be a rather abstract concept, but the theory we develop with this channel model turns out to be what we need.

- *A transmitter:* Mathematically, the transmitter is a mapping from the message set $\mathcal{H} = \{0, 1, \ldots, m-1\}$ to the *signal set* $\mathcal{C} = \{c_0, c_1, \ldots, c_{m-1}\}$ (also called *signal constellation*), where $c_i \in \mathcal{X}^n$ for some n. From an engineering point of view, the transmitter is needed because the signals that represent the message are not suitable to excite the channel. From this viewpoint, the transmitter is just a sort of sophisticated connector. The elements of the signal set $\mathcal{C}$ are chosen in such a way that a well-designed receiver that observes the channel output can tell (with high probability) which signal from $\mathcal{C}$ has excited the channel input.

- *A receiver:* The receiver's task is to "guess" the realization of the hypothesis H from the realization of the channel output sequence. We use $\hat{\imath}$ to represent the guess made by the receiver. Like the message, the guess of the message is the outcome of a random experiment. The corresponding random variable

2.1. Introduction

is denoted by $\hat{H} \in \mathcal{H}$. Unless specified otherwise, the receiver will always be designed to minimize the probability of error, denoted P_e and defined as the probability that $\hat{H}$ differs from H. Guessing the value of a discrete random variable H from the realization of another random variable (or random vector) is called a *hypothesis testing* problem. We are interested in hypothesis testing in order to design communication systems, but it can also be used in other applications, for instance to develop a smoke detector.

EXAMPLE 2.1 (Source) *As a first example of a source, consider $\mathcal{H} = \{0, 1\}$ and $P_H(0) = P_H(1) = 1/2$. H could model individual bits of, say, a file. Alternatively, one could model an entire file of, say, 1 Mbit by saying that $\mathcal{H} = \{0, 1, \ldots, (2^{10^6} - 1)\}$ and $P_H(i) = \frac{1}{2^{10^6}}, i \in \mathcal{H}$.* □

EXAMPLE 2.2 (Transmitter) *A transmitter for a binary source could be a map from $\mathcal{H} = \{0, 1\}$ to $\mathcal{C} = \{-a, a\}$ for some real-valued constant a. If the source is 4-ary, the transmitter could be any one-to-one map from $\mathcal{H} = \{0, 1, 2, 3\}$ to $\mathcal{C} = \{-3a, -a, a, 3a\}$. Alternatively, the map could be from $\mathcal{H} = \{0, 1, 2, 3\}$ to $\mathcal{C} = \{a, \mathrm{j}a, -a, -\mathrm{j}a\}$, where $\mathrm{j} = \sqrt{-1}$. The latter is a valid choice if $\mathcal{X} \subseteq \mathbb{C}$. All three cases have real-world applications, but we have to wait until Chapter 7 to fully understand the utility of complex-valued signal sets.* □

EXAMPLE 2.3 (Channel) *The channel model that we will use frequently in this chapter is the one that maps a signal $c \in \mathbb{R}^n$ into $Y = c + Z$, where Z is a Gaussian random vector of independent and identically distributed components. As we will see later, this is the discrete-time equivalent of the continuous-time additive white Gaussian noise (AWGN) channel.* □

The chapter is organized as follows. We first learn the basic ideas behind hypothesis testing, the field that deals with the problem of guessing the outcome of a random variable based on the observation of another random variable (or random vector). Then we study the Q function as it is a very valuable tool in dealing with communication problems that involve Gaussian noise. At that point, we will be ready to consider the problem of communicating across the discrete-time additive white Gaussian noise channel. We will first consider the case that involves two messages and scalar signals, then the case of two messages and n-tuple signals, and finally the case of an arbitrary number m of messages and n-tuple signals. Then we study techniques that we use, for instance, to tell if we can reduce the dimensionality of the channel output signals without undermining the receiver performance. The last part of the chapter deals with techniques to bound the error probability when an exact expression is unknown or too difficult to evaluate.

A point about terminology and symbolism needs to be clarified. We are using c_i (and not s_i) to denote the signal used for message i because the signals of this chapter will become codewords in subsequent chapters.

2.2 Hypothesis testing

Hypothesis testing refers to the problem of guessing the outcome of a random variable H that takes values in a finite alphabet $\mathcal{H} = \{0, 1, \ldots, m-1\}$, based on the outcome of a random variable Y called *observable*.

This problem comes up in various applications under different names. Hypothesis testing is the terminology used in statistics where the problem is studied from a fundamental point of view. A receiver does hypothesis testing, but communication people call it *decoding*. An alarm system such as a smoke detector also does hypothesis testing, but people would call it *detection*. A more appealing name for hypothesis testing is *decision making*. Hypothesis testing, decoding, detection, and decision making are all synonyms.

In communication, the hypothesis H is the message to be transmitted and the observable Y is the channel output (or a sequence of channel outputs). The receiver guesses the realization of H based on the realization of Y. Unless stated otherwise, we assume that, for all $i \in \mathcal{H}$, the system designer knows $P_H(i)$ and $f_{Y|H}(\cdot|i)$.[1]

The receiver's decision will be denoted by $\hat{\imath}$ and the corresponding random variable by $\hat{H} \in \mathcal{H}$. If we could, we would ensure that $\hat{H} = H$, but this is generally not possible. The goal is to devise a decision strategy that maximizes the probability $P_c = Pr\{\hat{H} = H\}$ that the decision is correct.[2] An equivalent goal is to minimize the *error probability* $P_e = Pr\{\hat{H} \neq H\} = 1 - P_c$.

Hypothesis testing is at the heart of the communication problem. As described by Claude Shannon in the introduction to what is arguably the most influential paper ever written on the subject [24], "The fundamental problem of communication is that of reproducing at one point either exactly or approximately a message selected at another point".

EXAMPLE 2.4 *As a typical example of a hypothesis testing problem, consider the problem of communicating one bit of information across an optical fiber. The bit being transmitted is modeled by the random variable $H \in \{0,1\}$, $P_H(0) = 1/2$. If $H = 1$, we switch on a light-emitting diode (LED) and its light is carried across the optical fiber to a photodetector. The photodetector outputs the number of photons $Y \in \mathbb{N}$ it detects. The problem is to decide whether $H = 0$ (the LED is off) or $H = 1$ (the LED is on). Our decision can only be based on whatever prior information we have about the model and on the actual observation $Y = y$. What makes the problem interesting is that it is impossible to determine H from Y with certainty. Even if the LED is off, the detector is likely to detect some photons (e.g. due to "ambient light"). A good assumption is that Y is Poisson distributed with intensity λ, which depends on whether the LED is on or off. Mathematically, the situation is as follows:*

---

[1] We assume that Y is a continuous random variable (or continuous random vector). If it is discrete, then we use $P_{Y|H}(\cdot|i)$ instead of $f_{Y|H}(\cdot|i)$.
[2] $Pr\{\cdot\}$ is a short-hand for *probability of the enclosed event*.

2.2. Hypothesis testing

$$\text{when } H = 0, \quad Y \sim P_{Y|H}(y|0) = \frac{\lambda_0^y}{y!}e^{-\lambda_0},$$

$$\text{when } H = 1, \quad Y \sim P_{Y|H}(y|1) = \frac{\lambda_1^y}{y!}e^{-\lambda_1},$$

where $0 \leq \lambda_0 < \lambda_1$. We read the above as follows: "When $H = 0$, the observable Y is Poisson distributed with intensity λ_0. When $H = 1$, Y is Poisson distributed with intensity λ_1". Once again, the problem of deciding the value of H from the observable Y is a standard hypothesis testing problem. □

From P_H and $f_{Y|H}$, via Bayes' rule, we obtain

$$P_{H|Y}(i|y) = \frac{P_H(i)f_{Y|H}(y|i)}{f_Y(y)},$$

where $f_Y(y) = \sum_i P_H(i)f_{Y|H}(y|i)$. In the above expression $P_{H|Y}(i|y)$ is the *posterior* (also called *a posteriori probability* of H given Y). By observing $Y = y$, the probability that $H = i$ goes from the prior $P_H(i)$ to the posterior $P_{H|Y}(i|y)$.

If the decision is $\hat{H} = i$, the probability that it is the correct decision is the probability that $H = i$, i.e. $P_{H|Y}(i|y)$. As our goal is to maximize the probability of being correct, the optimum decision rule is

$$\hat{H}(y) = \arg\max_{i \in \mathcal{H}} P_{H|Y}(i|y) \qquad \text{(MAP decision rule)}, \qquad (2.1)$$

where $\arg\max_i g(i)$ stands for "one of the arguments i for which the function $g(i)$ achieves its maximum". The above is called the *maximum a posteriori (MAP) decision rule*. In case of ties, i.e. if $P_{H|Y}(j|y)$ equals $P_{H|Y}(k|y)$ equals $\max_i P_{H|Y}(i|y)$, then it does not matter if we decide for $\hat{H} = k$ or for $\hat{H} = j$. In either case, the probability that we have decided correctly is the same.

Because the MAP rule maximizes the probability of being correct for each observation y, it also maximizes the unconditional probability P_c of being correct. The former is $P_{H|Y}(\hat{H}(y)|y)$. If we plug in the random variable Y instead of y, then we obtain a random variable. (A real-valued function of a random variable is a random variable.) The expected value of this random variable is the (unconditional) probability of being correct, i.e.

$$P_c = \mathbb{E}[P_{H|Y}(\hat{H}(Y)|Y)] = \int_y P_{H|Y}(\hat{H}(y)|y)f_Y(y)dy. \qquad (2.2)$$

There is an important special case, namely when H is uniformly distributed. In this case $P_{H|Y}(i|y)$, as a function of i, is proportional to $f_{Y|H}(y|i)$. Therefore, the argument that maximizes $P_{H|Y}(i|y)$ also maximizes $f_{Y|H}(y|i)$. Then the MAP decision rule is equivalent to the decision

$$\hat{H}(y) = \arg\max_{i \in \mathcal{H}} f_{Y|H}(y|i) \qquad \text{(ML decision rule)}, \qquad (2.3)$$

called the *maximum likelihood (ML) decision rule*. The name stems from the fact that $f_{Y|H}(y|i)$, as a function of i, is called the *likelihood function*.

Notice that the ML decision rule is defined even if we do not know P_H. Hence it is the solution of choice when the prior is not known. (The MAP and the ML decision rules are equivalent only when the prior is uniform.)

2.2.1 Binary hypothesis testing

The special case in which we have to make a binary decision, i.e. $\mathcal{H} = \{0, 1\}$, is both instructive and of practical relevance. We begin with it and generalize in the next section.

As there are only two alternatives to be tested, the MAP test may now be written as

$$\frac{f_{Y|H}(y|1)P_H(1)}{f_Y(y)} \underset{\hat{H}=0}{\overset{\hat{H}=1}{\gtrless}} \frac{f_{Y|H}(y|0)P_H(0)}{f_Y(y)}.$$

The above notation means that the MAP test decides for $\hat{H} = 1$ when the left is bigger than or equal to the right, and decides for $\hat{H} = 0$ otherwise. Observe that the denominator is irrelevant because $f_Y(y)$ is a positive constant – hence it will not affect the decision. Thus an equivalent decision rule is

$$f_{Y|H}(y|1)P_H(1) \underset{\hat{H}=0}{\overset{\hat{H}=1}{\gtrless}} f_{Y|H}(y|0)P_H(0).$$

The above test is depicted in Figure 2.2 assuming $y \in \mathbb{R}$. This is a very important figure that helps us visualize what goes on and, as we will see, will be helpful to compute the probability of error.

The above test is insightful as it shows that we are comparing posteriors after rescaling them by canceling the positive number $f_Y(y)$ from the denominator. However, there are alternative forms of the test that, depending on the details, can be computationally more convenient. An equivalent test is obtained by dividing

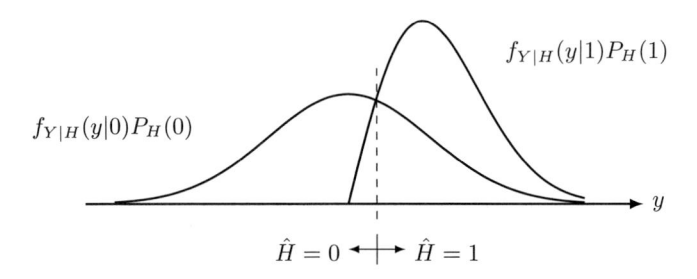

Figure 2.2. Binary MAP decision.

2.2. Hypothesis testing

both sides with the non-negative quantity $f_{Y|H}(y|0)P_H(1)$. This results in the following *binary MAP test*:

$$\Lambda(y) = \frac{f_{Y|H}(y|1)}{f_{Y|H}(y|0)} \begin{array}{c} \hat{H}=1 \\ \geq \\ < \\ \hat{H}=0 \end{array} \frac{P_H(0)}{P_H(1)} = \eta. \qquad (2.4)$$

The left side of the above test is called the *likelihood ratio*, denoted by $\Lambda(y)$, whereas the right side is the *threshold* η. Notice that if $P_H(0)$ increases, so does the threshold. In turn, as we would expect, the region $\{y : \hat{H}(y) = 0\}$ becomes larger.

When $P_H(0) = P_H(1) = 1/2$ the threshold η is unity and the MAP test becomes a *binary ML test*:

$$f_{Y|H}(y|1) \begin{array}{c} \hat{H}=1 \\ \geq \\ < \\ \hat{H}=0 \end{array} f_{Y|H}(y|0).$$

A function $\hat{H} : \mathcal{Y} \to \mathcal{H} = \{0, \ldots, m-1\}$ is called a *decision function* (also called *decoding function*). One way to describe a decision function is by means of the *decision regions* $\mathcal{R}_i = \{y \in \mathcal{Y} : \hat{H}(y) = i\}$, $i \in \mathcal{H}$. Hence $\mathcal{R}_i$ is the set of $y \in \mathcal{Y}$ for which $\hat{H}(y) = i$. We continue this subsection with $\mathcal{H} = \{0, 1\}$.

To compute the error probability, it is often convenient to compute the error probability for each hypothesis and then take the average. When $H = 0$, the decision is incorrect if $Y \in \mathcal{R}_1$ or, equivalently, if $\Lambda(y) \geq \eta$. Hence, denoting by $P_e(i)$ the error probability when $H = i$,

$$P_e(0) = Pr\{Y \in \mathcal{R}_1 | H = 0\} = \int_{\mathcal{R}_1} f_{Y|H}(y|0) dy \qquad (2.5)$$

or, equivalently,

$$P_e(0) = Pr\{\Lambda(Y) \geq \eta | H = 0\}. \qquad (2.6)$$

Whether it is easier to work with the right side of (2.5) or that of (2.6) depends on whether it is easier to work with the conditional density of Y or of $\Lambda(Y)$. We will see examples of both cases.

Similar expressions hold for the probability of error conditioned on $H = 1$, denoted by $P_e(1)$. Using the law of total probability, we obtain the (unconditional) error probability

$$P_e = P_e(1)P_H(1) + P_e(0)P_H(0).$$

In deriving the probability of error we have tacitly used an important technique that we use all the time in probability: conditioning as an intermediate step. Conditioning as an intermediate step may be seen as a divide-and-conquer strategy. The idea is to solve a problem that seems hard by breaking it up into subproblems that (i) we know how to solve and (ii) once we have the solution to the sub-problems we also have the solution to the original problem. Here is how it works in probability. We want to compute the expected value of a random variable Z. Assume that it is not immediately clear how to compute the expected value of Z, but we know that Z is related to another random variable W that tells us something useful about Z: useful in the sense that for every value w we are able to compute the expected value of Z given $W = w$. Then, via the *law of total expectation*, we compute: $\mathbb{E}[Z] = \sum_w \mathbb{E}[Z|W=w] P_W(w)$. The same principle applies for probabilities. (This is not a coincidence: The probability of an event is the expected value of the indicator function of that event.) For probabilities, the expression is $Pr(Z \in \mathcal{A}) = \sum_w Pr(Z \in \mathcal{A}|W=w) P_W(w)$. It is called the *law of total probability*.

Let us revisit what we have done in light of the above comments and what else we could have done. The computation of the probability of error involves two random variables, H and Y, as well as an event $\{H \neq \hat{H}\}$. To compute the probability of error (2.5) we have first conditioned on all possible values of H. Alternatively, we could have conditioned on all possible values of Y. This is indeed a viable alternative. In fact we have already done so (without saying it) in (2.2). Between the two, we use the one that seems more promising for the problem at hand. We will see examples of both.

2.2.2 m-ary hypothesis testing

Now we go back to the m-ary hypothesis testing problem. This means that $\mathcal{H} = \{0, 1, \ldots, m-1\}$.

Recall that the MAP decision rule, which minimizes the probability of making an error, is

$$\hat{H}_{MAP}(y) = \arg\max_{i \in \mathcal{H}} P_{H|Y}(i|y)$$

$$= \arg\max_{i \in \mathcal{H}} \frac{f_{Y|H}(y|i) P_H(i)}{f_Y(y)}$$

$$= \arg\max_{i \in \mathcal{H}} f_{Y|H}(y|i) P_H(i),$$

where $f_{Y|H}(\cdot|i)$ is the probability density function of the observable Y when the hypothesis is i and $P_H(i)$ is the probability of the ith hypothesis. This rule is well defined up to ties. If there is more than one i that achieves the maximum on the right side of one (and thus all) of the above expressions, then we may decide for any such i without affecting the probability of error. If we want the decision rule to be unambiguous, we can for instance agree that in case of ties we choose the largest i that achieves the maximum.

When all hypotheses have the same probability, then the MAP rule specializes to the ML rule, i.e.

$$\hat{H}_{ML}(y) = \arg\max_{i \in \mathcal{H}} f_{Y|H}(y|i).$$

2.3. The Q function

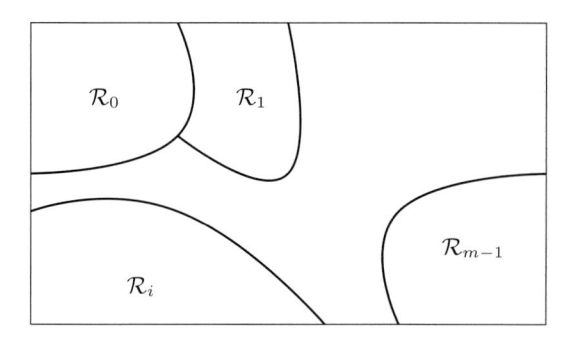

Figure 2.3. Decision regions.

We will always assume that $f_{Y|H}$ is either given as part of the problem formulation or that it can be figured out from the setup. In communication, we typically know the transmitter and the channel. In this chapter, the transmitter is the map from $\mathcal{H}$ to $\mathcal{C} \subset \mathcal{X}^n$ and the channel is described by the pdf $f_{Y|X}(y|x)$ known for all $x \in \mathcal{X}^n$ and all $y \in \mathcal{Y}^n$. From these two, we immediately obtain $f_{Y|H}(y|i) = f_{Y|X}(y|c_i)$, where c_i is the signal assigned to i.

Note that the decision (or decoding) function $\hat{H}$ assigns an $i \in \mathcal{H}$ to each $y \in \mathbb{R}^n$. As already mentioned, it can be described by the decision (or decoding) regions $\mathcal{R}_i$, $i \in \mathcal{H}$, where $\mathcal{R}_i$ consists of those y for which $\hat{H}(y) = i$. It is convenient to think of $\mathbb{R}^n$ as being partitioned by decoding regions as depicted in Figure 2.3.

We use the decoding regions to express the error probability P_e or, equivalently, the probability $P_c = 1 - P_e$ of deciding correctly. Conditioned on $H = i$ we have

$$P_e(i) = 1 - P_c(i)$$
$$= 1 - \int_{\mathcal{R}_i} f_{Y|H}(y|i) dy.$$

2.3 The Q function

The Q function plays a very important role in communication. It will come up frequently throughout this text. It is defined as

$$Q(x) := \frac{1}{\sqrt{2\pi}} \int_x^\infty e^{-\frac{\xi^2}{2}} d\xi,$$

where the symbol $:=$ means that the left side is defined by the expression on the right. Hence if Z is a normally distributed zero-mean random variable of unit variance, denoted by $Z \sim \mathcal{N}(0,1)$, then $Pr\{Z \geq x\} = Q(x)$. (The Q function has been defined specifically to make this true.)

If Z is normally distributed with mean m and variance σ^2, denoted by $Z \sim \mathcal{N}(m, \sigma^2)$, the probability $Pr\{Z \geq x\}$ can also be written using the Q function. In fact the event $\{Z \geq x\}$ is equivalent to $\{\frac{Z-m}{\sigma} \geq \frac{x-m}{\sigma}\}$. But $\frac{Z-m}{\sigma} \sim \mathcal{N}(0,1)$.

Hence $Pr\{Z \geq x\} = Q(\frac{x-m}{\sigma})$. This result will be used frequently. It should be memorized.

We now describe some of the key properties of the Q function.

(a) If $Z \sim \mathcal{N}(0,1)$, $F_Z(z) := Pr\{Z \leq z\} = 1 - Q(z)$. (The reader is advised to draw a picture that expresses this relationship in terms of areas under the probability density function of Z.)
(b) $Q(0) = 1/2$, $Q(-\infty) = 1$, $Q(\infty) = 0$.
(c) $Q(-x) + Q(x) = 1$. (Again, it is advisable to draw a picture.)
(d) $\frac{1}{\sqrt{2\pi}\alpha} e^{-\frac{\alpha^2}{2}} (\frac{\alpha^2}{1+\alpha^2}) < Q(\alpha) < \frac{1}{\sqrt{2\pi}\alpha} e^{-\frac{\alpha^2}{2}}$, $\alpha > 0$.
(e) An alternative expression for the Q function with fixed integration limits is
$$Q(x) = \frac{1}{\pi} \int_0^{\frac{\pi}{2}} e^{-\frac{x^2}{2\sin^2\theta}} d\theta.$$ It holds for $x \geq 0$.
(f) $Q(\alpha) \leq \frac{1}{2} e^{-\frac{\alpha^2}{2}}$, $\alpha \geq 0$.

Proofs The proofs of (a), (b), and (c) are immediate (a picture suffices). The proof of part (d) is left as an exercise (see Exercise 2.12). To prove (e), let $X \sim \mathcal{N}(0,1)$ and $Y \sim \mathcal{N}(0,1)$ be independent random variables and let $\xi \geq 0$. Hence $Pr\{X \geq 0, Y \geq \xi\} = Q(0)Q(\xi) = \frac{Q(\xi)}{2}$. Using polar coordinates to integrate over the region $x \geq 0$, $y \geq \xi$ (shaded region of the figure below), yields

$$\frac{Q(\xi)}{2} = Pr\{X \geq 0, Y \geq \xi\} = \int_0^{\frac{\pi}{2}} \int_{\frac{\xi}{\sin\theta}}^{\infty} \frac{e^{-\frac{r^2}{2}}}{2\pi} r \, dr d\theta$$

$$= \frac{1}{2\pi} \int_0^{\frac{\pi}{2}} \int_{\frac{\xi^2}{2\sin^2\theta}}^{\infty} e^{-t} \, dt d\theta = \frac{1}{2\pi} \int_0^{\frac{\pi}{2}} e^{-\frac{\xi^2}{2\sin^2\theta}} \, d\theta.$$

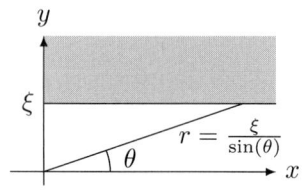

To prove (f), we use (e) and the fact that $e^{-\frac{\xi^2}{2\sin^2\theta}} \leq e^{-\frac{\xi^2}{2}}$ for $\theta \in [0, \frac{\pi}{2}]$. Hence

$$Q(\xi) \leq \frac{1}{\pi} \int_0^{\frac{\pi}{2}} e^{-\frac{\xi^2}{2}} d\theta = \frac{1}{2} e^{-\frac{\xi^2}{2}}.$$ □

A plot of the Q function and its bounds is given in Figure 2.4.

2.4 Receiver design for the discrete-time AWGN channel

The hypothesis testing problem discussed in this section is key in digital communication. It is the one performed by the top layer of Figure 1.7. The setup is depicted in Figure 2.5. The hypothesis $H \in \mathcal{H} = \{0, \ldots, m-1\}$ represents a randomly selected message. The transmitter maps $H = i$ to a signal n-tuple

2.4. Receiver design for the discrete-time AWGN channel

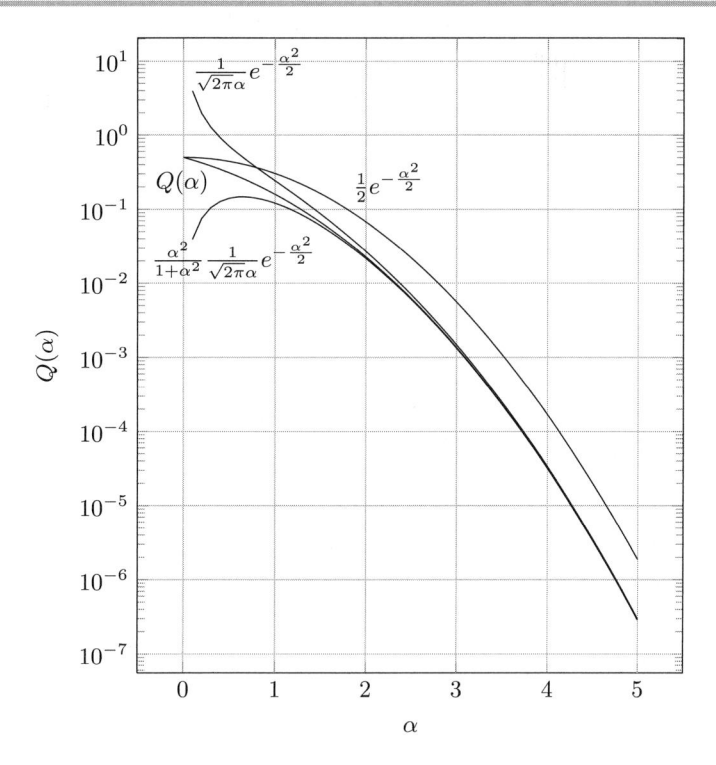

Figure 2.4. Q function with upper and lower bounds.

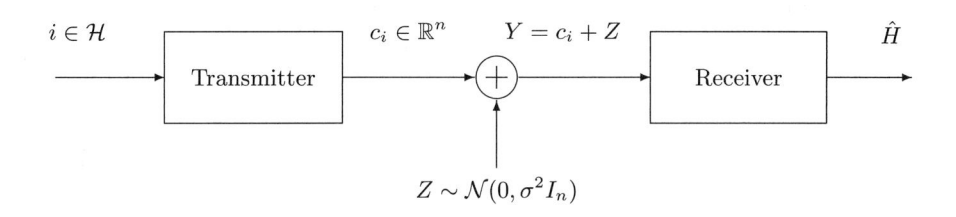

Figure 2.5. Communication over the discrete-time additive white Gaussian noise channel.

$c_i \in \mathbb{R}^n$. The channel adds a random (noise) vector Z which is zero-mean and has independent and identically distributed Gaussian components of variance σ^2. In short, $Z \sim \mathcal{N}(0, \sigma^2 I_n)$. The observable is $Y = c_i + Z$.

We begin with the simplest possible situation, specifically when there are only two equiprobable messages and the signals are scalar ($n = 1$). Then we generalize to arbitrary values for n and finally we consider arbitrary values also for the cardinality m of the message set.

2.4.1 Binary decision for scalar observations

Let the message $H \in \{0, 1\}$ be equiprobable and assume that the transmitter maps $H = 0$ into $c_0 \in \mathbb{R}$ and $H = 1$ into $c_1 \in \mathbb{R}$. The output statistic for the various hypotheses is as follows:

$$H = 0: \quad Y \sim \mathcal{N}(c_0, \sigma^2)$$
$$H = 1: \quad Y \sim \mathcal{N}(c_1, \sigma^2).$$

An equivalent way to express the output statistic for each hypothesis is

$$f_{Y|H}(y|0) = \frac{1}{\sqrt{2\pi\sigma^2}} \exp\left\{-\frac{(y-c_0)^2}{2\sigma^2}\right\}$$
$$f_{Y|H}(y|1) = \frac{1}{\sqrt{2\pi\sigma^2}} \exp\left\{-\frac{(y-c_1)^2}{2\sigma^2}\right\}.$$

We compute the likelihood ratio

$$\Lambda(y) = \frac{f_{Y|H}(y|1)}{f_{Y|H}(y|0)} = \exp\left\{-\frac{(y-c_1)^2 - (y-c_0)^2}{2\sigma^2}\right\} = \exp\left\{y\frac{c_1 - c_0}{\sigma^2} + \frac{c_0^2 - c_1^2}{2\sigma^2}\right\}. \tag{2.7}$$

The threshold is

$$\eta = \frac{P_H(0)}{P_H(1)}.$$

Now we have all the ingredients for the MAP rule.

Instead of comparing $\Lambda(y)$ to the threshold η, we can compare $\ln \Lambda(y)$ to $\ln \eta$. The function $\ln \Lambda(y)$ is called *log likelihood ratio*. Hence the MAP decision rule can be expressed as

$$y\frac{c_1 - c_0}{\sigma^2} + \frac{c_0^2 - c_1^2}{2\sigma^2} \underset{\hat{H}=0}{\overset{\hat{H}=1}{\gtrless}} \ln \eta.$$

The progress consists of the fact that the receiver no longer computes an exponential function of the observation. It has to compute $\ln(\eta)$, but this is done once and for all.

Without loss of essential generality, assume $c_1 > c_0$. Then we can divide both sides by $\frac{c_1 - c_0}{\sigma^2}$ (which is positive) without changing the outcome of the above comparison. We can further simplify by moving the constants to the right. The result is the simple test

$$\hat{H}_{\text{MAP}}(y) = \begin{cases} 1, & y \geq \theta \\ 0, & \text{otherwise,} \end{cases}$$

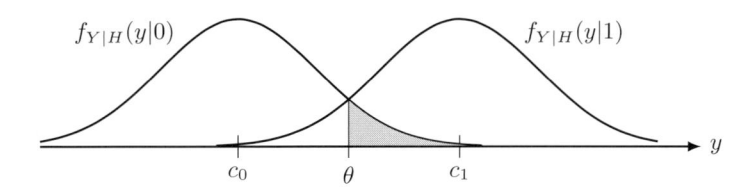

2.4. Receiver design for the discrete-time AWGN channel

Figure 2.6. When $P_H(0) = P_H(1)$, the decision threshold θ is the midpoint between c_0 and c_1. The shaded area represents the probability of error conditioned on $H = 0$.

where

$$\theta = \frac{\sigma^2}{c_1 - c_0} \ln \eta + \frac{c_0 + c_1}{2}.$$

Notice that if $P_H(0) = P_H(1)$, then $\ln \eta = 0$ and the threshold θ becomes the midpoint $\frac{c_0+c_1}{2}$ (Figure 2.6).

We now determine the error probability.

$$P_e(0) = Pr\{Y > \theta | H = 0\} = \int_\theta^\infty f_{Y|H}(y|0) dy.$$

This is the probability that a Gaussian random variable with mean c_0 and variance σ^2 exceeds the threshold θ. From our review on the Q function we know immediately that $P_e(0) = Q\left(\frac{\theta - c_0}{\sigma}\right)$. Similarly, $P_e(1) = Q\left(\frac{c_1 - \theta}{\sigma}\right)$. Finally, $P_e = P_H(0) Q\left(\frac{\theta - c_0}{\sigma}\right) + P_H(1) Q\left(\frac{c_1 - \theta}{\sigma}\right)$.

The most common case is when $P_H(0) = P_H(1) = 1/2$. Then $\frac{\theta - c_0}{\sigma} = \frac{c_1 - \theta}{\sigma} = \frac{c_1 - c_0}{2\sigma} = \frac{d}{2\sigma}$, where d is the distance between c_0 and c_1. In this case, $P_e(0) = P_e(1) = P_e$, where

$$P_e = Q\left(\frac{d}{2\sigma}\right).$$

This result can be obtained straightforwardly without side calculations. As shown in Figure 2.6, the threshold is the middle point between c_0 and c_1 and $P_e = P_e(0) = Q(\frac{d}{2\sigma})$. This result should be known by heart.

2.4.2 Binary decision for n-tuple observations

As in the previous subsection, we assume that H takes values in $\{0, 1\}$. What is new is that the signals are now n-tuples for $n \geq 1$. So when $H = 0$, the transmitter sends some $c_0 \in \mathbb{R}^n$ and when $H = 1$, it sends $c_1 \in \mathbb{R}^n$. The noise added by the channel is $Z \sim \mathcal{N}(0, \sigma^2 I_n)$ and independent of H.

From here on, we assume that the reader is familiar with the definitions and basic results related to Gaussian random vectors. (See Appendix 2.10 for a review.) We also assume familiarity with the notions of *inner product*, *norm*, and *affine plane*.

(See Appendix 2.12 for a review.) The inner product between the vectors u and v will be denoted by $\langle u, v \rangle$, whereas $\|u\| = \sqrt{\langle u, u \rangle}$ denotes the norm of u. We will make extensive use of these notations.

Even though for now the vector space is over the reals, in Chapter 7 we will encounter complex vector spaces. Whether the vector space is over $\mathbb{R}$ or over $\mathbb{C}$, the notation is almost identical. For instance, if a and b are (column) n-tuples in $\mathbb{C}^n$, then $\langle a, b \rangle = b^\dagger a$, where $\dagger$ denotes conjugate transpose. The equality holds even if a and b are in $\mathbb{R}^n$, but in this case the conjugation is inconsequential and we could write $\langle a, b \rangle = b^\mathsf{T} a$, where T denotes transpose. By default, we will use the more general notation for complex vector spaces. An equality that we will use frequently, therefore should be memorized, is

$$\|a \pm b\|^2 = \|a\|^2 + \|b\|^2 \pm 2\Re\{\langle a, b \rangle\}, \tag{2.8}$$

where $\Re\{\cdot\}$ denotes the real part of the enclosed complex number. Of course we can drop the $\Re\{\cdot\}$ for elements of a real vector space.

As done earlier, to derive a MAP decision rule, we start by writing down the output statistic for each hypothesis

$$H = 0: \quad Y = c_0 + Z \sim \mathcal{N}(c_0, \sigma^2 I_n)$$
$$H = 1: \quad Y = c_1 + Z \sim \mathcal{N}(c_1, \sigma^2 I_n),$$

or, equivalently,

$$H = 0: \quad Y \sim f_{Y|H}(y|0) = \frac{1}{(2\pi\sigma^2)^{n/2}} \exp\left\{-\frac{\|y - c_0\|^2}{2\sigma^2}\right\}$$
$$H = 1: \quad Y \sim f_{Y|H}(y|1) = \frac{1}{(2\pi\sigma^2)^{n/2}} \exp\left\{-\frac{\|y - c_1\|^2}{2\sigma^2}\right\}.$$

Like in the scalar case, we compute the likelihood ratio

$$\Lambda(y) = \frac{f_{Y|H}(y|1)}{f_{Y|H}(y|0)} = \exp\left\{\frac{\|y - c_0\|^2 - \|y - c_1\|^2}{2\sigma^2}\right\}.$$

Taking the logarithm on both sides and using (2.8), we obtain

$$\ln \Lambda(y) = \frac{\|y - c_0\|^2 - \|y - c_1\|^2}{2\sigma^2} \tag{2.9}$$
$$= \left\langle y, \frac{c_1 - c_0}{\sigma^2} \right\rangle + \frac{\|c_0\|^2 - \|c_1\|^2}{2\sigma^2}. \tag{2.10}$$

From (2.10), the MAP rule can be written as

$$\left\langle y, \frac{c_1 - c_0}{\sigma^2} \right\rangle + \frac{\|c_0\|^2 - \|c_1\|^2}{2\sigma^2} \underset{\hat{H} = 0}{\overset{\hat{H} = 1}{\gtrless}} \ln \eta. \tag{2.11}$$

2.4. Receiver design for the discrete-time AWGN channel

Notice the similarity with the corresponding expression of the scalar case. As for the scalar case, we move the constants to the right and normalize to obtain

$$\langle y, \psi \rangle \underset{\hat{H}=0}{\overset{\hat{H}=1}{\gtrless}} \theta, \qquad (2.12)$$

where

$$\psi = \frac{c_1 - c_0}{d}$$

is the unit-length vector that points in the direction $c_1 - c_0$, $d = \|c_1 - c_0\|$ is the distance between the signals, and

$$\theta = \frac{\sigma^2}{d} \ln \eta + \frac{\|c_1\|^2 - \|c_0\|^2}{2d}$$

is the decision threshold. Hence the decision regions $\mathcal{R}_0$ and $\mathcal{R}_1$ are delimited by the affine plane

$$\left\{ y \in \mathbb{R}^n : \langle y, \psi \rangle = \theta \right\}.$$

For definiteness, we are assigning the points of the delimiting affine plane to $\mathcal{R}_1$, but this is an arbitrary decision that has no effect on the error probability because the probability that Y is on any given affine plane is zero.

We obtain additional geometrical insight by considering those y for which (2.9) is constant. The situation is depicted in Figure 2.7, where the signed distance p is positive if the delimiting affine plane lies in the direction pointed by ψ with respect to c_0 and q is positive if the affine plane lies in the direction pointed by $-\psi$ with respect to c_1. (In the figure, both p and q are positive.) By Pythagoras' theorem applied to the two right triangles with common edge, for all y on the affine plane, $\|y - c_0\|^2 - \|y - c_1\|^2$ equals $p^2 - q^2$.

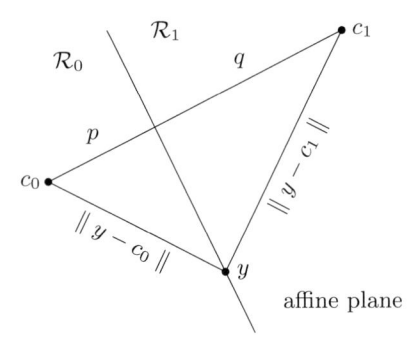

Figure 2.7. Affine plane delimiting $\mathcal{R}_0$ and $\mathcal{R}_1$.

Note that p and q are related to η, σ^2, and d via

$$\|y - c_0\|^2 - \|y - c_1\|^2 = p^2 - q^2$$
$$\|y - c_0\|^2 - \|y - c_1\|^2 = 2\sigma^2 \ln \eta.$$

Hence $p^2 - q^2 = 2\sigma^2 \ln \eta$. Using this and $d = p + q$, we obtain

$$p = \frac{d}{2} + \frac{\sigma^2 \ln \eta}{d}$$
$$q = \frac{d}{2} - \frac{\sigma^2 \ln \eta}{d}.$$

When $P_H(0) = P_H(1) = \frac{1}{2}$, the delimiting affine plane is the set of $y \in \mathbb{R}^n$ for which (2.9) equals 0. These are the points y that are at the same distance from c_0 and from c_1. Hence, $\mathcal{R}_0$ contains all the points $y \in \mathbb{R}^n$ that are closer to c_0 than to c_1.

A few additional observations are in order.

- The vector ψ is not affected by the prior but the threshold θ is. Hence the prior affects the position but not the orientation of the delimiting affine plane. As one would expect, the plane moves away from c_0 when $P_H(0)$ increases. This is consistent with our intuition that the decoding region for a hypothesis becomes larger as the probability of that hypothesis increases.
- The above-mentioned effect of the prior is amplified when σ^2 increases. This is also consistent with our intuition that the decoder relies less on the observation and more on the prior when the observation becomes noisier.
- Notice the similarity of (2.9) and (2.10) with (2.7). This suggests a tight relationship between the scalar and the vector case. We can gain additional insight by placing the origin of a new coordinate system at $\frac{c_0+c_1}{2}$ and by letting the first coordinate be in the direction of $\psi = \frac{c_1-c_0}{d}$, where again $d = \|c_1 - c_0\|$. In this new coordinate system, $H = 0$ is mapped into the vector $\tilde{c}_0 = (-\frac{d}{2}, 0, \ldots, 0)^\mathsf{T}$ and $H = 1$ is mapped into $\tilde{c}_1 = (\frac{d}{2}, 0, \ldots, 0)^\mathsf{T}$. If $\tilde{y} = (\tilde{y}_1, \ldots, \tilde{y}_n)^\mathsf{T}$ is the channel output in this new coordinate system, $\langle \tilde{y}, \psi \rangle = \tilde{y}_1$. This shows that for a binary decision, the vector case is essentially the scalar case embedded in an n-dimensional space.

As for the scalar case, we compute the probability of error by conditioning on $H = 0$ and $H = 1$ and then remove the conditioning by averaging: $P_e = P_e(0)P_H(0) + P_e(1)P_H(1)$.

When $H = 0$, $Y = c_0 + Z$ and the MAP decoder makes the wrong decision when $\langle Z, \psi \rangle \geq p$, i.e. when the projection of Z onto the directional unit vector ψ has (signed) length that is equal to or greater than p. That this is the condition for an error should be clear from Figure 2.7, but it can also be derived by inserting $Y = c_0 + Z$ into $\langle Y, \psi \rangle \geq \theta$ and using (2.8). Since $\langle Z, \psi \rangle$ is a zero-mean Gaussian random variable of variance σ^2 (see Appendix 2.10), we obtain

$$P_e(0) = Q\left(\frac{p}{\sigma}\right) = Q\left(\frac{d}{2\sigma} + \frac{\sigma \ln \eta}{d}\right).$$

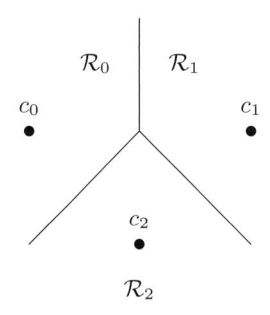

2.4. Receiver design for the discrete-time AWGN channel

Figure 2.8. Example of Voronoi regions in $\mathbb{R}^2$.

Proceeding similarly, we find

$$P_e(1) = Q\Big(\frac{q}{\sigma}\Big) = Q\Big(\frac{d}{2\sigma} - \frac{\sigma \ln \eta}{d}\Big).$$

The case $P_H(0) = P_H(1) = 0.5$ is the most common one. Because $p = q = \frac{d}{2}$, determining the error probability for this special case is straightforward:

$$P_e = P_e(1) = P_e(0) = Pr\left\{\langle Z, \psi \rangle \geq \frac{d}{2}\right\} = Q\Big(\frac{d}{2\sigma}\Big).$$

2.4.3 m-ary decision for n-tuple observations

When $H = i$, $i \in \mathcal{H} = \{0, 1, \ldots, m-1\}$, the channel input is $c_i \in \mathbb{R}^n$. For now we make the simplifying assumption that $P_H(i) = \frac{1}{m}$, which is a common assumption in communication. Later on we will see that generalizing is straightforward.

When $Y = y \in \mathbb{R}^n$, the ML decision rule is

$$\hat{H}_{ML}(y) = \arg \max_{i \in \mathcal{H}} f_{Y|H}(y|i)$$

$$= \arg \max_{i \in \mathcal{H}} \frac{1}{(2\pi\sigma^2)^{n/2}} \exp\left\{-\frac{\|y - c_i\|^2}{2\sigma^2}\right\} \qquad (2.13)$$

$$= \arg \min_{i} \|y - c_i\|.$$

Hence *an ML decision rule for the AWGN channel is a minimum-distance decision rule* as shown in Figure 2.8. Up to ties, $\mathcal{R}_i$ corresponds to the *Voronoi region* of c_i, defined as the set of points in $\mathbb{R}^n$ that are at least as close to c_i as to any other c_j.

EXAMPLE 2.5 *(m-PAM) Figure 2.9 shows the signal set $\{c_0, c_1, \ldots, c_5\} \subset \mathbb{R}$ for 6-ary PAM (pulse amplitude modulation) (the appropriateness of the name will become clear in the next chapter). The figure also shows the decoding regions of an ML decoder, assuming that the channel is AWGN. The signal points are elements of $\mathbb{R}$, and the ML decoder chooses according to the minimum-distance rule. When the hypothesis is $H = 0$, the receiver makes the wrong decision if the observation*

$y \in \mathbb{R}$ falls outside the decoding region $\mathcal{R}_0$. This is the case if the noise $Z \in \mathbb{R}$ is larger than $d/2$, where $d = c_i - c_{i-1}$, $i = 1, \ldots, 5$. Thus

$$P_e(0) = Pr\left\{Z > \frac{d}{2}\right\} = Q\left(\frac{d}{2\sigma}\right).$$

By symmetry, $P_e(5) = P_e(0)$. For $i \in \{1, 2, 3, 4\}$, the probability of error when $H = i$ is the probability that the event $\{Z \geq \frac{d}{2}\} \cup \{Z < -\frac{d}{2}\}$ occurs. This event is the union of disjoint events. Its probability is the sum of the probabilities of the individual events. Hence

$$P_e(i) = Pr\left\{\left\{Z \geq \frac{d}{2}\right\} \cup \left\{Z < -\frac{d}{2}\right\}\right\} = 2Pr\left\{Z \geq \frac{d}{2}\right\} = 2Q\left(\frac{d}{2\sigma}\right), \; i \in \{1,2,3,4\}.$$

Finally, $P_e = \frac{2}{6}Q\left(\frac{d}{2\sigma}\right) + \frac{4}{6}2Q\left(\frac{d}{2\sigma}\right) = \frac{5}{3}Q\left(\frac{d}{2\sigma}\right)$. We see immediately how to generalize. For a PAM constellation of m points (m positive integer), the error probability is

$$P_e = \left(2 - \frac{2}{m}\right)Q\left(\frac{d}{2\sigma}\right).$$

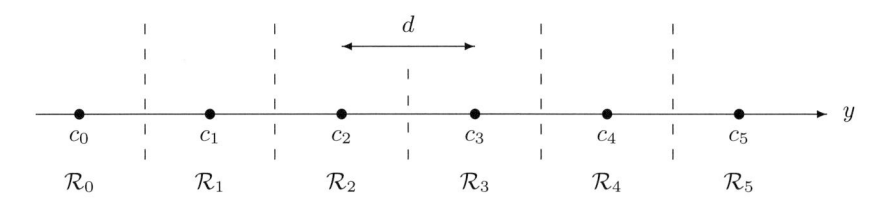

Figure 2.9. 6-ary PAM constellation in $\mathbb{R}$.

□

EXAMPLE 2.6 (*m-QAM*) *Figure 2.10 shows the signal set* $\{c_0, c_1, c_2, c_3\} \subset \mathbb{R}^2$ *for 4-ary quadrature amplitude modulation (QAM). We consider signals as points in $\mathbb{R}^2$. (We could choose to consider signals as points in $\mathbb{C}$, but we have to postpone this view until we know how to deal with complex valued noise.) The noise is $Z \sim \mathcal{N}(0, \sigma^2 I_2)$ and the observable, when $H = i$, is $Y = c_i + Z$. We assume that the receiver implements an ML decision rule, which for the AWGN channel means minimum-distance decoding. The decoding region for c_0 is the first quadrant, for c_1 the second quadrant, etc. When $H = 0$, the decoder makes the correct decision if $\{Z_1 > -\frac{d}{2}\} \cap \{Z_2 \geq -\frac{d}{2}\}$, where d is the minimum distance among signal points. This is the intersection of independent events. Hence the probability of the intersection is the product of the probability of each event, i.e.*

$$P_c(0) = Pr\left\{\left\{Z_1 \geq -\frac{d}{2}\right\} \cap \left\{Z_2 \geq -\frac{d}{2}\right\}\right\} = Q^2\left(-\frac{d}{2\sigma}\right) = \left[1 - Q\left(\frac{d}{2\sigma}\right)\right]^2.$$

2.5. Irrelevance and sufficient statistic

By symmetry, for all i, $P_c(i) = P_c(0)$. Hence,

$$P_e = P_e(0) = 1 - P_c(0) = 2Q\left(\frac{d}{2\sigma}\right) - Q^2\left(\frac{d}{2\sigma}\right).$$

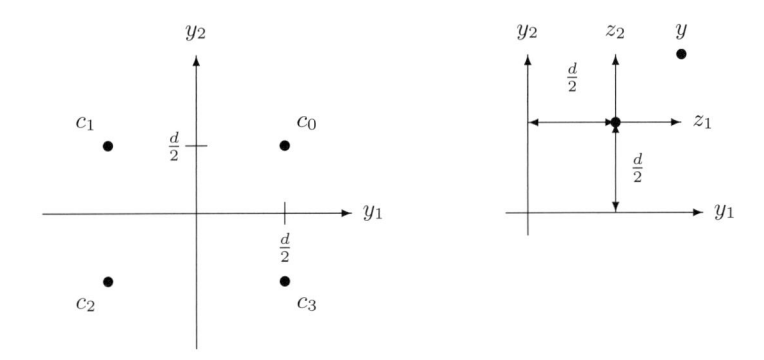

Figure 2.10. 4-ary QAM constellation in $\mathbb{R}^2$.

m-QAM is defined for all m of the form $(2j)^2$, for $j = 1, 2, \ldots$. An example of 16-QAM is given later in Figure 2.22. □

EXAMPLE 2.7 In Example 2.6 we have computed $P_e(0)$ via $P_c(0)$, but we could have opted for computing $P_e(0)$ directly. Here is how:

$$P_e(0) = \Pr\left\{\left\{Z_1 \leq -\frac{d}{2}\right\} \cup \left\{Z_2 \leq -\frac{d}{2}\right\}\right\}$$

$$= \Pr\left\{Z_1 \leq -\frac{d}{2}\right\} + \Pr\left\{Z_2 \leq -\frac{d}{2}\right\} - \Pr\left\{\left\{Z_1 \leq -\frac{d}{2}\right\} \cap \left\{Z_2 \leq -\frac{d}{2}\right\}\right\}$$

$$= 2Q\left(\frac{d}{2\sigma}\right) - Q^2\left(\frac{d}{2\sigma}\right).$$

Notice that, in determining $P_c(0)$ (Example 2.6), we compute the probability of the intersection of independent events (which is the product of the probability of the individual events) whereas in determining $P_e(0)$ without passing through $P_c(0)$ (this example), we compute the probability of the union of events that are not disjoint (which is not the sum of the probability of the individual events). □

2.5 Irrelevance and sufficient statistic

Have you ever tried to drink from a fire hydrant? There are situations in which the observable Y contains excessive data, but if we reduce the amount of data, how to be sure that we are not throwing away anything useful for a MAP decision? In this section we derive a criterion to answer that question. We begin by recalling the notion of Markov chain.

DEFINITION 2.8 *Three random variables U, V, W are said to form a Markov chain in that order, symbolized by $U \to V \to W$, if the distribution of W given both U and V is independent of U, i.e. $P_{W|V,U}(w|v,u) = P_{W|V}(w|v)$.* □

The reader should verify the correctness of the following two statements, which are straightforward consequences of the above Markov chain definition.

(i) $U \to V \to W$ if and only if $P_{U,W|V}(u, w|v) = P_{U|V}(u|v)P_{W|V}(w|v)$. In words, U, V, W form a Markov chain (in that order) if and only if U and W are independent when conditioned on V.
(ii) $U \to V \to W$ if and only if $W \to V \to U$, i.e. Markovity in one direction implies Markovity in the other direction.

Let Y be the observable and $T(Y)$ be a function (either stochastic or deterministic) of Y. Observe that $H \to Y \to T(Y)$ is always true, but in general it is *not* true that $H \to T(Y) \to Y$.

DEFINITION 2.9 *Let $T(Y)$ be a random variable obtained from processing an observable Y. If $H \to T(Y) \to Y$, then we say that $T(Y)$ is a sufficient statistic (for the hypothesis H).* □

If $T(Y)$ is a sufficient statistic, then the error probability of a MAP decoder that observes $T(Y)$ is identical to that of a MAP decoder that observes Y. Indeed for all $i \in \mathcal{H}$ and all $y \in \mathcal{Y}$, $P_{H|Y}(i|y) = P_{H|Y,T}(i|y,t) = P_{H|T}(i|t)$, where $t = T(y)$. (The first equality holds because the random variable T is a function of Y and the second equality holds because $H \to T(Y) \to Y$.) Hence if i maximizes $P_{H|Y}(\cdot|y)$ then it also maximizes $P_{H|T}(\cdot|t)$. We state this important result as a theorem.

THEOREM 2.10 *If $T(Y)$ is a sufficient statistic for H, then a MAP decoder that estimates H from $T(Y)$ achieves the exact same error probability as one that estimates H from Y.* □

EXAMPLE 2.11 *Consider the communication system depicted in Figure 2.11 where H, Z_1, Z_2 are independent random variables. Let $Y = (Y_1, Y_2)$. Then $H \to Y_1 \to Y$. Hence a MAP receiver that observes $Y_1 = T(Y)$ achieves the same error probability as a MAP receiver that observes Y. Note that the independence assumption is essential here. For instance, if $Z_2 = Z_1$, we can obtain Z_2 (and Z_1) from the difference $Y_2 - Y_1$. We can then remove Z_1 from Y_1 and obtain H. In this case from (Y_1, Y_2) we can make an error-free decision about H.* □

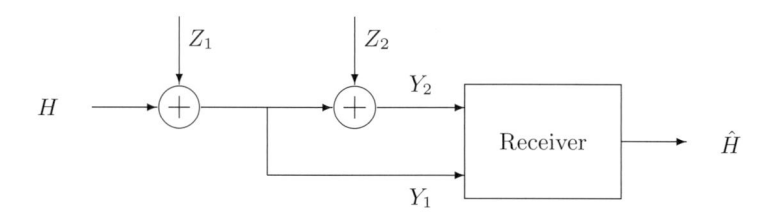

Figure 2.11. Example of irrelevant measurement.

2.5. Irrelevance and sufficient statistic

In some situations, like in the above example, we make multiple measurements and want to prove that some of the measurements are relevant for the detection problem and some are not. Specifically, the observable Y may consist of two components $Y = (Y_1, Y_2)$ where Y_1 and Y_2 may be m and n tuples, respectively. If $T(Y) = Y_1$ is a sufficient statistic, then we say that Y_2 is *irrelevant*.

We have seen that $H \to T(Y) \to Y$ implies that Y cannot be used to reduce the error probability of a MAP decoder that observes $T(Y)$. Is the contrary also true? Specifically, assume that a MAP decoder that observes Y always makes the same decision as one that observes only $T(Y)$. Does this imply $H \to T(Y) \to Y$?

The answer is "yes and no". We may expect the answer to be "no" because, for $H \to U \to V$ to hold, it has to be true that $P_{H|U,V}(i|u,v)$ equals $P_{H|U}(i|u)$ for all values of i, u, v, whereas for v to have no effect on a MAP decision it is "sufficient" that for all u, v the *maximum* of $P_{H|U}(\cdot|u)$ and that of $P_{H|U,V}(\cdot|u,v)$ be achieved for the same i. It seems clear that the former requirement is stronger. Indeed in Exercise 2.21 we give an example to show that the answer to the above question is "no" in general.

The choice of distribution on H is relevant for the example in Exercise 2.21. That we can construct artificial examples by "playing" with the distribution on H becomes clear if we choose, say, $P_H(0) = 1$. Now the decoder that chooses $\hat{H} = 0$ all the time is always correct. Yet one should not conclude that Y is irrelevant. However, if for every choice of P_H, the MAP decoder that observes Y and the MAP decoder that observes $T(Y)$ make the same decision, then $H \to T(Y) \to Y$ must hold. We prove this in Exercise 2.23. The following example is an application of this result.

EXAMPLE 2.12 *Regardless of the distribution on H, the binary test (2.4) depends on Y only through the likelihood ratio $\Lambda(Y)$. Hence $H \to \Lambda(Y) \to Y$ must hold, which makes the likelihood ratio a sufficient statistic. Notice that $\Lambda(y)$ is a scalar even when y is an n-tuple.* □

The following result is a useful tool in verifying that a function $T(y)$ is a sufficient statistic. It is proved in Exercise 2.22.

THEOREM 2.13 (Fisher–Neyman factorization theorem) *Suppose that $g_0, g_1, \ldots, g_{m-1}$ and h are functions such that for each $i \in \mathcal{H}$ one can write*

$$f_{Y|H}(y|i) = g_i(T(y))h(y). \tag{2.14}$$

Then T is a sufficient statistic. □

We will often use the notion of indicator function. Recall that if $\mathcal{A}$ is an arbitrary set, the indicator function $\mathbb{1}\{x \in \mathcal{A}\}$ is defined as

$$\mathbb{1}\{x \in \mathcal{A}\} = \begin{cases} 1, & x \in \mathcal{A} \\ 0, & \text{otherwise.} \end{cases}$$

EXAMPLE 2.14 *Let $H \in \mathcal{H} = \{0, 1, \ldots, m-1\}$ be the hypothesis and when $H = i$ let the components of $Y = (Y_1, \ldots, Y_n)^\mathsf{T}$ be iid uniformly distributed in*

$[0, i]$. We use the Fisher–Neyman factorization theorem to show that $T(Y) = \max\{Y_1, \ldots, Y_n\}$ is a sufficient statistic. In fact

$$f_{Y|H}(y|i) = \frac{1}{i}\mathbb{1}\{y_1 \in [0,i]\}\frac{1}{i}\mathbb{1}\{y_2 \in [0,i]\} \cdots \frac{1}{i}\mathbb{1}\{y_n \in [0,i]\}$$

$$= \begin{cases} \frac{1}{i^n}, & \text{if } 0 \leq \min\{y_1, \ldots, y_n\} \text{ and } \max\{y_1, \ldots, y_n\} \leq i \\ 0, & \text{otherwise} \end{cases}$$

$$= \frac{1}{i^n}\mathbb{1}\{0 \leq \min\{y_1, \ldots, y_n\}\}\mathbb{1}\{\max\{y_1, \ldots, y_n\} \leq i\}.$$

In this case, the Fisher–Neyman factorization theorem applies with $g_i(T) = \frac{1}{i^n}\mathbb{1}\{T \leq i\}$, where $T(y) = \max\{y_1, \ldots, y_n\}$, and $h(y) = \mathbb{1}\{0 \leq \min\{y_1, \ldots, y_n\}\}$. $\square$

2.6 Error probability bounds

2.6.1 Union bound

Here is a simple and extremely useful bound. Recall that for general events $\mathcal{A}, \mathcal{B}$

$$P(\mathcal{A} \cup \mathcal{B}) = P(\mathcal{A}) + P(\mathcal{B}) - P(\mathcal{A} \cap \mathcal{B})$$
$$\leq P(\mathcal{A}) + P(\mathcal{B}).$$

More generally, using induction, we obtain the *union bound*

$$P\left(\bigcup_{i=1}^m \mathcal{A}_i\right) \leq \sum_{i=1}^m P(\mathcal{A}_i), \qquad (UB)$$

that applies to any collection of events $\mathcal{A}_i$, $i = 1, \ldots, m$. We now apply the union bound to approximate the probability of error in multi-hypothesis testing. Recall that

$$P_e(i) = Pr\{Y \in \mathcal{R}_i^c | H = i\} = \int_{\mathcal{R}_i^c} f_{Y|H}(y|i) dy,$$

where $\mathcal{R}_i^c$ denotes the complement of $\mathcal{R}_i$. If we are able to evaluate the above integral for every i, then we are able to determine the probability of error exactly. The bound that we derive is useful if we are unable to evaluate the above integral.

For $i \neq j$ define

$$\mathcal{B}_{i,j} = \{y : P_H(j) f_{Y|H}(y|j) \geq P_H(i) f_{Y|H}(y|i)\}.$$

$\mathcal{B}_{i,j}$ is the set of y for which the a posteriori probability of H given $Y = y$ is at least as high for j as it is for i. Roughly speaking,[3] it contains the ys for which a MAP decision rule would choose j over i.

---

[3] A y for which the a posteriori probability is the same for i and for j is contained in both $\mathcal{B}_{i,j}$ and $\mathcal{B}_{j,i}$.

2.6. Error probability bounds

The following fact is very useful:

$$\mathcal{R}_i^c \subseteq \bigcup_{j:j\neq i} \mathcal{B}_{i,j}. \quad (2.15)$$

To see that the above inclusion holds, consider an arbitrary $y \in \mathcal{R}_i^c$. By definition, there is at least one $k \in \mathcal{H}$ such that $P_H(k)f_{Y|H}(y|k) \geq P_H(i)f_{Y|H}(y|i)$. Hence $y \in \mathcal{B}_{i,k}$.

The reader may wonder why we do not have equality in (2.15). To see that equality may or may not apply, consider a y that belongs to $\mathcal{B}_{i,l}$ for some l. It could be so because $P_H(l)f_{Y|H}(y|l) = P_H(i)f_{Y|H}(y|i)$ (notice the equality sign). To simplify the argument, let us assume that for the chosen y there is only one such l. The MAP decoding rule does not prescribe whether y should be in the decoding region of i or l. If it is in that of i, then equality in (2.15) does not hold. If none of the y for which $P_H(l)f_{Y|H}(y|l) = P_H(i)f_{Y|H}(y|i)$ for some l has been assigned to $\mathcal{R}_i$ then we have equality in (2.15). In one sentence, we have equality if all the ties have been resolved against i.

We are now in the position to upper bound $P_e(i)$. Using (2.15) and the union bound we obtain

$$P_e(i) = Pr\{Y \in \mathcal{R}_i^c | H = i\} \leq Pr\Big\{Y \in \bigcup_{j:j\neq i} \mathcal{B}_{i,j} | H = i\Big\}$$
$$\leq \sum_{j:j\neq i} Pr\{Y \in \mathcal{B}_{i,j} | H = i\} \quad (2.16)$$
$$= \sum_{j:j\neq i} \int_{\mathcal{B}_{i,j}} f_{Y|H}(y|i)dy.$$

The gain is that it is typically easier to integrate over $\mathcal{B}_{i,j}$ than over $\mathcal{R}_i^c$.

For instance, when the channel is AWGN and the decision rule is ML, $\mathcal{B}_{i,j}$ is the set of points in $\mathbb{R}^n$ that are at least as close to c_j as they are to c_i. Figure 2.12 depicts this situation.

In this case,

$$\int_{\mathcal{B}_{i,j}} f_{Y|H}(y|i)dy = Q\left(\frac{\|c_j - c_i\|}{2\sigma}\right),$$

and the union bound yields the simple expression

$$P_e(i) \leq \sum_{j:j\neq i} Q\left(\frac{\|c_j - c_i\|}{2\sigma}\right).$$

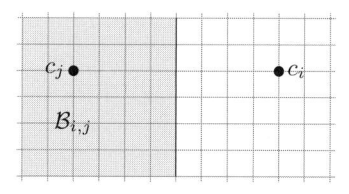

Figure 2.12. The shape of $\mathcal{B}_{i,j}$ for AWGN channels and ML decision.

In the next section we derive an easy-to-compute tight upper bound on

$$\int_{\mathcal{B}_{i,j}} f_{Y|H}(y|i)dy$$

for a general $f_{Y|H}$. Notice that the above integral is the probability of error under $H = i$ when there are only two hypotheses, the other hypothesis is $H = j$, and the priors are proportional to $P_H(i)$ and $P_H(j)$.

EXAMPLE 2.15 *(m-PSK) Figure 2.13 shows a signal set for 8-ary PSK (phase-shift keying). m-PSK is defined for all integers $m \geq 2$. Formally, the signal transmitted when $H = i$, $i \in \mathcal{H} = \{0, 1, \ldots, m-1\}$, is*

$$c_i = \sqrt{\mathcal{E}_s} \begin{pmatrix} \cos\left(\frac{2\pi i}{m}\right) \\ \sin\left(\frac{2\pi i}{m}\right) \end{pmatrix}.$$

For now $\sqrt{\mathcal{E}_s}$ is just the radius of the PSK constellation. As we will see, $\mathcal{E}_s = \|c_i\|^2$ is (proportional to) the energy required to generate c_i.

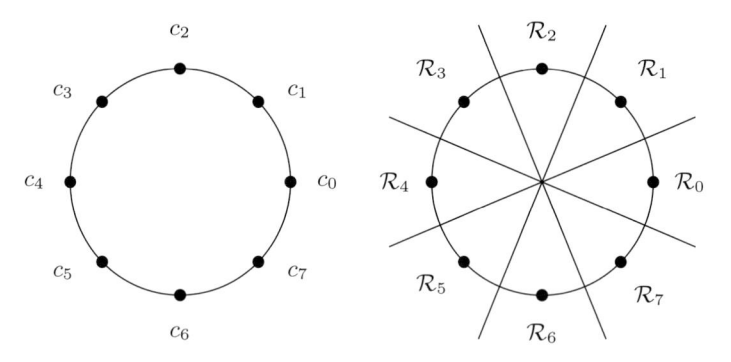

Figure 2.13. 8-ary PSK constellation in $\mathbb{R}^2$ and decoding regions.

Assuming the AWGN channel, the hypothesis testing problem is specified by

$$H = i: \quad Y \sim \mathcal{N}(c_i, \sigma^2 I_2)$$

and the prior $P_H(i)$ is assumed to be uniform. Because we have a uniform prior, the MAP and the ML decision rules are identical. Furthermore, since the channel is the AWGN channel, the ML decoder is a minimum-distance decoder. The decoding regions are also shown in Figure 2.13.

By symmetry, $P_e = P_e(i)$. Using polar coordinates to integrate the density of the noise, it is not hard to show that

$$P_e(i) = \frac{1}{\pi} \int_0^{\pi - \frac{\pi}{m}} \exp\left\{-\frac{\sin^2 \frac{\pi}{m}}{\sin^2(\theta)} \frac{\mathcal{E}_s}{2\sigma^2}\right\} d\theta.$$

The above expression is rather complicated. Let us see what we obtain through the union bound.

2.6. Error probability bounds

With reference to Figure 2.14 we have:

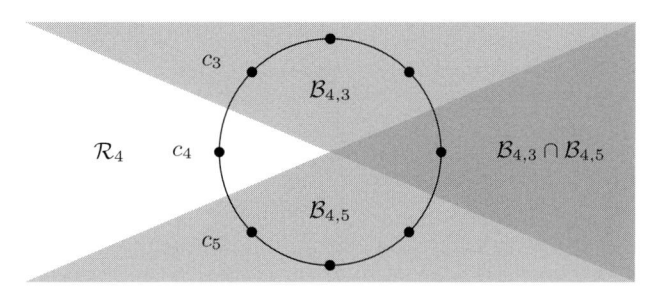

Figure 2.14. Bounding the error probability of PSK by means of the union bound.

$$P_e(i) = Pr\{Y \in \mathcal{B}_{i,i-1} \cup \mathcal{B}_{i,i+1} | H = i\}$$
$$\leq Pr\{Y \in \mathcal{B}_{i,i-1} | H = i\} + Pr\{Y \in \mathcal{B}_{i,i+1} | H = i\}$$
$$= 2Pr\{Y \in \mathcal{B}_{i,i-1} | H = i\}$$
$$= 2Q\left(\frac{\|c_i - c_{i-1}\|}{2\sigma}\right)$$
$$= 2Q\left(\frac{\sqrt{\mathcal{E}_s}}{\sigma} \sin \frac{\pi}{m}\right).$$

Notice that we have been using a version of the union bound adapted to the problem: we get a tighter bound by using the fact that $\mathcal{R}_i^c \subseteq \mathcal{B}_{i,i-1} \cup \mathcal{B}_{i,i+1}$ rather than $\mathcal{R}_i^c \subseteq \cup_{j \neq i} \mathcal{B}_{i,j}$.

How good is the upper bound? We know that it is good if we can find a lower bound which is close enough to the upper bound. From Figure 2.14 with $i = 4$ in mind we see that

$$P_e(i) = Pr\{Y \in \mathcal{B}_{i,i-1} | H = i\} + Pr\{Y \in \mathcal{B}_{i,i+1} | H = i\}$$
$$- Pr\{Y \in \mathcal{B}_{i,i-1} \cap \mathcal{B}_{i,i+1} | H = i\}.$$

The above expression can be used to upper and lower bound $P_e(i)$. In fact, if we lower bound the last term by setting it to zero, we obtain the upper bound that we have just derived. To the contrary, if we upper bound the last term, we obtain a lower bound to $P_e(i)$. To do so, observe that $\mathcal{R}_i^c$ is the union of $(m-1)$ disjoint cones, one of which is $\mathcal{B}_{i,i-1} \cap \mathcal{B}_{i,i+1}$ (see again Figure 2.14). The integral of $f_{Y|H}(\cdot|i)$ over those cones is P_e. If all those integrals gave the same result (which is not the case) the result would be $\frac{P_e(i)}{m-1}$. From the figure, the integral of $f_{Y|H}(\cdot|i)$ over $\mathcal{B}_{i,i-1} \cap \mathcal{B}_{i,i+1}$ is clearly smaller than that over the other cones. Hence its value must be less than $\frac{P_e(i)}{m-1}$. Mathematically,

$$Pr\{Y \in (\mathcal{B}_{i,i-1} \cap \mathcal{B}_{i,i+1}) | H = i\} \leq \frac{P_e(i)}{m-1}.$$

Inserting in the previous expression, solving for $P_e(i)$ and using the fact that $P_e(i) = P_e$ yields the desired lower bound

$$P_e \geq 2Q\left(\sqrt{\frac{\mathcal{E}_s}{\sigma^2}}\sin\frac{\pi}{m}\right)\frac{m-1}{m}.$$

The ratio between the upper and the lower bound is the constant $\frac{m}{m-1}$. For m large, the bounds become very tight.

The way we upper-bounded $Pr\{Y \in \mathcal{B}_{i,i-1} \cap \mathcal{B}_{i,i+1}|H=i\}$ is not the only way to proceed. Alternatively, we could use the fact that $\mathcal{B}_{i,i-1} \cap \mathcal{B}_{i,i+1}$ is included in $\mathcal{B}_{i,k}$ where k is the index of the codeword opposed to c_i. (In Figure 2.14, $\mathcal{B}_{4,3} \cap \mathcal{B}_{4,5} \subset \mathcal{B}_{4,0}$.) Hence $Pr\{Y \in \mathcal{B}_{i,i-1} \cap \mathcal{B}_{i,i+1}|H=i\} \leq Pr\{Y \in \mathcal{B}_{i,k}|H=i\} = Q\left(\sqrt{\mathcal{E}_s}/\sigma\right)$. This goes to zero as $\mathcal{E}_s/\sigma^2 \to \infty$. It implies that the lower bound obtained this way becomes tight as $\mathcal{E}_s/\sigma^2$ becomes large.

It is not surprising that the upper bound to $P_e(i)$ becomes tighter as m or $\mathcal{E}_s/\sigma^2$ (or both) become large. In fact it should be clear that under those conditions $Pr\{Y \in \mathcal{B}_{i,i-1} \cap \mathcal{B}_{i,i+1}|H=i\}$ becomes smaller. □

PAM, QAM, and PSK are widely used in modern communications systems. See Section 2.7 for examples of standards using these constellations.

2.6.2 Union Bhattacharyya bound

Let us summarize. From the union bound applied to $\mathcal{R}_i^c \subseteq \bigcup_{j:j\neq i} \mathcal{B}_{i,j}$ we have obtained the upper bound

$$P_e(i) = Pr\{Y \in \mathcal{R}_i^c|H=i\}$$
$$\leq \sum_{j:j\neq i} Pr\{Y \in \mathcal{B}_{i,j}|H=i\}$$

and we have used this bound for the AWGN channel. With the bound, instead of having to compute

$$Pr\{Y \in \mathcal{R}_i^c|H=i\} = \int_{\mathcal{R}_i^c} f_{Y|H}(y|i)dy,$$

which requires integrating over a possibly complicated region $\mathcal{R}_i^c$, we have only to compute

$$Pr\{Y \in \mathcal{B}_{i,j}|H=i\} = \int_{\mathcal{B}_{i,j}} f_{Y|H}(y|i)dy.$$

The latter integral is simply $Q(d_{i,j}/\sigma)$, where $d_{i,j}$ is the distance between c_i and the affine plane bounding $\mathcal{B}_{i,j}$. For an ML decision rule, $d_{i,j} = \frac{\|c_i-c_j\|}{2}$.

What if the channel is *not* AWGN? Is there a relatively simple expression for $Pr\{Y \in \mathcal{B}_{i,j}|H=i\}$ that applies for general channels? Such an expression does

2.6. Error probability bounds

exist. It is the *Bhattacharyya bound* that we now derive.[4] We will need it only for those i for which $P_H(i) > 0$. Hence, for the derivation that follows, we assume that it is the case.

The definition of $\mathcal{B}_{i,j}$ may be rewritten in either of the following two forms

$$\left\{y : \frac{P_H(j)f_{Y|H}(y|j)}{P_H(i)f_{Y|H}(y|i)} \geq 1\right\} = \left\{y : \sqrt{\frac{P_H(j)f_{Y|H}(y|j)}{P_H(i)f_{Y|H}(y|i)}} \geq 1\right\}$$

except that the above fraction is not defined when $f_{Y|H}(y|i)$ vanishes. This exception apart, we see that

$$\mathbb{1}\{y \in \mathcal{B}_{i,j}\} \leq \sqrt{\frac{P_H(j)f_{Y|H}(y|j)}{P_H(i)f_{Y|H}(y|i)}}$$

is true when y is inside $\mathcal{B}_{i,j}$; it is also true when outside because the left side vanishes and the right is never negative. We do not have to worry about the exception because we will use

$$f_{Y|H}(y|i)\mathbb{1}\{y \in \mathcal{B}_{i,j}\} \leq f_{Y|H}(y|i)\sqrt{\frac{P_H(j)f_{Y|H}(y|j)}{P_H(i)f_{Y|H}(y|i)}}$$

$$= \sqrt{\frac{P_H(j)}{P_H(i)}}\sqrt{f_{Y|H}(y|i)f_{Y|H}(y|j)},$$

which is obviously true when $f_{Y|H}(y|i)$ vanishes.

We are now ready to derive the Bhattacharyya bound:

$$Pr\{Y \in \mathcal{B}_{i,j}|H = i\} = \int_{y \in \mathcal{B}_{i,j}} f_{Y|H}(y|i)dy$$

$$= \int_{y \in \mathbb{R}^n} f_{Y|H}(y|i)\mathbb{1}\{y \in \mathcal{B}_{i,j}\}dy$$

$$\leq \sqrt{\frac{P_H(j)}{P_H(i)}} \int_{y \in \mathbb{R}^n} \sqrt{f_{Y|H}(y|i)f_{Y|H}(y|j)}\, dy. \qquad (2.17)$$

What makes the last integral appealing is that we integrate over the entire $\mathbb{R}^n$. The above bound takes a particularly simple form when there are only two hypotheses of equal probability. In this case,

$$P_e(0) = P_e(1) = P_e \leq \int_{y \in \mathbb{R}^n} \sqrt{f_{Y|H}(y|0)f_{Y|H}(y|1)}\, dy. \qquad (2.18)$$

As shown in Exercise 2.32, for *discrete memoryless channels* the bound further simplifies.

---

[4] There are two versions of the Bhattacharyya bound. Here we derive the one that has the simpler derivation. The other version, which is tighter by a factor 2, is derived in Exercises 2.29 and 2.30.

As the name indicates, the *union Bhattacharyya bound* combines (2.16) and (2.17), namely

$$P_e(i) \leq \sum_{j:j\neq i} Pr\{Y \in \mathcal{B}_{i,j} | H = i\} \leq \sum_{j:j\neq i} \sqrt{\frac{P_H(j)}{P_H(i)}} \int_{y \in \mathbb{R}^n} \sqrt{f_{Y|H}(y|i) f_{Y|H}(y|j)} \, dy.$$

We can now remove the conditioning on $H = i$ and obtain

$$P_e \leq \sum_i \sum_{j:j\neq i} \sqrt{P_H(i) P_H(j)} \int_{y \in \mathbb{R}^n} \sqrt{f_{Y|H}(y|i) f_{Y|H}(y|j)} \, dy. \tag{2.19}$$

EXAMPLE 2.16 (Tightness of the Bhattacharyya bound) *Let the message $H \in \{0, 1\}$ be equiprobable, let the channel be the binary erasure channel described in Figure 2.15, and let $c_i = (i, i, \ldots, i)^\mathsf{T}$, $i \in \{0, 1\}$.*

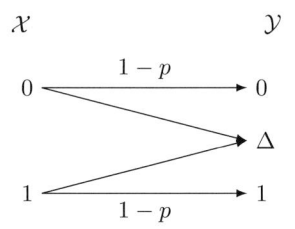

Figure 2.15. Binary erasure channel.

The Bhattacharyya bound for this case yields

$$Pr\{Y \in \mathcal{B}_{0,1} | H = 0\} \leq \sum_{y \in \{0,1,\Delta\}^n} \sqrt{P_{Y|H}(y|1) P_{Y|H}(y|0)}$$

$$= \sum_{y \in \{0,1,\Delta\}^n} \sqrt{P_{Y|X}(y|c_1) P_{Y|X}(y|c_0)}$$

$$\stackrel{(a)}{=} p^n,$$

where in (a) *we used the fact that the first factor under the square root vanishes if y contains 0s and the second vanishes if y contains 1s. Hence the only non-vanishing term in the sum is the one for which $y_i = \Delta$ for all i. The same bound applies for $H = 1$. Hence $P_e \leq \frac{1}{2}p^n + \frac{1}{2}p^n = p^n$.*

If we use the tighter version of the union Bhattacharyya bound, which as mentioned earlier is tighter by a factor of 2, then we obtain

$$P_e \leq \frac{1}{2} p^n.$$

For the binary erasure channel and the two codewords c_0 and c_1 we can actually compute the exact probability of error. An error can occur only if $Y = (\Delta, \Delta, \ldots, \Delta)^\mathsf{T}$, and in this case it occurs with probability $\frac{1}{2}$. Hence,

$$P_e = \frac{1}{2} Pr\{Y = (\Delta, \Delta, \ldots, \Delta)^\mathsf{T}\} = \frac{1}{2} p^n.$$

The Bhattacharyya bound is tight for the scenario considered in this example! □

2.7 Summary

The maximum a posteriori probability (MAP) rule is a decision rule that does exactly what the name implies – it maximizes the a posteriori probability – and in so doing it maximizes the probability that the decision is correct. With hindsight, the key idea is quite simple and it applies even when there is no observable. Let us review it.

Assume that a coin is flipped and we have to guess the outcome. We model the coin by the random variable $H \in \{0, 1\}$. All we know is $P_H(0)$ and $P_H(1)$. Suppose that $P_H(0) \leq P_H(1)$. Clearly we have the highest chance of being correct if we guess $\hat{H} = 1$ every time we perform the experiment of flipping the coin. We will be correct if indeed $H = 1$, and this has probability $P_H(1)$. More generally, for an arbitrary number m of hypotheses, we choose (one of) the i that maximizes $P_H(\cdot)$ and the probability of being correct is $P_H(i)$.

It is more interesting when there is some "side information". The side information is obtained when we observe the outcome of a related random variable Y. Once we have made the observation $Y = y$, our knowledge about the distribution of H gets updated from the prior distribution $P_H(\cdot)$ to the posterior distribution $P_{H|Y}(\cdot|y)$. What we have said in the previous paragraphs applies with the posterior instead of the prior.

In a typical example $P_H(\cdot)$ is constant whereas for the observed y, $P_{H|Y}(\cdot|y)$ is strongly biased in favor of one hypothesis. If it is strongly biased, the observable has been very informative, which is what we hope of course.

Often $P_{H|Y}$ is not given to us, but we can find it from P_H and $f_{Y|H}$ via Bayes' rule. Although $P_{H|Y}$ is the most fundamental quantity associated to a MAP test and therefore it would make sense to write the test in terms of $P_{H|Y}$, the test is typically written in terms of P_H and $f_{Y|H}$ because these are the quantities that are specified as part of the model.

Ideally a receiver performs a MAP decision. We have emphasized the case in which all hypotheses have the same probability as this is a common assumption in digital communication. Then the MAP and the ML rule are identical.

The following is an example of how the posterior becomes more and more selective as the number of observations increases. The example also shows that the posterior becomes less selective if the observations are more "noisy".

EXAMPLE 2.17 *Assume $H \in \{0, 1\}$ and $P_H(0) = P_H(1) = 1/2$. The outcome of H is communicated across a binary symmetric channel (BSC) of crossover probability $p < \frac{1}{2}$ via a transmitter that sends n 0s when $H = 0$ and n 1s when $H = 1$. The BSC has input alphabet $\mathcal{X} = \{0, 1\}$, output alphabet $\mathcal{Y} = \mathcal{X}$, and transition probability $p_{Y|X}(y|x) = \prod_{i=1}^{n} p_{Y|X}(y_i|x_i)$ where $p_{Y|X}(y_i|x_i)$ equals $1 - p$ if $y_i = x_i$ and p otherwise. (We obtain a BSC, for instance, if we place an appropriately chosen 1-bit quantizer at the output of the AWGN channel used with a binary* input alphabet.) Letting k be the number of 1s in the observed channel output y we have

$$P_{Y|H}(y|i) = \begin{cases} p^k(1-p)^{n-k}, & H=0 \\ p^{n-k}(1-p)^k, & H=1. \end{cases}$$

Using Bayes' rule,

$$P_{H|Y}(i|y) = \frac{P_{H,Y}(i,y)}{P_Y(y)} = \frac{P_H(i)P_{Y|H}(y|i)}{P_Y(y)},$$

where $P_Y(y) = \sum_i P_{Y|H}(y|i)P_H(i)$ is the normalization that ensures $\sum_i P_{H|Y}(i|y)$ equals 1.

Hence

$$P_{H|Y}(0|y) = \frac{p^k(1-p)^{n-k}}{2P_Y(y)} = \left(\frac{p}{1-p}\right)^k \frac{(1-p)^n}{2P_Y(y)}$$

$$P_{H|Y}(1|y) = \frac{p^{n-k}(1-p)^k}{2P_Y(y)} = \left(\frac{1-p}{p}\right)^k \frac{p^n}{2P_Y(y)}.$$

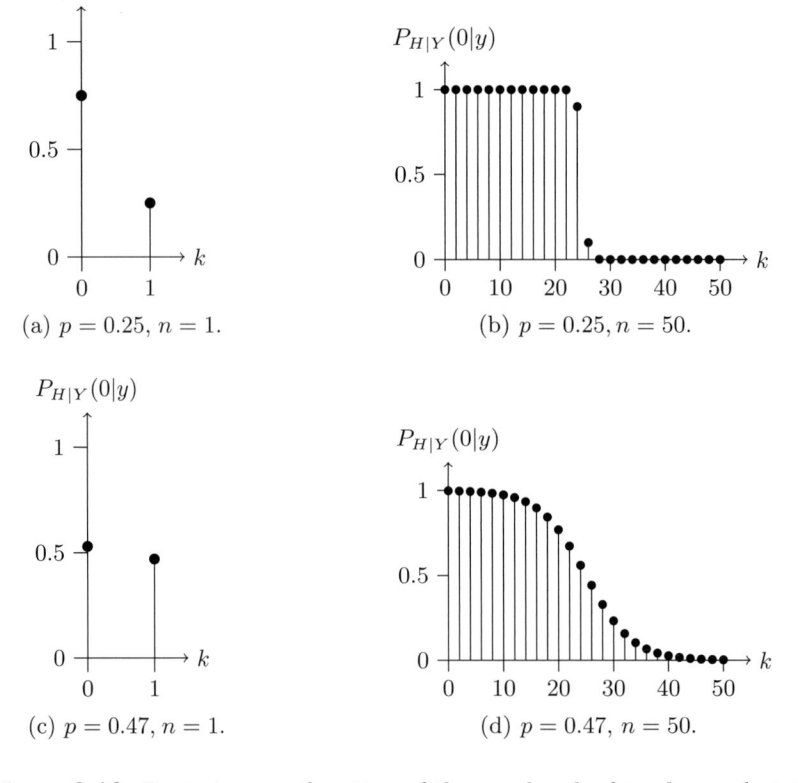

(a) $p = 0.25$, $n = 1$.

(b) $p = 0.25$, $n = 50$.

(c) $p = 0.47$, $n = 1$.

(d) $p = 0.47$, $n = 50$.

Figure 2.16. Posterior as a function of the number k of 1s observed at the output of a BSC of crossover probability p. The channel input consists of n 0s when $H = 0$ and of n 1s when $H = 1$.

Figure 2.16 depicts the behavior of $P_{H|Y}(0|y)$ as a function of the number k of 1s in y. For the top two figures, $p = 0.25$. We see that when $n = 50$ (top right figure), the prior is very biased in favor of one or the other hypothesis, unless the number k of observed 1s is nearly $n/2 = 25$. Comparing to $n = 1$ (top left figure), we see that many observations allow the receiver to make a more confident decision. This is true also for $p = 0.47$ (bottom row), but we see that with the crossover probability p close to $1/2$, there is a smoother transition between the region in favor of one hypothesis and the region in favor of the other. If we make only one observation (bottom left figure), then there is only a slight difference between the posterior for $H = 0$ and that for $H = 1$. This is the worse of the four cases (fewer observations through noisier channel). The best situation is of course the one of figure (b) (more observations through a more reliable channel). □

We have paid particular attention to the discrete-time AWGN channel as it will play an important role in subsequent chapters. The ML receiver for the AWGN channel is a minimum-distance decision rule. For signal constellations of the PAM and QAM family, the error probability can be computed exactly by means of the Q function. For other constellations, it can be upper bounded by means of the union bound and the Q function.

A quite general and useful technique to upper bound the probability of error is the union Bhattacharyya bound. Notice that it applies to MAP decisions associated to general hypothesis testing problems, not only to communication problems. The union Bhattacharyya bound just depends on $f_{Y|H}$ and P_H (no need to know the decoding regions).

Most current communications standards use PAM, QAM, or PSK to form their codewords. The basic idea is to form long codewords with components that belong to a PAM, QAM, or PSK set of points, called *alphabet*. Only a subset of the sequences that can be obtained this way are used as codewords. Chapter 6 is dedicated to a comprehensive case study. Here are a few concrete applications: 5-PAM is used as the underlying constellation for the Ethernet; QAM is used in telephone modems and in various digital video broadcasting standards (DVB-C2, DVB-T2); depending on the data rate, PSK and QAM are used in the wireless LAN (local area network) IEEE 802.11 standard, as well as in the third-generation partnership project (3GPP) and long-term evolution (LTE) standards; and PSK (with variations) is used in the Bluetooth 2, ZigBee, and EDGE standards.

2.8 Appendix: Facts about matrices

In this appendix we provide a summary of useful definitions and facts about matrices over $\mathbb{C}$. An excellent text about matrices is [12]. Hereafter $H^\dagger$ is the conjugate transpose of the matrix H. It is also called the *Hermitian adjoint* of H.

DEFINITION 2.18 *A matrix $U \in \mathbb{C}^{n \times n}$ is said to be* unitary *if $U^\dagger U = I$. If U is unitary and has real-valued entries, then it is said to be* orthogonal. □

The following theorem lists a number of handy facts about unitary matrices. Most of them are straightforward. Proofs can be found in [12, page 67].

THEOREM 2.19 *If $U \in \mathbb{C}^{n \times n}$, the following are equivalent:*

(a) *U is unitary;*
(b) *U is nonsingular and $U^\dagger = U^{-1}$;*
(c) *$UU^\dagger = I$;*
(d) *$U^\dagger$ is unitary;*
(e) *the columns of U form an orthonormal set;*
(f) *the rows of U form an orthonormal set; and*
(g) *for all $x \in \mathbb{C}^n$ the Euclidean length of $y = Ux$ is the same as that of x; that is, $y^\dagger y = x^\dagger x$.* □

The following is an important result that we use to prove Lemma 2.22.

THEOREM 2.20 (Schur) *Any square matrix A can be written as $A = URU^\dagger$ where U is unitary and R is an upper triangular matrix whose diagonal entries are the eigenvalues of A.* □

Proof Let us use induction on the size n of the matrix. The theorem is clearly true for $n = 1$. Let us now show that if it is true for $n - 1$, it follows that it is true for n. Given a matrix $A \in \mathbb{C}^{n \times n}$, let v be an eigenvector of unit norm, and β the corresponding eigenvalue. Let V be a unitary matrix whose first column is v. Consider the matrix $V^\dagger A V$. The first column of this matrix is given by $V^\dagger A v = \beta V^\dagger v = \beta e_1$, where e_1 is the unit vector along the first coordinate. Thus

$$V^\dagger A V = \begin{pmatrix} \beta & * \\ 0 & B \end{pmatrix},$$

where B is square and of dimension $n-1$. By the induction hypothesis $B = WSW^\dagger$, where W is unitary and S is upper triangular. Thus,

$$V^\dagger A V = \begin{pmatrix} \beta & * \\ 0 & WSW^\dagger \end{pmatrix} = \begin{pmatrix} 1 & 0 \\ 0 & W \end{pmatrix} \begin{pmatrix} \beta & * \\ 0 & S \end{pmatrix} \begin{pmatrix} 1 & 0 \\ 0 & W^\dagger \end{pmatrix} \qquad (2.20)$$

and putting

$$U = V \begin{pmatrix} 1 & 0 \\ 0 & W \end{pmatrix} \quad \text{and} \quad R = \begin{pmatrix} \beta & * \\ 0 & S \end{pmatrix},$$

we see that U is unitary, R is upper triangular, and $A = URU^\dagger$, completing the induction step. The eigenvalues of a matrix are the roots of the characteristic polynomial. To see that the diagonal entries of R are indeed the eigenvalues of A, it suffices to bring the characteristic polynomial of A in the following form: $\det(\lambda I - A) = \det \left[U(\lambda I - R)U^\dagger \right] = \det(\lambda I - R) = \prod_i (\lambda - r_{ii})$. □

DEFINITION 2.21 *A matrix $H \in \mathbb{C}^{n \times n}$ is said to be* Hermitian *if $H = H^\dagger$. It is said to be* skew-Hermitian *if $H = -H^\dagger$. If H is Hermitian and has real-valued entries, then it is said to be* symmetric. □

2.8. Appendix: Facts about matrices

LEMMA 2.22 *A Hermitian matrix $H \in \mathbb{C}^{n \times n}$ can be written as*

$$H = U\Lambda U^\dagger = \sum_i \lambda_i u_i u_i^\dagger,$$

where U is unitary and $\Lambda = \mathrm{diag}(\lambda_1, \ldots, \lambda_n)$ is a diagonal matrix that consists of the eigenvalues of H. Moreover, the eigenvalues are real and the ith column of U, u_i, is an eigenvector associated to λ_i. □

Proof By Theorem 2.20 (Schur) we can write $H = URU^\dagger$, where U is unitary and R is upper triangular with the diagonal elements consisting of the eigenvalues of H. From $R = U^\dagger H U$ we immediately see that R is Hermitian. Hence it must be a diagonal matrix and the diagonal elements must be real. If u_i is the ith column of U, then

$$Hu_i = U\Lambda U^\dagger u_i = U\Lambda e_i = U\lambda_i e_i = \lambda_i u_i,$$

showing that it is indeed an eigenvector associated to the ith eigenvalue λ_i. □

If $H \in \mathbb{C}^{n \times n}$ is Hermitian, then $u^\dagger H u$ is real for all $u \in \mathbb{C}^n$. Indeed, $[u^\dagger H u]^\dagger = u^\dagger H^\dagger u = u^\dagger H u$. Hence, if we compare the set $\mathbb{C}^{n \times n}$ of square matrices to the set $\mathbb{C}$ of complex numbers, then the subset of Hermitian matrices is analogous to the real numbers.

A class of Hermitian matrices with a special positivity property arises naturally in many applications, including communication theory. They are the analogs of the positive numbers.

DEFINITION 2.23 *A Hermitian matrix $H \in \mathbb{C}^{n \times n}$ is said to be* positive definite *if*

$$u^\dagger H u > 0 \quad \textit{for all non-zero} \quad u \in \mathbb{C}^n.$$

If H satisfies the weaker inequality $u^\dagger H u \geq 0$, then H is said to be positive semidefinite. □

For any matrix $A \in \mathbb{C}^{m \times n}$, the matrix $AA^\dagger$ is positive semidefinite. Indeed, for any vector $v \in \mathbb{C}^m$, $v^\dagger AA^\dagger v = \|A^\dagger v\|^2 \geq 0$. Similarly, a covariance matrix is positive semidefinite. To see why, let $X \in \mathbb{C}^m$ be a zero-mean random vector and let $K = \mathbb{E}\left[XX^\dagger\right]$ be its covariance matrix. For any $v \in \mathbb{C}^m$,

$$\begin{aligned} v^\dagger K v &= v^\dagger \mathbb{E}\left[XX^\dagger\right] v \\ &= \mathbb{E}\left[v^\dagger X X^\dagger v\right] \\ &= \mathbb{E}\left[v^\dagger X (v^\dagger X)^\dagger\right] \\ &= \mathbb{E}\left[|v^\dagger X|^2\right] \geq 0. \end{aligned}$$

LEMMA 2.24 (Eigenvalues of positive (semi)definite matrices) *The eigenvalues of a positive definite matrix are positive and those of a positive semidefinite matrix are non-negative.* □

Proof Let A be a positive definite matrix, let u be an eigenvector of A associated to the eigenvalue λ, and calculate $u^\dagger A u = u^\dagger \lambda u = \lambda u^\dagger u$. Hence $\lambda = u^\dagger A u / u^\dagger u$, proving that λ must be positive, because it is the ratio of two positive numbers. If A is positive semidefinite, then the numerator of $\lambda = u^\dagger A u / u^\dagger u$ can vanish. □

THEOREM 2.25 (SVD) *Any matrix $A \in \mathbb{C}^{m \times n}$ can be written as a product*

$$A = UDV^\dagger,$$

where U and V are unitary (of dimension $m \times m$ and $n \times n$, respectively) and $D \in \mathbb{R}^{m \times n}$ is non-negative and diagonal. This is called the singular value decomposition (SVD) *of A. Moreover, by letting k be the rank of A, the following statements are true.*

(a) *The columns of V are the eigenvectors of $A^\dagger A$. The last $n - k$ columns span the null space of A.*
(b) *The columns of U are eigenvectors of $AA^\dagger$. The first k columns span the range of A.*
(c) *If $m \geq n$ then*

$$D = \begin{pmatrix} \mathrm{diag}(\sqrt{\lambda_1}, \ldots, \sqrt{\lambda_n}) \\ \cdots\cdots\cdots\cdots\cdots\cdots \\ 0_{(n-m)\times n} \end{pmatrix},$$

where $\lambda_1 \geq \lambda_2 \geq \cdots \geq \lambda_k > \lambda_{k+1} = \cdots = \lambda_n = 0$ are the eigenvalues of $A^\dagger A \in \mathbb{C}^{n \times n}$ which are non-negative because $A^\dagger A$ is positive semidefinite.
(d) *If $m \leq n$ then*

$$D = \left(\mathrm{diag}(\sqrt{\lambda_1}, \ldots, \sqrt{\lambda_m}) : 0_{m \times (n-m)} \right),$$

where $\lambda_1 \geq \lambda_2 \geq \cdots \geq \lambda_k > \lambda_{k+1} = \cdots = \lambda_m = 0$ are the eigenvalues of $AA^\dagger$. □

Note 1: Recall that the set of non-zero eigenvalues of AB equals the set of non-zero eigenvalues of BA, see e.g. [12, Theorem 1.3.29]. Hence the non-zero eigenvalues in (c) and (d) are the same.

Note 2: To remember that V contains the eigenvectors of $A^\dagger A$ (as opposed to containing those of $AA^\dagger$) it suffices to look at the dimensions: V has to be an $n \times n$ matrix, and so is $A^\dagger A$.

Proof It is sufficient to consider the case with $m \geq n$ because if $m < n$, we can apply the result to $A^\dagger = UDV^\dagger$ and obtain $A = VD^\dagger U^\dagger$. Hence let $m \geq n$, and consider the matrix $A^\dagger A \in \mathbb{C}^{n \times n}$. This matrix is positive semidefinite. Hence its eigenvalues $\lambda_1 \geq \lambda_2 \geq \cdots \lambda_n \geq 0$ are real and non-negative and we can choose the eigenvectors $v_1, v_2, \ldots, v_n$ to form an orthonormal basis for $\mathbb{C}^n$. Let $V = (v_1, \ldots, v_n)$. Let k be the number of positive eigenvalues and choose

$$u_i = \frac{1}{\sqrt{\lambda_i}} A v_i, \quad i = 1, 2, \ldots, k. \tag{2.21}$$

2.8. Appendix: Facts about matrices

Observe that

$$u_i^\dagger u_j = \frac{1}{\sqrt{\lambda_i \lambda_j}} v_i^\dagger A^\dagger A v_j = \sqrt{\frac{\lambda_j}{\lambda_i}} v_i^\dagger v_j = \delta_{ij}, \quad 1 \leq i, j \leq k.$$

Hence $\{u_i : i = 1, \ldots, k\}$ is a set of orthonormal vectors. Complete this set of vectors to an orthonormal basis for $\mathbb{C}^m$ by choosing $\{u_i : i = k+1, \ldots, m\}$ and let $U = (u_1, u_2, \ldots, u_m)$. Note that (2.21) implies

$$u_i \sqrt{\lambda_i} = A v_i, \quad i = 1, 2, \ldots, k, k+1, \ldots, n,$$

where for $i = k+1, \ldots, n$ the above relationship holds since $\lambda_i = 0$ and v_i is a corresponding eigenvector. Using matrix notation we obtain

$$U \begin{pmatrix} \sqrt{\lambda_1} & & 0 \\ & \ddots & \\ 0 & & \sqrt{\lambda_n} \\ \multicolumn{3}{c}{\dotfill} \\ & 0 & \end{pmatrix} = AV, \tag{2.22}$$

i.e. $A = UDV^\dagger$. For $i = 1, 2, \ldots, m$,

$$AA^\dagger u_i = UDV^\dagger V D^\dagger U^\dagger u_i$$
$$= UDD^\dagger U^\dagger u_i = u_i \lambda_i,$$

where in the last equality we use the fact that $U^\dagger u_i$ contains 1 at position i and 0 elsewhere, and $DD^\dagger = \operatorname{diag}(\lambda_1, \lambda_2, \ldots, \lambda_k, 0, \ldots, 0)$. This shows that λ_i is also an eigenvalue of $AA^\dagger$. We have also shown that $\{v_i : i = k+1, \ldots, n\}$ spans the null space of A and from (2.22) we see that $\{u_i : i = 1, \ldots, k\}$ spans the range of A. □

The following key result is a simple application of the SVD.

LEMMA 2.26 *The linear transformation described by a matrix $A \in \mathbb{R}^{n \times n}$ maps the unit cube into a parallelepiped of volume $|\det A|$.* □

Proof From the singular value decomposition, we can write $A = UDV^\dagger$, where D is diagonal and U and V are unitary matrices. A transformation described by a unitary matrix is volume preserving. (In fact if we apply an orthogonal matrix to an object, we obtain the same object described in a new coordinate system.) Hence we can focus our attention on the effect of D. But D maps the unit vectors $e_1, e_2, \ldots, e_n$ into $\lambda_1 e_1, \lambda_2 e_2, \ldots, \lambda_n e_n$, respectively. Therefore it maps the unit square into a rectangular parallelepiped of sides $\lambda_1, \lambda_2, \ldots, \lambda_n$ and of volume $|\prod \lambda_i| = |\det D| = |\det A|$, where the last equality holds because the determinant of a product (of matrices) is the product of the determinants and the determinant of a unitary matrix has unit magnitude. □

2.9 Appendix: Densities after one-to-one differentiable transformations

In this appendix we outline how to determine the density of a random vector Y when we know the density of a random vector X and $Y = g(X)$ for some differentiable and one-to-one function g.

We begin with the scalar case. Generalizing to the vector case is conceptually straightforward. Let $X \in \mathcal{X}$ be a random variable of density f_X and define $Y = g(X)$ for a given one-to-one differentiable function $g : \mathcal{X} \to \mathcal{Y}$. The density becomes useful when we integrate it over some set $\mathcal{A}$ to obtain the probability that $X \in \mathcal{A}$. In Figure 2.17 the shaded area "under" f_X equals $Pr\{X \in \mathcal{A}\}$. Now assume that g maps the interval $\mathcal{A}$ into the interval $\mathcal{B}$. Then $X \in \mathcal{A}$ if and only if $Y \in \mathcal{B}$. Hence $Pr\{X \in \mathcal{A}\} = Pr\{Y \in \mathcal{B}\}$, which means that the two shaded areas in the figure must be identical. This requirement completely specifies f_Y.

For the mathematical details we need to consider an infinitesimally small interval $\mathcal{A}$. Then $Pr\{X \in \mathcal{A}\} = f_X(\bar{x})l(\mathcal{A})$, where $l(\mathcal{A})$ denotes the length of $\mathcal{A}$ and $\bar{x}$ is any point in $\mathcal{A}$. Similarly, $Pr\{Y \in \mathcal{B}\} = f_Y(\bar{y})l(\mathcal{B})$ where $\bar{y} = g(\bar{x})$. Hence f_Y fulfills $f_X(\bar{x})l(\mathcal{A}) = f_Y(\bar{y})l(\mathcal{B})$.

The last ingredient is the fact that the absolute value of the slope of g at $\bar{x}$ is the ratio $\frac{l(\mathcal{B})}{l(\mathcal{A})}$. (We are still assuming infinitesimally small intervals.) Hence $f_Y(y)|g'(x)| = f_X(x)$ and after solving for $f_Y(y)$ and using $x = g^{-1}(y)$ we obtain the desired result

$$f_Y(y) = \frac{f_X(g^{-1}(y))}{|g'(g^{-1}(y))|}. \qquad (2.23)$$

EXAMPLE 2.27 *If $g(x) = ax + b$ then $f_Y(y) = \frac{f_X(\frac{y-b}{a})}{|a|}.$* □

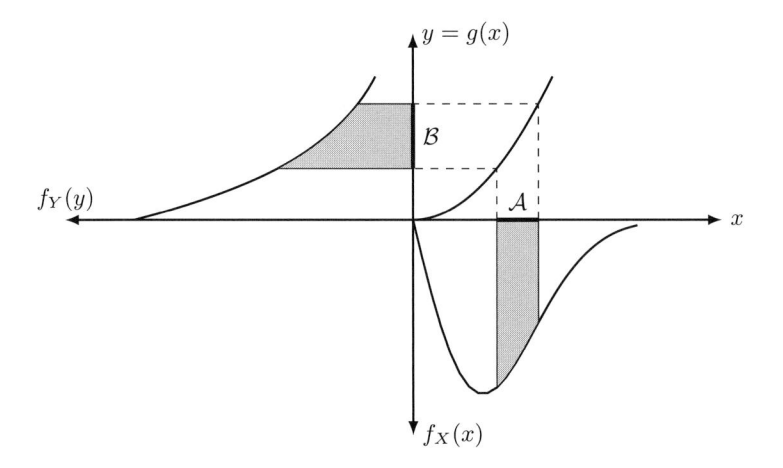

Figure 2.17. Finding the density of $Y = g(X)$ from that of X. Shaded surfaces have the same area.

2.9. Appendix: Densities after one-to-one differentiable transformations

EXAMPLE 2.28   Let f_X be Rayleigh, specifically

$$f_X(x) = \begin{cases} xe^{-\frac{x^2}{2}}, & x \geq 0 \\ 0, & \text{otherwise} \end{cases}$$

and let $Y = g(X) = X^2$. Then

$$f_Y(y) = \begin{cases} 0.5e^{-\frac{y}{2}}, & y \geq 0 \\ 0, & \text{otherwise.} \end{cases} \qquad \square$$

Next we consider the multidimensional case, starting with two dimensions. Let $X = (X_1, X_2)^\mathsf{T}$ have pdf $f_X(x)$ and consider first the random vector Y obtained from the affine transformation

$$Y = AX + b$$

for some nonsingular matrix A and vector b. The procedure to determine f_Y parallels that for the scalar case. If $\mathcal{A}$ is a small rectangle, small enough that $f_X(x)$ can be considered as constant for all $x \in \mathcal{A}$, then $Pr\{X \in \mathcal{A}\}$ is approximated by $f_X(\bar{x})a(\mathcal{A})$, where $a(\mathcal{A})$ is the area of $\mathcal{A}$ and $\bar{x} \in \mathcal{A}$. If $\mathcal{B}$ is the image of $\mathcal{A}$, then

$$f_Y(\bar{y})a(\mathcal{B}) \to f_X(\bar{x})a(\mathcal{A}) \text{ as } a(\mathcal{A}) \to 0.$$

Hence

$$f_Y(\bar{y}) \to f_X(\bar{x})\frac{a(\mathcal{A})}{a(\mathcal{B})} \text{ as } a(\mathcal{A}) \to 0.$$

For the next and final step, we need to know that A maps $\mathcal{A}$ of area $a(\mathcal{A})$ into $\mathcal{B}$ of area $a(\mathcal{B}) = a(\mathcal{A})|\det A|$. So the absolute value of the determinant of a matrix is the amount by which areas scale through the affine transformation associated to the matrix. This is true in any dimension n, but for $n = 1$ we speak of length rather than area and for $n \geq 3$ we speak of volume. (For the one-dimensional case, observe that the determinant of the scalar a is a). See Lemma 2.26 (in Appendix 2.8) for an outline of the proof of this important geometrical interpretation of the determinant of a matrix. Hence

$$f_Y(y) = \frac{f_X(A^{-1}(y - b))}{|\det A|}.$$

We are ready to generalize to a function $g : \mathbb{R}^n \to \mathbb{R}^n$ which is one-to-one and differentiable. Write $g(x) = (g_1(x), \ldots, g_n(x))$ and define its *Jacobian* $J(x)$ to be the matrix that has $\frac{\partial g_i}{\partial x_j}$ at position (i, j). In the neighborhood of x the relationship $y = g(x)$ may be approximated by means of an affine expression of the form

$$y = Ax + b,$$

where A is precisely the Jacobian $J(x)$. Hence, leveraging on the affine case, we can immediately conclude that

$$f_Y(y) = \frac{f_X(g^{-1}(y))}{|\det J(g^{-1}(y))|}, \qquad (2.24)$$

which holds for any n.

Sometimes the new random vector Y is described by the inverse function, namely $X = g^{-1}(Y)$ (rather than the other way around, as assumed so far). In this case there is no need to find g. The determinant of the Jacobian of g at x is one over the determinant of the Jacobian of g^{-1} at $y = g(x)$.

As a final note we mention that if g is a many-to-one map, then for a specific y the pull-back $g^{-1}(y)$ will be a set $\{x_1, \ldots, x_k\}$ for some k. In this case the right side of (2.24) will be $\sum_i \frac{f_X(x_i)}{|\det J(x_i)|}$.

EXAMPLE 2.29 (Rayleigh distribution) *Let X_1 and X_2 be two independent, zero-mean, unit-variance, Gaussian random variables. Let R and Θ be the corresponding polar coordinates, i.e. $X_1 = R\cos\Theta$ and $X_2 = R\sin\Theta$. We are interested in the probability density functions $f_{R,\Theta}$, f_R, and f_Θ. Because we are given the map g from (r,θ) to (x_1, x_2), we pretend that we know $f_{R,\Theta}$ and that we want to find f_{X_1, X_2}. Thus*

$$f_{X_1, X_2}(x_1, x_2) = \frac{1}{|\det J|} f_{R,\Theta}(r, \theta),$$

where J is the Jacobian of g, namely

$$J = \begin{pmatrix} \cos\theta & -r\sin\theta \\ \sin\theta & r\cos\theta \end{pmatrix}.$$

Hence $\det J = r$ and

$$f_{X_1, X_2}(x_1, x_2) = \frac{1}{r} f_{R,\theta}(r, \theta).$$

Using $f_{X_1, X_2}(x_1, x_2) = \frac{1}{2\pi} \exp\left\{-\frac{x_1^2 + x_2^2}{2}\right\}$ and $x_1^2 + x_2^2 = r^2$, we obtain

$$f_{R,\theta}(r, \theta) = \frac{r}{2\pi} \exp\left\{-\frac{r^2}{2}\right\}.$$

Since $f_{R,\Theta}(r, \theta)$ depends only on r, we infer that R and Θ are independent random variables and that Θ is uniformly distributed in $[0, 2\pi)$. Hence

$$f_\Theta(\theta) = \begin{cases} \frac{1}{2\pi} & \theta \in [0, 2\pi) \\ 0 & \text{otherwise} \end{cases}$$

and

$$f_R(r) = \begin{cases} re^{-\frac{r^2}{2}} & r \geq 0 \\ 0 & \text{otherwise.} \end{cases}$$

We come to the same conclusion by integrating $f_{R,\Theta}$ over θ to obtain f_R and by integrating over r to obtain f_Θ. Notice that f_R is a Rayleigh probability density. It is often used to evaluate the error probability of a wireless link subject to fading. □

2.10 Appendix: Gaussian random vectors

A Gaussian random vector is a collection of jointly Gaussian random variables. We learn to use vector notation as it simplifies matters significantly.

Recall that a random variable $W : \Omega \to \mathbb{R}$ is a mapping from the sample space to the reals. W is a Gaussian random variable with mean m and variance σ^2 if and only if its probability density function (pdf) is

$$f_W(w) = \frac{1}{\sqrt{2\pi\sigma^2}} \exp\left\{-\frac{(w-m)^2}{2\sigma^2}\right\}.$$

Because a Gaussian random variable is completely specified by its mean m and variance σ^2, we use the short-hand notation $\mathcal{N}(m, \sigma^2)$ to denote its pdf. Hence $W \sim \mathcal{N}(m, \sigma^2)$.

An *n-dimensional random vector* X is a mapping $X : \Omega \to \mathbb{R}^n$. It can be seen as a collection $X = (X_1, X_2, \ldots, X_n)^\mathsf{T}$ of n random variables. The pdf of X is the joint pdf of $X_1, X_2, \ldots, X_n$. The expected value of X, denoted by $\mathbb{E}[X]$, is the n-tuple $(\mathbb{E}[X_1], \mathbb{E}[X_2], \ldots, \mathbb{E}[X_n])^\mathsf{T}$. The *covariance matrix* of X is $K_X = \mathbb{E}[(X - \mathbb{E}[X])(X - \mathbb{E}[X])^\mathsf{T}]$. Notice that XX^T is an $n \times n$ random matrix, i.e. a matrix of random variables, and the expected value of such a matrix is, by definition, the matrix whose components are the expected values of those random variables. A covariance matrix is always Hermitian. This follows immediately from the definitions.

The pdf of a vector $W = (W_1, W_2, \ldots, W_n)^\mathsf{T}$ that consists of independent and identically distributed (iid) $\sim \mathcal{N}(0, 1)$ components is

$$f_W(w) = \prod_{i=1}^{n} \frac{1}{\sqrt{2\pi}} \exp\left(-\frac{w_i^2}{2}\right) \tag{2.25}$$

$$= \frac{1}{(2\pi)^{n/2}} \exp\left(-\frac{w^\mathsf{T} w}{2}\right). \tag{2.26}$$

DEFINITION 2.30 *The random vector $Z = (Z_1, \ldots, Z_m)^\mathsf{T} \in \mathbb{R}^m$ is a zero-mean Gaussian random vector and $Z_1, \ldots, Z_m$ are zero-mean* jointly *Gaussian random variables if and only if there exists a matrix $A \in \mathbb{R}^{m \times n}$ such that Z can be expressed as*

$$Z = AW, \tag{2.27}$$

where $W \in \mathbb{R}^n$ is a random vector of iid $\sim \mathcal{N}(0, 1)$ components. □

More generally, $Z = AW + m$, $m \in \mathbb{R}^m$, is a Gaussian random vector of mean m. We focus on zero-mean Gaussian random vectors since we can always add or remove the mean as needed.

It follows immediately from the above definition that linear combinations of zero-mean jointly Gaussian random variables are zero-mean jointly Gaussian random variables. Indeed, $BZ = BAW$, where the right-hand side is as in (2.27) with the matrix BA instead of A.

Recall from Appendix 2.9 that, if $Z = AW$ for some nonsingular matrix $A \in \mathbb{R}^{n \times n}$, then

$$f_Z(z) = \frac{f_W(A^{-1}z)}{|\det A|}.$$

Using (2.26) we obtain

$$f_Z(z) = \frac{\exp\left(-\frac{(A^{-1}z)^\mathsf{T}(A^{-1}z)}{2}\right)}{(2\pi)^{n/2}|\det A|}.$$

The above expression can be written using K_Z instead of A. Indeed from

$$K_Z = \mathbb{E}[AW(AW)^\mathsf{T}] = \mathbb{E}[AWW^\mathsf{T}A^\mathsf{T}] = AI_nA^\mathsf{T} = AA^\mathsf{T}$$

we obtain

$$\begin{aligned}(A^{-1}z)^\mathsf{T}(A^{-1}z) &= z^\mathsf{T}(A^{-1})^\mathsf{T}A^{-1}z \\ &= z^\mathsf{T}(AA^\mathsf{T})^{-1}z \\ &= z^\mathsf{T}K_Z^{-1}z\end{aligned}$$

and

$$\sqrt{\det K_Z} = \sqrt{\det AA^\mathsf{T}} = \sqrt{\det A \det A} = |\det A|.$$

Thus

$$f_Z(z) = \frac{1}{\sqrt{(2\pi)^n \det K_Z}} \exp\left(-\frac{1}{2}z^\mathsf{T}K_Z^{-1}z\right) \tag{2.28a}$$

$$= \frac{1}{\sqrt{\det(2\pi K_Z)}} \exp\left(-\frac{1}{2}z^\mathsf{T}K_Z^{-1}z\right). \tag{2.28b}$$

The above densities specialize to (2.26) when $K_Z = I_n$.

EXAMPLE 2.31 *Let $W \in \mathbb{R}^n$ have iid $\sim \mathcal{N}(0,1)$ components and define $Z = UW$ for some orthogonal matrix U. Then Z is zero-mean, Gaussian, and its covariance matrix is $K_Z = UU^\mathsf{T} = I_n$. Hence Z has the same distribution as W.* □

If $A \in \mathbb{R}^{m \times n}$ (hence not necessarily square) *with linearly independent rows* and $Z = AW$, then we can find an $m \times m$ nonsingular matrix $\tilde{A}$ and write $Z = \tilde{A}\tilde{W}$ for a Gaussian random vector $\tilde{W} \in \mathbb{R}^m$ with iid $\sim \mathcal{N}(0,1)$ components. To see this, we use the SVD (Appendix 2.8) to write $A = UDV^\mathsf{T}$. Now

$$Z = UDV^\mathsf{T}W = UDW,$$

where (with a slight abuse of notation) we have substituted W for $V^\mathsf{T}W$ because they have the same distribution (Example 2.31). The $m \times n$ diagonal matrix D can be written as $[\tilde{D} : 0]$, where $\tilde{D}$ is an $m \times m$ diagonal matrix with positive diagonal elements and 0 is the $m \times (n-m)$ zero-matrix. Hence $DW = \tilde{D}\tilde{W}$, where $\tilde{W}$ consists of the first m components of W. Thus $Z = U\tilde{D}\tilde{W} = \tilde{A}\tilde{W}$ with $\tilde{A} = U\tilde{D}$ nonsingular because U is orthogonal and $\tilde{D}$ is nonsingular.

2.10. Appendix: Gaussian random vectors

If A has *linearly dependent* rows, then $K_Z = AA^\mathsf{T}$ is singular. In fact, A^T has linearly dependent columns and we cannot recover $x \in \mathbb{R}^m$ from $A^\mathsf{T} x$, let alone from $K_Z x = AA^\mathsf{T} x$. In this case the random vector Z is still Gaussian but it is not possible to write its pdf as in (2.28), and we say that its pdf does not exist.[5] This is not a problem because on the rare occasions when we encounter such a case, we can generally find a workaround that goes as follows. Typically we want to know the density of a random vector Z so that we can determine the probability of an event such as $Z \in \mathcal{B}$. If A has linearly dependent rows, then (with probability 1) some of the components of Z are linear combinations of some of the other components. There is always a way to write the event $Z \in \mathcal{B}$ in terms of an event that involves only a linearly independent subset of the $Z_1, \ldots, Z_m$. This subset forms a Gaussian random vector of nonsingular covariance matrix, for which the density is defined as in (2.28). An example follows. (For more on degenerate cases see e.g. [3, Section 23.4.3].)

EXAMPLE 2.32 (Degenerate case) *Let $W \sim \mathcal{N}(0,1)$, $A = (1,1)^\mathsf{T}$, and $Z = AW$. By our definition, Z is a Gaussian random vector. However, A is a matrix of linearly dependent rows implying that Z has linearly dependent components. Indeed $Z_1 = Z_2$. We can easily check that K_Z is the 2×2 matrix with 1 in each position, hence it is singular and f_Z is not defined. How do we compute the probability of events involving Z if we do not know its pdf? Any event involving Z can be rewritten as an event involving Z_1 only (or equivalently involving Z_2 only). For instance, the event $\{Z_1 \in [1,3]\} \cap \{Z_2 \in [2,5]\}$ occurs if and only if $\{Z_1 \in [2,3]\}$ (see Figure 2.18). Hence*

$$Pr\{\{Z_1 \in [1,3]\} \cap \{Z_2 \in [2,5]\}\} = Pr\{Z_1 \in [2,3]\} = Q(2) - Q(3).$$

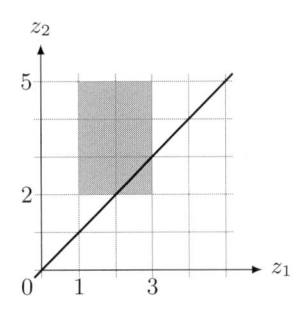

Figure 2.18.

□

Next we show that if a random vector has density as in (2.28), then it can be written as in (2.27). Let $Z \in \mathbb{R}^m$ be such a random vector and let K_Z be its

---

[5] It is possible to play tricks and define a function that can be considered as being the density of a Gaussian random vector of singular covariance matrix. But what we gain in doing so is not worth the trouble.

nonsingular covariance matrix. As a covariance matrix is Hermitian, we can write (see Appendix 2.8)

$$K_Z = U\Lambda U^\dagger, \tag{2.29}$$

where U is unitary and Λ is diagonal. Because K_Z is nonsingular, all diagonal elements of Λ must be positive. Define $W = \Lambda^{-\frac{1}{2}}U^\dagger Z$, where for $\alpha \in \mathbb{R}$, Λ^α is the diagonal matrix obtained by raising the diagonal elements of Λ to the power α. Then $Z = U\Lambda^{\frac{1}{2}}W = AW$ with $A = U\Lambda^{\frac{1}{2}}$ nonsingular. It remains to be shown that $f_W(w)$ is as on the right-hand side of (2.26). It must be, because the transformation from $\mathbb{R}^n$ to $\mathbb{R}^n$ that sends W to $Z = AW$ is one-to-one. Hence the density of $f_W(w)$ that leads to $f_Z(z)$ is unique. It must be (2.26), because (2.28) was obtained from (2.26) assuming $Z = AW$.

Many authors use (2.28) to define a Gaussian random vector. We favor (2.27) because it is more general (it does not depend on the covariance being nonsingular), and because from this definition it is straightforward to prove a number of key results associated to Gaussian random vectors. Some of these are dealt with in the examples that follow.

EXAMPLE 2.33 *The ith component Z_i of a Gaussian random vector $Z = (Z_1, \ldots, Z_m)^\mathsf{T}$ is a Gaussian random variable. This is an immediate consequence of Definition 2.30. In fact, $Z_i = AZ$, where A is the row vector that has 1 at position i and 0 elsewhere. To appreciate the convenience of working with (2.27) instead of (2.28), compare this answer with the tedious derivation that consists of integrating over f_Z to obtain f_{Z_i} (see Exercise 2.37).* □

EXAMPLE 2.34 (Gaussian random variables are not necessarily jointly Gaussian) *Let $Y_1 \sim \mathcal{N}(0,1)$, let $X \in \{\pm 1\}$ be uniformly distributed, and let $Y_2 = Y_1 X$. Notice that Y_2 has the same pdf as Y_1. This follows from the fact that the pdf of Y_1 is an even function. Hence Y_1 and Y_2 are both Gaussian. However, they are not jointly Gaussian. We come to this conclusion by observing that $Y = Y_1 + Y_2 = Y_1(1+X)$ is 0 with probability 1/2. Hence Y cannot be Gaussian.* □

EXAMPLE 2.35 *Two random variables are said to be* uncorrelated *if their covariance is 0. Uncorrelated Gaussian random variables are not necessarily independent. For instance, Y_1 and Y_2 of Example 2.34 are uncorrelated Gaussian random variables, yet they are not independent. However, uncorrelated jointly Gaussian random variables are always independent. This follows immediately from the pdf (2.28). The contrary is always true: random variables (not necessarily Gaussian) that are independent are always uncorrelated. The proof is straightforward.* □

The shorthand $Z \sim \mathcal{N}(m, K_Z)$ means that Z is a Gaussian random vector of mean m and covariance matrix K_Z. If K_Z is nonsingular, then

$$f_Z(z) = \frac{1}{\sqrt{(2\pi)^n \det K_Z}} \exp\left(-\frac{1}{2}(z - \mathbb{E}[Z])^\mathsf{T} K_Z^{-1}(z - \mathbb{E}[Z])\right).$$

2.11 Appendix: A fact about triangles

In Example 2.15 we derive the error probability for PSK by using the fact that for a triangle with edges a, b, c and angles α, β, γ as shown in the figure below, the following relationship holds:

$$\frac{a}{\sin\alpha} = \frac{b}{\sin\beta} = \frac{c}{\sin\gamma}. \qquad (2.30)$$

$$a\sin\beta = b\sin(\pi - \alpha)$$

To prove the first equality relating a and b we consider the distance between the vertex γ (common to a and b) and its projection onto the extension of c. As shown in the figure, this distance may be computed in two ways obtaining $a\sin\beta$ and $b\sin(\pi - \alpha)$, respectively. The latter may be written as $b\sin(\alpha)$. Hence $a\sin\beta = b\sin(\alpha)$, which is the first equality. The second equality is proved similarly. □

2.12 Appendix: Inner product spaces

2.12.1 Vector space

Most readers are familiar with the notion of vector space from a linear algebra course. Unfortunately, some linear algebra courses for engineers associate vectors to n-tuples rather than taking the axiomatic point of view – which is what we need. A *vector space* (or linear space) consists of the following (see e.g. [10, 11] for more).

(1) A field $\mathbb{F}$ of scalars.[6]
(2) A set $\mathcal{V}$ of objects called *vectors*.[7]
(3) An operation called vector addition, which associates with each pair of vectors α and β in $\mathcal{V}$ a vector $\alpha + \beta$ in $\mathcal{V}$, in such a way that
  (i) it is commutative: $\alpha + \beta = \beta + \alpha$;
  (ii) it is associative: $\alpha + (\beta + \gamma) = (\alpha + \beta) + \gamma$ for every α, β, γ in $\mathcal{V}$;
  (iii) there is a unique vector, called the zero vector and denoted by 0, such that $\alpha + 0 = \alpha$ for all α in $\mathcal{V}$;
  (iv) for each α in $\mathcal{V}$, there is a β in $\mathcal{V}$ such that $\alpha + \beta = 0$.
(4) An operation called scalar multiplication, which associates with each vector α in $\mathcal{V}$ and each scalar a in $\mathbb{F}$ a vector $a\alpha$ in $\mathcal{V}$, in such a way that
  (i) $1\alpha = \alpha$ for every α in $\mathcal{V}$;
  (ii) $(a_1 a_2)\alpha = a_1(a_2\alpha)$ for every a_1, a_2 in $\mathbb{F}$;

---

[6] In this book the field is almost exclusively $\mathbb{R}$ (the field of real numbers) or $\mathbb{C}$ (the field of complex numbers). In Chapter 6, where we talk about coding, we also work with the field $\mathbb{F}_2$ of binary numbers.
[7] We are concerned with two families of vectors: n-tuples and functions.

(iii) $a(\alpha + \beta) = a\alpha + a\beta$ for every α, β in $\mathcal{V}$;
(iv) $(a + b)\alpha = a\alpha + b\alpha$ for every $a, b \in \mathbb{F}$.

In this appendix we consider general vector spaces for which the scalar field is $\mathbb{C}$. They are commonly called *complex vector spaces*. Vector spaces for which the scalar field is $\mathbb{R}$ are called *real vector spaces*.

2.12.2 Inner product space

Given a vector space and nothing more, one can introduce the notion of a basis for the vector space, but one does not have the tool needed to define an orthonormal basis. Indeed the axioms of a vector space say nothing about geometric ideas such as "length" or "angle". To remedy this, one endows the vector space with the notion of inner product.

DEFINITION 2.36 *Let $\mathcal{V}$ be a vector space over $\mathbb{C}$. An* inner product *on $\mathcal{V}$ is a function that assigns to each ordered pair of vectors α, β in $\mathcal{V}$ a scalar $\langle \alpha, \beta \rangle$ in $\mathbb{C}$ in such a way that, for all α, β, γ in $\mathcal{V}$ and all scalars c in $\mathbb{C}$,*

(a) $\langle \alpha + \beta, \gamma \rangle = \langle \alpha, \gamma \rangle + \langle \beta, \gamma \rangle$
$\langle c\alpha, \beta \rangle = c \langle \alpha, \beta \rangle$;
(b) $\langle \beta, \alpha \rangle = \langle \alpha, \beta \rangle^*$; *(Hermitian symmetry)*
(c) $\langle \alpha, \alpha \rangle \geq 0$ *with equality if and only if $\alpha = 0$.*

It is implicit in (c) that $\langle \alpha, \alpha \rangle$ is real for all $\alpha \in \mathcal{V}$. From (a) and (b), we obtain the additional properties

(d) $\langle \alpha, \beta + \gamma \rangle = \langle \alpha, \beta \rangle + \langle \alpha, \gamma \rangle$
$\langle \alpha, c\beta \rangle = c^* \langle \alpha, \beta \rangle$.

□

Notice that the above definition is also valid for a vector space over the field of real numbers, but in this case the complex conjugates appearing in (b) and (2.36) are superfluous. However, over the field of complex numbers they are necessary for any $\alpha \neq 0$, otherwise we could write

$$0 < \langle j\alpha, j\alpha \rangle = -1 \langle \alpha, \alpha \rangle < 0,$$

where the first inequality follows from condition (c) and the fact that $j\alpha$ is a valid vector ($j = \sqrt{-1}$), and the equality follows from (a) and (2.36) without the complex conjugate. We see that the complex conjugate is necessary or else we can create the contradictory statement $0 < 0$.

On $\mathbb{C}^n$ there is an inner product that is sometimes called the *standard inner product*. It is defined on $a = (a_1, \ldots, a_n)$ and $b = (b_1, \ldots, b_n)$ by

$$\langle a, b \rangle = \sum_j a_j b_j^*.$$

On $\mathbb{R}^n$, the standard inner product is often called the dot or scalar product and denoted by $a \cdot b$. Unless explicitly stated otherwise, over $\mathbb{R}^n$ and over $\mathbb{C}^n$ we will always assume the standard inner product.

2.12. Appendix: Inner product spaces

An *inner product space* is a real or complex vector space, together with a specified inner product on that space. We will use the letter $\mathcal{V}$ to denote a generic inner product space.

EXAMPLE 2.37 *The vector space $\mathbb{R}^n$ equipped with the dot product is an inner product space and so is the vector space $\mathbb{C}^n$ equipped with the standard inner product.* □

By means of the inner product, we introduce the notion of length, called *norm*, of a vector α, via

$$\|\alpha\| = \sqrt{\langle \alpha, \alpha \rangle}.$$

Using linearity, we immediately obtain that the *squared norm* satisfies

$$\|\alpha \pm \beta\|^2 = \langle \alpha \pm \beta, \alpha \pm \beta \rangle = \|\alpha\|^2 + \|\beta\|^2 \pm 2\Re\{\langle \alpha, \beta \rangle\}. \tag{2.31}$$

The above generalizes $(a \pm b)^2 = a^2 + b^2 \pm 2ab$, $a, b \in \mathbb{R}$, and $|a \pm b|^2 = |a|^2 + |b|^2 \pm 2\Re\{ab^*\}$, $a, b \in \mathbb{C}$.

We say that two vectors are *collinear* if one is a scalar multiple of the other.

THEOREM 2.38 *If $\mathcal{V}$ is an inner product space then, for any vectors α, β in $\mathcal{V}$ and any scalar c,*

(a) $\|c\alpha\| = |c|\|\alpha\|$;
(b) $\|\alpha\| \geq 0$ *with equality if and only if $\alpha = 0$;*
(c) $|\langle \alpha, \beta \rangle| \leq \|\alpha\|\|\beta\|$ *with equality if and only if α and β are collinear (Cauchy–Schwarz inequality);*
(d) $\|\alpha + \beta\| \leq \|\alpha\| + \|\beta\|$ *with equality if and only if one of α, β is a non-negative multiple of the other (triangle inequality);*
(e) $\|\alpha + \beta\|^2 + \|\alpha - \beta\|^2 = 2(\|\alpha\|^2 + \|\beta\|^2)$ *(parallelogram equality).*

Proof Statements (a) and (b) follow immediately from the definitions. We postpone the proof of the Cauchy–Schwarz inequality to Example 2.43 as at that time we will be able to make a more elegant proof based on the concept of a projection. To prove the triangle inequality we use (2.31) and the Cauchy–Schwarz inequality applied to $\Re\{\langle \alpha, \beta \rangle\} \leq |\langle \alpha, \beta \rangle|$ to prove that $\|\alpha + \beta\|^2 \leq (\|\alpha\| + \|\beta\|)^2$. Notice that $\Re\{\langle \alpha, \beta \rangle\} \leq |\langle \alpha, \beta \rangle|$ holds with equality if and only if $\langle \alpha, \beta \rangle$ is a non-negative real. The Cauchy–Schwarz inequality holds with equality if and only if α and β are collinear. Both conditions for equality are satisfied if and only if one of α, β is a non-negative multiple of the other. The parallelogram equality follows immediately from (2.31) used twice, once with each sign. □

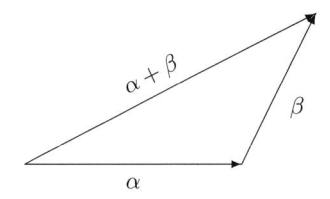

Triangle inequality. 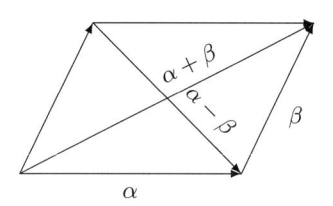   Parallelogram equality.

At this point we could use the inner product and the norm to define the angle between two vectors but we do not have any use for this. Instead, we will make frequent use of the notion of orthogonality. Two vectors α and β are defined to be *orthogonal* if $\langle \alpha, \beta \rangle = 0$.

EXAMPLE 2.39 *This example and the two that follow are relevant for what we do from Chapter 3 on. Let $\mathcal{W} = \{w_0(t), \ldots, w_{m-1}(t)\}$ be a finite collection of functions from $\mathbb{R}$ to $\mathbb{C}$ such that $\int_{-\infty}^{\infty} |w(t)|^2 dt < \infty$ for all elements of $\mathcal{W}$. Let $\mathcal{V}$ be the complex vector space spanned by the elements of $\mathcal{W}$, where the addition of two functions and the multiplication of a function by a scalar are defined in the obvious way. The reader should verify that the axioms of a vector space are fulfilled. A vector space of functions will be called a* signal space. *The standard inner product for functions from $\mathbb{R}$ to $\mathbb{C}$ is defined as*

$$\langle \alpha, \beta \rangle = \int \alpha(t) \beta^*(t) dt,$$

which implies the norm

$$\|\alpha\| = \sqrt{\int |\alpha(t)|^2 dt},$$

but it is not a given that $\mathcal{V}$ with the standard inner product forms an inner product space. It is straightforward to verify that the axioms (a), (c), and (d) of Definition 2.36 are fulfilled for all elements of $\mathcal{V}$ but axiom (b) is not necessarily fulfilled (see Example 2.40). If $\mathcal{V}$ is such that for all $\alpha \in \mathcal{V}$, $\langle \alpha, \alpha \rangle = 0$ implies that α is the zero vector, then $\mathcal{V}$ endowed with the standard inner product forms an inner product space. All we have said in this example applies also for the real vector spaces spanned by functions from $\mathbb{R}$ to $\mathbb{R}$. □

EXAMPLE 2.40 *Let $\mathcal{V}$ be the set of functions from $\mathbb{R}$ to $\mathbb{R}$ spanned by the function that is zero everywhere, except at 0 where it takes value 1. It can easily be checked that this is a vector space. It contains all the functions that are zero everywhere, except at 0 where they can take on any value in $\mathbb{R}$. Its zero vector is the function that is 0 everywhere, including at 0. For all α in $\mathcal{V}$, the standard inner product $\langle \alpha, \alpha \rangle$ equals 0. Hence $\mathcal{V}$ with the standard inner product is* not *an inner product space.* □

The problem highlighted with Example 2.40 is that for a general function $\alpha : \mathcal{I} \to \mathbb{C}$, $\int |\alpha(t)|^2 dt = 0$ does not necessarily imply $\alpha(t) = 0$ for all $t \in \mathcal{I}$. It

2.12. Appendix: Inner product spaces

is important to be aware of this fact. However, this potential problem will never arise in practice because all electrical signals are continuous. Sometimes we work out examples using signals that have discontinuities (e.g. rectangles) but even then the problem will not arise unless we use rather bizarre signals.

EXAMPLE 2.41 *Let $p(t)$ be a complex-valued square-integrable function (i.e. $\int |p(t)|^2 dt < \infty$) and let $\int |p(t)|^2 dt > 0$. For instance, $p(t)$ could be the rectangular pulse $\mathbb{1}\{t \in [0,T]\}$ for some $T > 0$. The set $\mathcal{V} = \{cp(t) : c \in \mathbb{C}\}$ with the standard inner product forms an inner product space. (In $\mathcal{V}$, only the zero-pulse has zero norm.)* □

THEOREM 2.42 *(Pythagoras' theorem) If α and β are orthogonal vectors in $\mathcal{V}$, then*

$$\|\alpha + \beta\|^2 = \|\alpha\|^2 + \|\beta\|^2.$$ □

Proof Pythagoras' theorem follows immediately from the equality $\|\alpha + \beta\|^2 = \|\alpha\|^2 + \|\beta\|^2 + 2\Re\{\langle \alpha, \beta \rangle\}$ and the fact that $\langle \alpha, \beta \rangle = 0$ by definition of orthogonality. □

Given two vectors $\alpha, \beta \in \mathcal{V}$, $\beta \neq 0$, we define the *projection of α on β* as the vector $\alpha_{|\beta}$ collinear to β (i.e. of the form $c\beta$ for some scalar c) such that $\alpha_{\perp\beta} = \alpha - \alpha_{|\beta}$ is orthogonal to β.

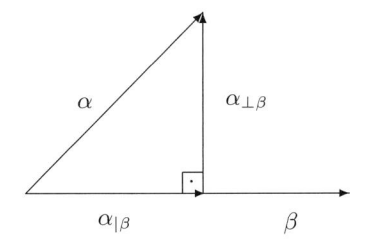

Projection of α on β.

Using the definition of orthogonality, what we want is

$$0 = \langle \alpha_{\perp\beta}, \beta \rangle = \langle \alpha - c\beta, \beta \rangle = \langle \alpha, \beta \rangle - c\|\beta\|^2.$$

Solving for c we obtain $c = \frac{\langle \alpha, \beta \rangle}{\|\beta\|^2}$. Hence

$$\alpha_{|\beta} = \frac{\langle \alpha, \beta \rangle}{\|\beta\|^2}\beta = \langle \alpha, \varphi \rangle \varphi \qquad \text{and} \qquad \alpha_{\perp\beta} = \alpha - \alpha_{|\beta},$$

where $\varphi = \frac{\beta}{\|\beta\|}$ is β scaled to unit norm. Notice that the projection of α on β does not depend on the norm of β. In fact, the norm of $\alpha_{|\beta}$ is $|\langle \alpha, \varphi \rangle|$.

Any non-zero vector $\beta \in \mathcal{V}$ defines a *hyperplane* by the relationship

$$\{\alpha \in \mathcal{V} : \langle \alpha, \beta \rangle = 0\}.$$

Hyperplane defined by β.

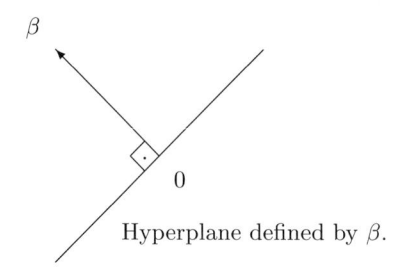

The hyperplane is the set of vectors in $\mathcal{V}$ that are orthogonal to β. A hyperplane always contains the zero vector.

An *affine plane*, defined by a vector β and a scalar c, is an object of the form

$$\{\alpha \in \mathcal{V} : \langle \alpha, \beta \rangle = c\}.$$

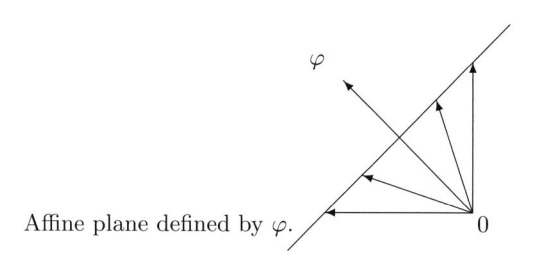

Affine plane defined by φ.

The vector β and scalar c that define a hyperplane are not unique, unless we agree that we use only normalized vectors to define hyperplanes. By letting $\varphi = \frac{\beta}{\|\beta\|}$, the above definition of affine plane may equivalently be written as $\{\alpha \in \mathcal{V} : \langle \alpha, \varphi \rangle = \frac{c}{\|\beta\|}\}$ or even as $\{\alpha \in \mathcal{V} : \langle \alpha - \frac{c}{\|\beta\|}\varphi, \varphi \rangle = 0\}$. The first shows that at an affine plane is the set of vectors that have the same projection $\frac{c}{\|\beta\|}\varphi$ on φ. The second form shows that the affine plane is a hyperplane translated by the vector $\frac{c}{\|\beta\|}\varphi$. Some authors make no distinction between affine planes and hyperplanes; in this case both are called hyperplane.

In the example that follows, we use the notion of projection to prove the Cauchy–Schwarz inequality stated in Theorem 2.38.

EXAMPLE 2.43 (Proof of the Cauchy–Schwarz inequality). *The Cauchy–Schwarz inequality states that, for any $\alpha, \beta \in \mathcal{V}$, $|\langle \alpha, \beta \rangle| \leq \|\alpha\|\|\beta\|$ with equality if and only if α and β are collinear. The statement is obviously true if $\beta = 0$. Assume $\beta \neq 0$ and write $\alpha = \alpha_{|\beta} + \alpha_{\perp \beta}$. (See the next figure.) Pythagoras' theorem states that $\|\alpha\|^2 = \|\alpha_{|\beta}\|^2 + \|\alpha_{\perp \beta}\|^2$. If we drop the second term, which is always non-negative, we obtain $\|\alpha\|^2 \geq \|\alpha_{|\beta}\|^2$ with equality if and only if α and β are collinear. From the definition of projection, $\|\alpha_{|\beta}\|^2 = \frac{|\langle \alpha, \beta \rangle|^2}{\|\beta\|^2}$. Hence $\|\alpha\|^2 \geq \frac{|\langle \alpha, \beta \rangle|^2}{\|\beta\|^2}$ with equality if and only if α and β are collinear. This is the Cauchy–Schwarz inequality.* □

2.12. Appendix: Inner product spaces

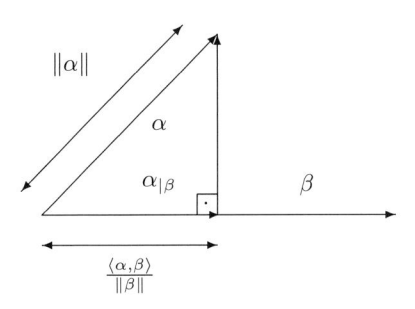

The Cauchy–Schwarz inequality.

A *basis* of $\mathcal{V}$ is a list of vectors in $\mathcal{V}$ that is linearly independent and spans $\mathcal{V}$. Every finite-dimensional vector space has a basis. If $\beta_1, \beta_2, \ldots, \beta_n$ is a basis for the inner product space $\mathcal{V}$ and $\alpha \in \mathcal{V}$ is an arbitrary vector, then there are unique scalars (coefficients) $a_1, \ldots, a_n$ such that $\alpha = \sum a_i \beta_i$ but finding them may be difficult. However, finding the coefficients of a vector is particularly easy when the basis is orthonormal.

A basis $\psi_1, \psi_2, \ldots, \psi_n$ for an inner product space $\mathcal{V}$ is orthonormal if

$$\langle \psi_i, \psi_j \rangle = \begin{cases} 0, & i \neq j \\ 1, & i = j. \end{cases}$$

Finding the ith coefficient a_i of an *orthonormal expansion* $\alpha = \sum a_i \psi_i$ is immediate. It suffices to observe that all but the ith term of $\sum a_i \psi_i$ are orthogonal to ψ_i and that the inner product of the ith term with ψ_i yields a_i. Hence if $\alpha = \sum a_i \psi_i$ then

$$a_i = \langle \alpha, \psi_i \rangle.$$

Observe that $|a_i|$ is the norm of the projection of α on ψ_i. This should not be surprising given that the ith term of the orthonormal expansion of α is collinear to ψ_i and the sum of all the other terms are orthogonal to ψ_i.

There is another major advantage to working with an orthonormal basis. If a and b are the n-tuples of coefficients of the expansion of α and β with respect to the same orthonormal basis, then

$$\langle \alpha, \beta \rangle = \langle a, b \rangle, \tag{2.32}$$

where the right-hand side inner product is given with respect to the standard inner product. Indeed

$$\langle \alpha, \beta \rangle = \left\langle \sum_i a_i \psi_i, \sum_j b_j \psi_j \right\rangle = \sum_i a_i \left\langle \psi_i, \sum_j b_j \psi_j \right\rangle$$
$$= \sum_i a_i \langle \psi_i, b_i \psi_i \rangle = \sum_i a_i b_i^* = \langle a, b \rangle.$$

Letting $\beta = \alpha$, the above also implies

$$\|\alpha\| = \|a\|,$$

where $\|a\| = \sqrt{\sum_i |a_i|^2}$.

An orthonormal set of vectors $\psi_1, \ldots, \psi_n$ of an inner product space $\mathcal{V}$ is a linearly independent set. Indeed $0 = \sum a_i \psi_i$ implies $a_i = \langle 0, \psi_i \rangle = 0$. By normalizing the vectors and recomputing the coefficients, we can easily extend this reasoning to a set of orthogonal (but not necessarily orthonormal) vectors $\alpha_1, \ldots, \alpha_n$. They too must be linearly independent.

The idea of a projection on a vector generalizes to a projection on a subspace. If $\mathcal{U}$ is a subspace of an inner product space $\mathcal{V}$, and $\alpha \in \mathcal{V}$, the projection of α on $\mathcal{U}$ is defined to be a vector $\alpha_{|\mathcal{U}} \in \mathcal{U}$ such that $\alpha - \alpha_{|\mathcal{U}}$ is orthogonal to all vectors in $\mathcal{U}$. If $\psi_1, \ldots, \psi_m$ is an orthonormal basis for $\mathcal{U}$, then the condition that $\alpha - \alpha_{|\mathcal{U}}$ is orthogonal to all vectors of $\mathcal{U}$ implies $0 = \langle \alpha - \alpha_{|\mathcal{U}}, \psi_i \rangle = \langle \alpha, \psi_i \rangle - \langle \alpha_{|\mathcal{U}}, \psi_i \rangle$. This shows that $\langle \alpha, \psi_i \rangle = \langle \alpha_{|\mathcal{U}}, \psi_i \rangle$. The right side of this equality is the ith coefficient of the orthonormal expansion of $\alpha_{|\mathcal{U}}$ with respect to the orthonormal basis. This proves that

$$\alpha_{|\mathcal{U}} = \sum_{i=1}^{m} \langle \alpha, \psi_i \rangle \psi_i$$

is the unique projection of α on $\mathcal{U}$. We summarize this important result and prove that the projection of α on $\mathcal{U}$ is the element of $\mathcal{U}$ that is closest to α.

THEOREM 2.44 *Let $\mathcal{U}$ be a subspace of an inner product space $\mathcal{V}$, and let $\alpha \in \mathcal{V}$. The projection of α on $\mathcal{U}$, denoted by $\alpha_{|\mathcal{U}}$, is the unique element of $\mathcal{U}$ that satisfies any (hence all) of the following conditions:*

(i) *$\alpha - \alpha_{|\mathcal{U}}$ is orthogonal to every element of $\mathcal{U}$;*
(ii) *$\alpha_{|\mathcal{U}} = \sum_{i=1}^{m} \langle \alpha, \psi_i \rangle \psi_i$;*
(iii) *for any $\beta \in \mathcal{U}$, $\|\alpha - \alpha_{|\mathcal{U}}\| \leq \|\alpha - \beta\|$.* □

Proof Statement (i) is the definition of projection and we have already proved that it is equivalent to statement (ii). Now consider any vector $\beta \in \mathcal{U}$. From Pythagoras' theorem and the fact that $\alpha - \alpha_{|\mathcal{U}}$ is orthogonal to $\alpha_{|\mathcal{U}} - \beta \in \mathcal{U}$, we obtain

$$\|\alpha - \beta\|^2 = \|\alpha - \alpha_{|\mathcal{U}} + \alpha_{|\mathcal{U}} - \beta\|^2 = \|\alpha - \alpha_{|\mathcal{U}}\|^2 + \|\alpha_{|\mathcal{U}} - \beta\|^2 \geq \|\alpha - \alpha_{|\mathcal{U}}\|^2.$$

Moreover, equality holds if and only if $\|\alpha_{|\mathcal{U}} - \beta\|^2 = 0$, i.e. if and only if $\beta = \alpha_{|\mathcal{U}}$. □

THEOREM 2.45 *Let $\mathcal{V}$ be an inner product space and let $\beta_1, \ldots, \beta_n$ be any collection of linearly independent vectors in $\mathcal{V}$. Then we may construct orthogonal vectors $\alpha_1, \ldots, \alpha_n$ in $\mathcal{V}$ such that they form a basis for the subspace spanned by $\beta_1, \ldots, \beta_n$.* □

Proof The proof is constructive via a procedure known as the Gram–Schmidt orthogonalization procedure. First let $\alpha_1 = \beta_1$. The other vectors are constructed inductively as follows. Suppose $\alpha_1, \ldots, \alpha_m$ have been chosen so that they form an orthogonal basis for the subspace $\mathcal{U}_m$ spanned by $\beta_1, \ldots, \beta_m$. We choose the next vector as

$$\alpha_{m+1} = \beta_{m+1} - \beta_{m+1|\mathcal{U}_m}, \tag{2.33}$$

where $\beta_{m+1|\mathcal{U}_m}$ is the projection of β_{m+1} on $\mathcal{U}_m$. By definition, α_{m+1} is orthogonal to every vector in $\mathcal{U}_m$, including $\alpha_1, \ldots, \alpha_m$. Also, $\alpha_{m+1} \neq 0$ for otherwise β_{m+1}

contradicts the hypothesis that it is linearly independent of $\beta_1, \ldots, \beta_m$. Therefore $\alpha_1, \ldots, \alpha_{m+1}$ is an orthogonal collection of non-zero vectors in the subspace $\mathcal{U}_{m+1}$ spanned by $\beta_1, \ldots, \beta_{m+1}$. Therefore it must be a basis for $\mathcal{U}_{m+1}$. Thus the vectors $\alpha_1, \ldots, \alpha_n$ may be constructed one after the other according to (2.33). □

COROLLARY 2.46 *Every finite-dimensional vector space has an orthonormal basis.* □

Proof Let $\beta_1, \ldots, \beta_n$ be a basis for the finite-dimensional inner product space $\mathcal{V}$. Apply the Gram–Schmidt procedure to find an orthogonal basis $\alpha_1, \ldots, \alpha_n$. Then $\psi_1, \ldots, \psi_n$, where $\psi_i = \frac{\alpha_i}{\|\alpha_i\|}$, is an orthonormal basis. □

Gram–Schmidt orthonormalization procedure

We summarize the Gram–Schmidt procedure, modified so as to produce orthonormal vectors. If $\beta_1, \ldots, \beta_n$ is a linearly independent collection of vectors in the inner product space $\mathcal{V}$, then we may construct a collection $\psi_1, \ldots, \psi_n$ that forms an orthonormal basis for the subspace spanned by $\beta_1, \ldots, \beta_n$ as follows. We let $\psi_1 = \frac{\beta_1}{\|\beta_1\|}$ and for $i = 2, \ldots, n$ we choose

$$\alpha_i = \beta_i - \sum_{j=1}^{i-1} \langle \beta_i, \psi_j \rangle \psi_j$$

$$\psi_i = \frac{\alpha_i}{\|\alpha_i\|}.$$

We have assumed that $\beta_1, \ldots, \beta_n$ is a linearly independent collection. Now assume that this is not the case. If β_j is linearly dependent of $\beta_1, \ldots, \beta_{j-1}$, then at step $i = j$ the procedure will produce $\alpha_i = \psi_i = 0$. Such vectors are simply disregarded.

Figure 2.19 gives an example of the Gram–Schmidt procedure applied to a set of signals.

i	β_i	$\langle \beta_i, \psi_j \rangle$ $j < i$	$\beta_{i\mid\mathcal{V}_{i-1}}$	$\alpha_i = \beta_i - \beta_{i\mid\mathcal{V}_{i-1}}$	$\|\alpha_i\|$	ψ_i	b_i
1	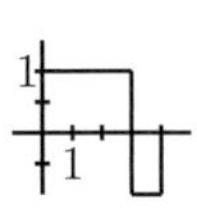	-	-	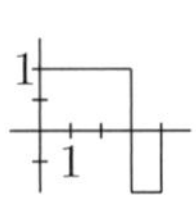	2	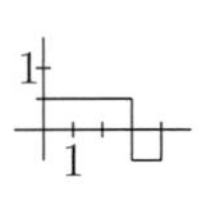	$\begin{pmatrix}2\\0\\0\end{pmatrix}$
2	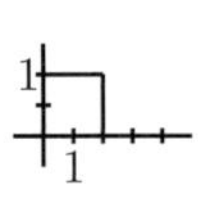	1	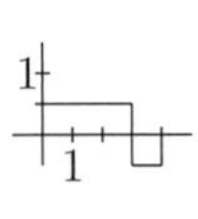	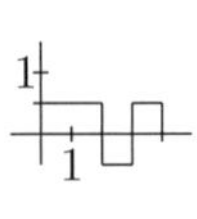	1	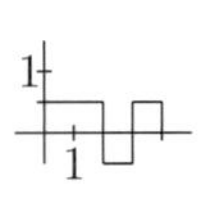	$\begin{pmatrix}1\\1\\0\end{pmatrix}$
3	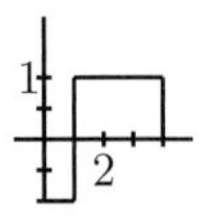	0, 0	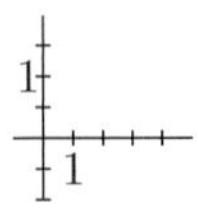	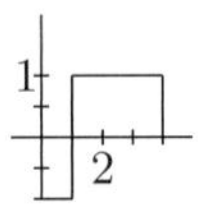	2	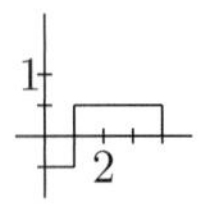	$\begin{pmatrix}0\\0\\2\end{pmatrix}$

Figure 2.19. Application of the Gram–Schmidt orthonormalization procedure starting with the waveforms given in the left column.

2.13 Exercises

Exercises for Section 2.2

EXERCISE 2.1 (Hypothesis testing: Uniform and uniform) *Consider a binary hypothesis testing problem in which the hypotheses $H = 0$ and $H = 1$ occur with probability $P_H(0)$ and $P_H(1) = 1 - P_H(0)$, respectively. The observable Y takes values in $\{0,1\}^{2k}$, where k is a fixed positive integer. When $H = 0$, each component of Y is 0 or 1 with probability $\frac{1}{2}$ and components are independent. When $H = 1$, Y is chosen uniformly at random from the set of all sequences of length $2k$ that have an equal number of ones and zeros. There are $\binom{2k}{k}$ such sequences.*

(a) *What is $P_{Y|H}(y|0)$? What is $P_{Y|H}(y|1)$?*
(b) *Find a maximum-likelihood decision rule for H based on y. What is the single number you need to know about y to implement this decision rule?*
(c) *Find a decision rule that minimizes the error probability.*
(d) *Are there values of $P_H(0)$ such that the decision rule that minimizes the error probability always chooses the same hypothesis regardless of y? If yes, what are these values, and what is the decision?*

EXERCISE 2.2 (The "Wetterfrosch") *Let us assume that a "weather frog" bases his forecast of tomorrow's weather entirely on today's air pressure. Determining a weather forecast is a hypothesis testing problem. For simplicity, let us assume that the weather frog only needs to tell us if the forecast for tomorrow's weather is "sunshine" or "rain". Hence we are dealing with binary hypothesis testing. Let $H = 0$ mean "sunshine" and $H = 1$ mean "rain". We will assume that both values of H are equally likely, i.e. $P_H(0) = P_H(1) = \frac{1}{2}$. For the sake of this exercise, suppose that on a day that precedes sunshine, the pressure may be modeled as a random variable Y with the following probability density function:*

$$f_{Y|H}(y|0) = \begin{cases} A - \frac{A}{2}y, & 0 \leq y \leq 1 \\ 0, & \text{otherwise.} \end{cases}$$

Similarly, the pressure on a day that precedes a rainy day is distributed according to

$$f_{Y|H}(y|1) = \begin{cases} B + \frac{B}{3}y, & 0 \leq y \leq 1 \\ 0, & \text{otherwise.} \end{cases}$$

The weather frog's purpose in life is to guess the value of H after measuring Y.

(a) *Determine A and B.*
(b) *Find the a posteriori probability $P_{H|Y}(0|y)$. Also find $P_{H|Y}(1|y)$.*
(c) *Show that the implementation of the decision rule $\hat{H}(y) = \arg\max_i P_{H|Y}(i|y)$ reduces to*

$$\hat{H}_\theta(y) = \begin{cases} 0, & \text{if } y \leq \theta \\ 1, & \text{otherwise,} \end{cases} \tag{2.34}$$

for some threshold θ and specify the threshold's value.

2.13. Exercises

(d) Now assume that the weather forecaster does not know about hypothesis testing and arbitrarily choses the decision rule $\hat{H}_\gamma(y)$ for some arbitrary $\gamma \in \mathbb{R}$. Determine, as a function of γ, the probability that the decision rule decides $\hat{H} = 1$ given that $H = 0$. This probability is denoted $Pr\{\hat{H}(Y) = 1 | H = 0\}$.

(e) For the same decision rule, determine the probability of error $P_e(\gamma)$ as a function of γ. Evaluate your expression at $\gamma = \theta$.

(f) Using calculus, find the γ that minimizes $P_e(\gamma)$ and compare your result to θ.

EXERCISE 2.3 (Hypothesis testing in Laplacian noise) *Consider the following hypothesis testing problem between two equally likely hypotheses. Under hypothesis $H = 0$, the observable Y is equal to $a + Z$ where Z is a random variable with Laplacian distribution*

$$f_Z(z) = \frac{1}{2}e^{-|z|}.$$

Under hypothesis $H = 1$, the observable is given by $-a + Z$. You may assume that a is positive.

(a) Find and draw the density $f_{Y|H}(y|0)$ of the observable under hypothesis $H = 0$, and the density $f_{Y|H}(y|1)$ of the observable under hypothesis $H = 1$.

(b) Find the decision rule that minimizes the probability of error.

(c) Compute the probability of error of the optimal decision rule.

EXERCISE 2.4 (Poisson parameter estimation) *In this example there are two hypotheses, $H = 0$ and $H = 1$, which occur with probabilities $P_H(0) = p_0$ and $P_H(1) = 1 - p_0$, respectively. The observable Y takes values in the set of non-negative integers. Under hypothesis $H = 0$, Y is distributed according to a Poisson law with parameter λ_0, i.e.*

$$P_{Y|H}(y|0) = \frac{\lambda_0^y}{y!}e^{-\lambda_0}. \quad (2.35)$$

Under hypothesis $H = 1$,

$$P_{Y|H}(y|1) = \frac{\lambda_1^y}{y!}e^{-\lambda_1}. \quad (2.36)$$

This is a model for the reception of photons in optical communication.

(a) Derive the MAP decision rule by indicating likelihood and log likelihood ratios. Hint: The direction of an inequality changes if both sides are multiplied by a negative number.

(b) Derive an expression for the probability of error of the MAP decision rule.

(c) For $p_0 = 1/3$, $\lambda_0 = 2$ and $\lambda_1 = 10$, compute the probability of error of the MAP decision rule. You may want to use a computer program to do this.

(d) Repeat (c) with $\lambda_1 = 20$ and comment.

EXERCISE 2.5 (Lie detector) *You are asked to develop a "lie detector" and analyze its performance. Based on the observation of brain-cell activity, your* detector has to decide if a person is telling the truth or is lying. For the purpose of this exercise, the brain cell produces a sequence of spikes. For your decision you may use only a sequence of n consecutive inter-arrival times $Y_1, Y_2, \ldots, Y_n$. Hence Y_1 is the time elapsed between the first and second spike, Y_2 the time between the second and third, etc. We assume that, a priori, a person lies with some known probability p. When the person is telling the truth, $Y_1, \ldots, Y_n$ is an iid sequence of exponentially distributed random variables with intensity α, $(\alpha > 0)$, i.e.

$$f_{Y_i}(y) = \alpha e^{-\alpha y}, \quad y \geq 0.$$

When the person lies, $Y_1, \ldots, Y_n$ is iid exponentially distributed with intensity β, $(\alpha < \beta)$.

(a) Describe the decision rule of your lie detector for the special case $n = 1$. Your detector should be designed so as to minimize the probability of error.
(b) What is the probability $P_{L|T}$ that your lie detector says that the person is lying when the person is telling the truth?
(c) What is the probability $P_{T|L}$ that your test says that the person is telling the truth when the person is lying.
(d) Repeat (a) and (b) for a general n. Hint: When $Y_1, \ldots, Y_n$ is a collection of iid random variables that are exponentially distributed with parameter $\alpha > 0$, then $Y_1 + \cdots + Y_n$ has the probability density function of the Erlang distribution, i.e.

$$f_{Y_1 + \cdots + Y_n}(y) = \frac{\alpha^n}{(n-1)!} y^{n-1} e^{-\alpha y}, \quad y \geq 0.$$

EXERCISE 2.6 (Fault detector) As an engineer, you are required to design the test performed by a fault detector for a "black-box" that produces a sequence of iid binary random variables $\ldots, X_1, X_2, X_3, \ldots$. Previous experience shows that this "black box" has an a priori failure probability of $\frac{1}{1025}$. When the "black box" works properly, $p_{X_i}(1) = p$. When it fails, the output symbols are equally likely to be 0 or 1. Your detector has to decide based on the observation of the past 16 symbols, i.e. at time k the decision will be based on $X_{k-16}, \ldots, X_{k-1}$.

(a) Describe your test.
(b) What does your test decide if it observes the sequence 0101010101010101? Assume that $p = 0.25$.

EXERCISE 2.7 (Multiple choice exam) You are taking a multiple choice exam. Question number 5 allows for two possible answers. According to your first impression, answer 1 is correct with probability $1/4$ and answer 2 is correct with probability $3/4$. You would like to maximize your chance of giving the correct answer and you decide to have a look at what your neighbors on the left and right have to say. The neighbor on the left has answered $\hat{H}_L = 1$. He is an excellent student who has a record of being correct 90% of the time when asked a binary question. The neighbor on the right has answered $\hat{H}_R = 2$. He is a weaker student who is correct 70% of the time.

2.13. Exercises

(a) You decide to use your first impression as a prior and to consider $\hat{H}_L$ and $\hat{H}_R$ as observations. Formulate the decision problem as a hypothesis testing problem.

(b) What is your answer $\hat{H}$?

EXERCISE 2.8 (MAP decoding rule: Alternative derivation) Consider the binary hypothesis testing problem where H takes values in $\{0, 1\}$ with probabilites $P_H(0)$ and $P_H(1)$. The conditional probability density function of the observation $Y \in \mathbb{R}$ given $H = i$, $i \in \{0, 1\}$ is given by $f_{Y|H}(\cdot|i)$. Let $\mathcal{R}_i$ be the decoding region for hypothesis i, i.e. the set of y for which the decision is $\hat{H} = i$, $i \in \{0, 1\}$.

(a) Show that the probability of error is given by

$$P_e = P_H(1) + \int_{\mathcal{R}_1} \left(P_H(0) f_{Y|H}(y|0) - P_H(1) f_{Y|H}(y|1) \right) dy.$$

Hint: Note that $\mathbb{R} = \mathcal{R}_0 \bigcup \mathcal{R}_1$ and $\int_{\mathbb{R}} f_{Y|H}(y|i) dy = 1$ for $i \in \{0, 1\}$.

(b) Argue that P_e is minimized when

$$\mathcal{R}_1 = \{ y \in \mathbb{R} : P_H(0) f_{Y|H}(y|0) < P_H(1) f_{Y|H}(y|1) \}$$

i.e. for the MAP rule.

EXERCISE 2.9 (Independent and identically distributed vs. first-order Markov) Consider testing two equally likely hypotheses $H = 0$ and $H = 1$. The observable $Y = (Y_1, \ldots, Y_k)^\mathsf{T}$ is a k-dimensional binary vector. Under $H = 0$ the components of the vector Y are independent uniform random variables (also called Bernoulli (1/2) random variables). Under $H = 1$, the component Y_1 is also uniform, but the components Y_i, $2 \leq i \leq k$, are distributed as follows:

$$P_{Y_i|Y_1,\ldots,Y_{i-1}}(y_i|y_1,\ldots,y_{i-1}) = \begin{cases} 3/4, & \text{if } y_i = y_{i-1} \\ 1/4, & \text{otherwise.} \end{cases} \qquad (2.37)$$

(a) Find the decision rule that minimizes the probability of error. Hint: Write down a short sample sequence $(y_1, \ldots, y_k)$ and determine its probability under each hypothesis. Then generalize.

(b) Give a simple sufficient statistic for this decision. (For the purpose of this question, a sufficient statistic is a function of y with the property that a decoder that observes y can not achieve a smaller error probability than a MAP decoder that observes this function of y.)

(c) Suppose that the observed sequence alternates between 0 and 1 except for one string of ones of length s, i.e. the observed sequence y looks something like

$$y = 0101010111111\ldots111111010101. \qquad (2.38)$$

What is the least s such that we decide for hypothesis $H = 1$?

EXERCISE 2.10 (SIMO channel with Laplacian noise, exercise from [1]) One of the two signals $c_0 = -1$, $c_1 = 1$ is transmitted over the channel shown in

Figure 2.20a. The two noise random variables Z_1 and Z_2 are statistically independent of the transmitted signal and of each other. Their density functions are

$$f_{Z_1}(\alpha) = f_{Z_2}(\alpha) = \frac{1}{2} e^{-|\alpha|}.$$

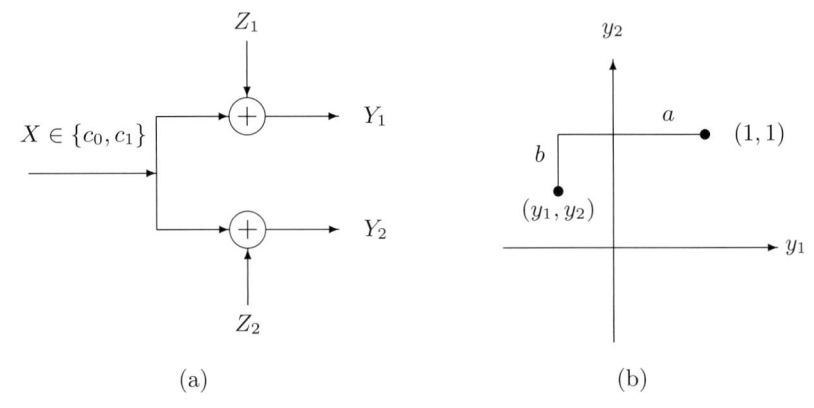

Figure 2.20.

(a) Derive a maximum likelihood decision rule.
(b) Describe the maximum likelihood decision regions in the (y_1, y_2) plane. Describe also the "either choice" regions, i.e. the regions where it does not matter if you decide for c_0 or for c_1. Hint: Use geometric reasoning and the fact that for a point (y_1, y_2) as shown in Figure 2.20b, $|y_1-1|+|y_2-1| = a+b$.
(c) A receiver decides that c_1 was transmitted if and only if $(y_1 + y_2) > 0$. Does this receiver minimize the error probability for equally likely messages?
(d) What is the error probability for the receiver in (c)? Hint: One way to do this is to use the fact that if $W = Z_1 + Z_2$ then $f_W(\omega) = \frac{e^{-\omega}}{4}(1+\omega)$ for $w > 0$ and $f_W(-\omega) = f_W(\omega)$.

Exercises for Section 2.3

EXERCISE 2.11 (Q function on regions, exercise from [1]) Let $X \sim \mathcal{N}(0, \sigma^2 I_2)$. For each of the three diagrams shown in Figure 2.21, express the probability that X lies in the shaded region. You may use the Q function when appropriate.

EXERCISE 2.12 (Properties of the Q function) Prove properties (a) through (d) of the Q function defined in Section 2.3. Hint: For property (d), multiply and divide inside the integral by the integration variable and integrate by parts. By upper- and lower-bounding the resulting integral, you will obtain the lower and upper bound.

2.13. Exercises

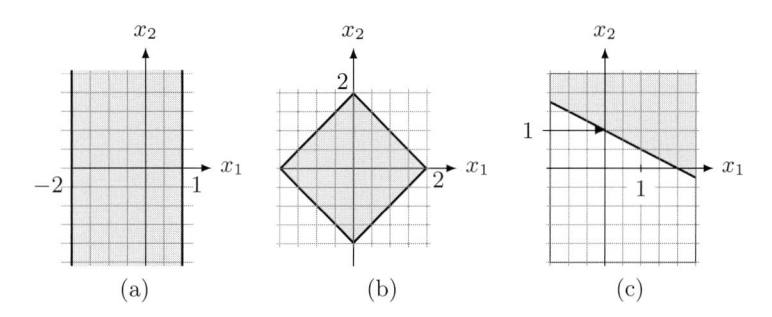

Figure 2.21.

Exercises for Section 2.4

EXERCISE 2.13 (16-PAM vs. 16-QAM) *The two signal constellations in Figure 2.22 are used to communicate across an additive white Gaussian noise channel. Let the noise variance be σ^2. Each point represents a codeword c_i for some i. Assume each codeword is used with the same probability.*

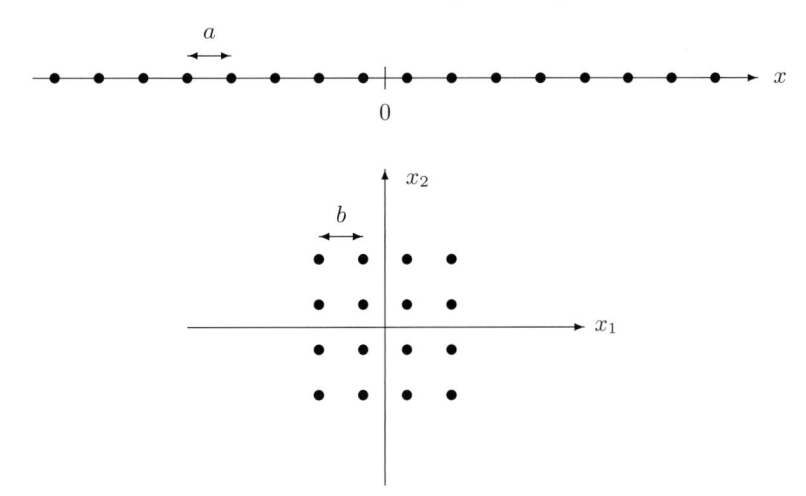

Figure 2.22.

(a) *For each signal constellation, compute the average probability of error P_e as a function of the parameters a and b, respectively.*

(b) *For each signal constellation, compute the average energy per symbol $\mathcal{E}$ as a function of the parameters a and b, respectively:*

$$\mathcal{E} = \sum_{i=1}^{16} P_H(i) \|c_i\|^2. \tag{2.39}$$

In the next chapter it will become clear in what sense $\mathcal{E}$ relates to the energy of the transmitted signal (see Example 3.2 and the discussion that follows).

(c) *Plot P_e versus $\frac{\mathcal{E}}{\sigma^2}$ for both signal constellations and comment.*

EXERCISE 2.14 (QPSK decision regions) *Let* $H \in \{0,1,2,3\}$ *and assume that when* $H = i$ *you transmit the codeword* c_i *shown in Figure 2.23. Under* $H = i$, *the receiver observes* $Y = c_i + Z$.

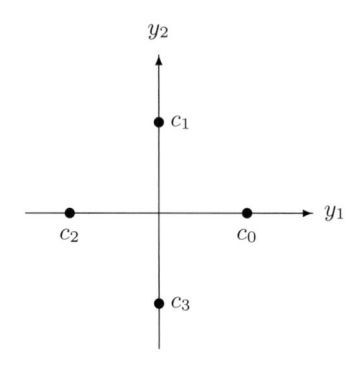

Figure 2.23.

(a) *Draw the decoding regions assuming that* $Z \sim \mathcal{N}(0, \sigma^2 I_2)$ *and that* $P_H(i) = 1/4$, $i \in \{0,1,2,3\}$.
(b) *Draw the decoding regions (qualitatively) assuming* $Z \sim \mathcal{N}(0, \sigma^2 I_2)$ *and* $P_H(0) = P_H(2) > P_H(1) = P_H(3)$. *Justify your answer.*
(c) *Assume again that* $P_H(i) = 1/4$, $i \in \{0,1,2,3\}$ *and that* $Z \sim \mathcal{N}(0, K)$, *where* $K = \begin{pmatrix} \sigma^2 & 0 \\ 0 & 4\sigma^2 \end{pmatrix}$. *How do you decode now?*

EXERCISE 2.15 (Antenna array) *The following problem relates to the design of multi-antenna systems. Consider the binary equiprobable hypothesis testing problem:*

$$H = 0 : Y_1 = A + Z_1, \quad Y_2 = A + Z_2$$
$$H = 1 : Y_1 = -A + Z_1, \quad Y_2 = -A + Z_2,$$

where Z_1, Z_2 *are independent Gaussian random variables with* **different** *variances* $\sigma_1^2 \neq \sigma_2^2$, *that is,* $Z_1 \sim \mathcal{N}(0, \sigma_1^2)$ *and* $Z_2 \sim \mathcal{N}(0, \sigma_2^2)$. $A > 0$ *is a constant.*

(a) *Show that the decision rule that minimizes the probability of error (based on the observable* Y_1 *and* Y_2) *can be stated as*

$$\sigma_2^2 y_1 + \sigma_1^2 y_2 \underset{1}{\overset{0}{\gtrless}} 0.$$

(b) *Draw the decision regions in the* (Y_1, Y_2) *plane for the special case where* $\sigma_1 = 2\sigma_2$.
(c) *Evaluate the probability of the error for the optimal detector as a function of* σ_1^2, σ_2^2 *and* A.

2.13. Exercises

EXERCISE 2.16 (Multi-antenna receiver) *Consider a communication system with one transmitter and n receiver antennas. The receiver observes the n-tuple $Y = (Y_1, \ldots, Y_n)^\mathsf{T}$ with*

$$Y_k = Bg_k + Z_k, \quad k = 1, 2, \ldots, n,$$

where $B \in \{\pm 1\}$ is a uniformly distributed source bit, g_k models the gain of antenna k and $Z_k \sim \mathcal{N}(0, \sigma^2)$. The random variables $B, Z_1, \ldots, Z_n$ are independent. Using n-tuple notation the model becomes

$$Y = Bg + Z,$$

where Y, g, and Z are n-tuples.

(a) *Suppose that the observation Y_k is weighted by an arbitrary real number w_k and combined with the other observations to form*

$$V = \sum_{k=1}^{n} Y_k w_k = \langle Y, w \rangle,$$

where w is an n-tuple. Describe the ML receiver for B given the observation V. (The receiver knows g and of course knows w.)
(b) *Give an expression for the probability of error P_e.*
(c) *Define $\beta = \frac{|\langle g, w \rangle|}{\|g\| \|w\|}$ and rewrite the expresson for P_e in a form that depends on w only through β.*
(d) *As a function of w, what are the maximum and minimum values for β and how do you choose w to achieve them?*
(e) *Minimize the probability of error over all possible choices of w. Could you reduce the error probability further by doing ML decision directly on Y rather than on V? Justify your answer.*
(f) *How would you choose w to minimize the error probability if Z_k had variance σ_k^2, $k = 1, \ldots, n$? Hint: With a simple operation at the receiver you can transform the new problem into the one you have already solved.*

EXERCISE 2.17 (Signal constellation) *The signal constellation of Figure 2.24 is used to communicate across the AWGN channel of noise variance σ^2. Assume that the six signals are used with equal probability.*

(a) *Draw the boundaries of the decision regions.*
(b) *Compute the average probability of error, P_e, for this signal constellation.*
(c) *Compute the average energy per symbol for this signal constellation.*

EXERCISE 2.18 (Hypothesis testing and fading) *Consider the following communication problem depicted in Figure 2.25. There are two equiprobable hypotheses. When $H = 0$, we transmit $c_0 = -b$, where b is an arbitrary but fixed positive number. When $H = 1$, we transmit $c_1 = b$. The channel is as shown in Figure 2.25, where $Z \sim \mathcal{N}(0, \sigma^2)$ represents the noise, $A \in \{0, 1\}$ represents a random attenuation (fading) with $P_A(0) = \frac{1}{2}$, and Y is the channel output. The random variables H, A, and Z are independent.*

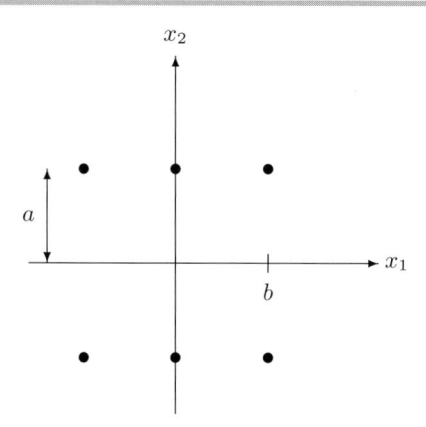

Figure 2.24.

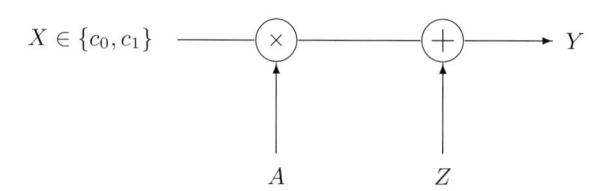

Figure 2.25.

(a) Find the decision rule that the receiver should implement to minimize the probability of error. Sketch the decision regions.
(b) Calculate the probability of error P_e, based on the above decision rule.

EXERCISE 2.19 (MAP decoding regions) *To communicate across an additive white Gaussian noise channel, an encoder uses the codewords c_i, $i = 0, 1, 2$, shown below:*

$$c_0 = (1, 0)^\mathsf{T}$$
$$c_1 = (-1, 0)^\mathsf{T}$$
$$c_2 = (-1, 1)^\mathsf{T}.$$

(a) Draw the decoding regions of an ML decoder.
(b) Now assume that codeword i is used with probability $P_H(i)$, where $P_H(0) = 0.25$, $P_H(1) = 0.25$, and $P_H(2) = 0.5$ and that the receiver performs a MAP decision. Adjust the decoding regions accordingly. (A qualitative illustration suffices.)
(c) Finally, assume that the noise variance increases (same variance in both components). Update the decoding regions of the MAP decision rule. (Again, a qualitative illustration suffices.)

2.13. Exercises

Exercises for Section 2.5

EXERCISE 2.20 (Sufficient statistic) *Consider a binary hypothesis testing problem specified by:*

$$H = 0 : \begin{cases} Y_1 = Z_1 \\ Y_2 = Z_1 Z_2 \end{cases}$$

$$H = 1 : \begin{cases} Y_1 = -Z_1 \\ Y_2 = -Z_1 Z_2, \end{cases}$$

where Z_1, Z_2, and H are independent random variables. Is Y_1 a sufficient statistic?

EXERCISE 2.21 (More on sufficient statistic) *We have seen that if $H \to T(Y) \to Y$, then the probability of error P_e of a MAP decoder that decides on the value of H upon observing both $T(Y)$ and Y is the same as that of a MAP decoder that observes only $T(Y)$. It is natural to wonder if the contrary is also true, specifically if the knowledge that Y does not help reduce the error probability that we can achieve with $T(Y)$ implies $H \to T(Y) \to Y$. Here is a counter-example. Let the hypothesis H be either 0 or 1 with equal probability (the choice of distribution on H is critical in this example). Let the observable Y take four values with the following conditional probabilities*

$$P_{Y|H}(y|0) = \begin{cases} 0.4 & \text{if } y = 0 \\ 0.3 & \text{if } y = 1 \\ 0.2 & \text{if } y = 2 \\ 0.1 & \text{if } y = 3 \end{cases} \qquad P_{Y|H}(y|1) = \begin{cases} 0.1 & \text{if } y = 0 \\ 0.2 & \text{if } y = 1 \\ 0.3 & \text{if } y = 2 \\ 0.4 & \text{if } y = 3 \end{cases}$$

and $T(Y)$ is the following function

$$T(y) = \begin{cases} 0 & \text{if } y = 0 \text{ or } y = 1 \\ 1 & \text{if } y = 2 \text{ or } y = 3. \end{cases}$$

(a) *Show that the MAP decoder $\hat{H}(T(y))$ that decides based on $T(y)$ is equivalent to the MAP decoder $\hat{H}(y)$ that operates based on y.*
(b) *Compute the probabilities $\Pr\{Y = 0 \mid T(Y) = 0, H = 0\}$ and $\Pr\{Y = 0 \mid T(Y) = 0, H = 1\}$. Is it true that $H \to T(Y) \to Y$?*

EXERCISE 2.22 (Fisher–Neyman factorization theorem) *Consider the hypothesis testing problem where the hypothesis is $H \in \{0, 1, \ldots, m-1\}$, the observable is Y, and $T(Y)$ is a function of the observable. Let $f_{Y|H}(y|i)$ be given for all $i \in \{0, 1, \ldots, m-1\}$. Suppose that there are positive functions $g_0, g_1, \ldots, g_{m-1}, h$ so that for each $i \in \{0, 1, \ldots, m-1\}$ one can write*

$$f_{Y|H}(y|i) = g_i(T(y))h(y). \tag{2.40}$$

(a) *Show that when the above conditions are satisfied, a MAP decision depends on the observable Y only through $T(Y)$. In other words, Y itself is not necessary. Hint: Work directly with the definition of a MAP decision rule.*

(b) *Show that $T(Y)$ is a sufficient statistic, that is $H \to T(Y) \to Y$. Hint: Start by observing the following fact. Given a random variable Y with probability density function $f_Y(y)$ and given an arbitrary event $\mathcal{B}$, we have*

$$f_{Y|Y \in \mathcal{B}} = \frac{f_Y(y)\mathbb{1}\{y \in \mathcal{B}\}}{\int_\mathcal{B} f_Y(y)dy}. \tag{2.41}$$

Proceed by defining $\mathcal{B}$ to be the event $\mathcal{B} = \{y : T(y) = t\}$ and make use of (2.41) applied to $f_{Y|H}(y|i)$ to prove that $f_{Y|H,T(Y)}(y|i,t)$ is independent of i.

(c) *(Example 1) Under hypothesis $H = i$, let $Y = (Y_1, Y_2, \ldots, Y_n)$, $Y_k \in \{0,1\}$, be an independent and identically distributed sequence of coin tosses such that $P_{Y_k|H}(1|i) = p_i$. Show that the function $T(y_1, y_2, \ldots, y_n) = \sum_{k=1}^n y_k$ fulfills the condition expressed in equation (2.40). Notice that $T(y_1, y_2, \ldots, y_n)$ is the number of 1s in $(y_1, y_2, \ldots, y_n)$.*

(d) *(Example 2) Under hypothesis $H = i$, let the observable Y_k be Gaussian distributed with mean m_i and variance 1; that is*

$$f_{Y_k|H}(y|i) = \frac{1}{\sqrt{2\pi}} e^{-\frac{(y-m_i)^2}{2}},$$

and $Y_1, Y_2, \ldots, Y_n$ be independently drawn according to this distribution. Show that the sample mean $T(y_1, y_2, \ldots, y_n) = \frac{1}{n}\sum_{k=1}^n y_k$ fulfills the condition expressed in equation (2.40).

EXERCISE 2.23 (Irrelevance and operational irrelevance) *Let the hypothesis H be related to the observables (U, V) via the channel $P_{U,V|H}$ and for simplicity assume that $P_{U|H}(u|h) > 0$ and $P_{V|U,H}(v|u, h) > 0$ for every $h \in \mathcal{H}$, $v \in \mathcal{V}$, and $u \in \mathcal{U}$. We say that V is operationally irrelevant if a MAP decoder that observes (U, V) achieves the same probability of error as one that observes only U, and this is true regardless of P_H. We now prove that irrelevance and operational irrelevance imply one another. We have already proved that irrelevance implies operational irrelevance. Hence it suffices to show that operational irrelevance implies irrelevance or, equivalently, that if V is not irrelevant, then it is not operationally irrelevant. We will prove the latter statement. We begin with a few observations that are instructive. By definition, V irrelevant means $H \to U \to V$. Hence V irrelevant is equivalent to the statement that, conditioned on U, the random variables H and V are independent. This gives us one intuitive explanation about why V is operationally irrelevant when $H \to U \to V$. Once we observe that $U = u$, we can restate the hypothesis testing problem in terms of a hypothesis H and an observable V that are independent (conditioned on $U = u$) and because of independence, from V we learn nothing about H. But if V is not irrelevant, then there is at least a u, call it $u^\star$, for which H and V are not independent conditioned on $U = u^\star$. It is when such a u is observed that we should be able to prove that V affects the decision. This suggests that the problem we are trying to solve is intimately related to the simpler problem that involves the hypothesis H and the observable V and the two are not independent. We begin with this problem and then we generalize.*

2.13. Exercises

(a) Let the hypothesis be $H \in \mathcal{H}$ (of yet unspecified distribution) and let the observable $V \in \mathcal{V}$ be related to H via an arbitrary but fixed channel $P_{V|H}$. Show that if V is not independent of H then there are distinct elements $i, j \in \mathcal{H}$ and distinct elements $k, l \in \mathcal{V}$ such that

$$P_{V|H}(k|i) > P_{V|H}(k|j) \\ P_{V|H}(l|i) < P_{V|H}(l|j). \tag{2.42}$$

Hint: For every $h \in \mathcal{H}$, $\sum_{v \in \mathcal{V}} P_{V|H}(v|h) = 1$.

(b) Under the condition of part (a), show that there is a distribution P_H for which the observable V affects the decision of a MAP decoder.

(c) Generalize to show that if the observables are U and V, and $P_{U,V|H}$ is fixed so that $H \to U \to V$ does not hold, then there is a distribution on H for which V is not operationally irrelevant. Hint: Argue as in parts (a) and (b) for the case $U = u^\star$, where $u^\star$ is as described above.

EXERCISE 2.24 (Antipodal signaling) *Consider the signal constellation shown in Figure 2.26.*

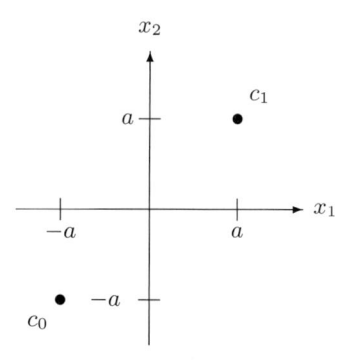

Figure 2.26.

Assume that the codewords c_0 and c_1 are used to communicate over the discrete-time AWGN channel. More precisely:

$$H = 0: \quad Y = c_0 + Z, \\ H = 1: \quad Y = c_1 + Z,$$

where $Z \sim \mathcal{N}(0, \sigma^2 I_2)$. Let $Y = (Y_1, Y_2)^\mathsf{T}$.

(a) Argue that Y_1 is not a sufficient statistic.
(b) Give a different signal constellation with two codewords $\tilde{c}_0$ and $\tilde{c}_1$ such that, when used in the above communication setting, Y_1 is a sufficient statistic.

EXERCISE 2.25 (Is it a sufficient statistic?) *Consider the following binary hypothesis testing problem*

$$H = 0: \quad Y = c_0 + Z \\ H = 1: \quad Y = c_1 + Z,$$

where $c_0 = (1,1)^\mathsf{T} = -c_1$ and $Z \sim \mathcal{N}(0, \sigma^2 I_2)$.

(a) Can the error probability of an ML decoder that observes $Y = (Y_1, Y_2)^\mathsf{T}$ be lower than that of an ML decoder that observes $Y_1 + Y_2$?

(b) Argue whether or not $H \to (Y_1 + Y_2) \to Y$ forms a Markov chain. Your argument should rely on first principles. Hint 1: Y is in a one-to-one relationship with $(Y_1 + Y_2, Y_1 - Y_2)$. Hint 2: argue that the random variables $Z_1 + Z_2$ and $Z_1 - Z_2$ are statistically independent.

Exercises for Section 2.6

EXERCISE 2.26 (Union bound) Let $Z \sim \mathcal{N}(c, \sigma^2 I_2)$ be a random vector that takes values in $\mathbb{R}^2$, where $c = (2,1)^\mathsf{T}$. Find a non-trivial upper bound to the probability that Z is in the shaded region of Figure 2.27.

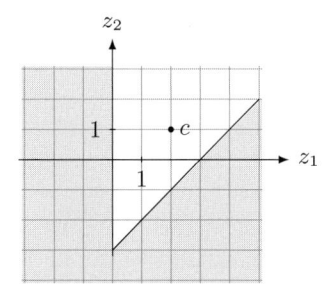

Figure 2.27.

EXERCISE 2.27 (QAM with erasure) Consider a QAM receiver that outputs a special symbol δ (called erasure) whenever the observation falls in the shaded area shown in Figure 2.28 and does minimum-distance decoding otherwise. (This is neither a MAP nor an ML receiver.) Assume that $c_0 \in \mathbb{R}^2$ is transmitted and that $Y = c_0 + N$ is received where $N \sim \mathcal{N}(0, \sigma^2 I_2)$. Let P_{0i}, $i = 0, 1, 2, 3$ be the probability that the receiver outputs $\hat{H} = i$ and let $P_{0\delta}$ be the probability that it outputs δ. Determine P_{00}, P_{01}, P_{02}, P_{03}, and $P_{0\delta}$.

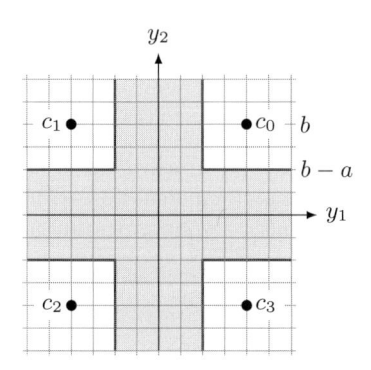

Figure 2.28.

2.13. Exercises

Comment: *If we choose $b - a$ large enough, we can make sure that the probability of the error is very small (we say that an error occurred if $\hat{H} = i$, $i \in \{0, 1, 2, 3\}$ and $H \neq \hat{H}$). When $\hat{H} = \delta$, the receiver can ask for a retransmission of H. This requires a feedback channel from the receiver to the transmitter. In most practical applications, such a feedback channel is available.*

EXERCISE 2.28 (Repeat codes and Bhattacharyya bound) *Consider two equally likely hypotheses. Under hypothesis $H = 0$, the transmitter sends $c_0 = (1, \ldots, 1)^\mathsf{T}$ and under $H = 1$ it sends $c_1 = (-1, \ldots, -1)^\mathsf{T}$, both of length n. The channel model is AWGN with variance σ^2 in each component. Recall that the probability of error for an ML receiver that observes the channel output $Y \in \mathbb{R}^n$ is*

$$P_e = Q\left(\frac{\sqrt{n}}{\sigma}\right).$$

Suppose now that the decoder has access only *to the sign of Y_i, $1 \leq i \leq n$, i.e. it observes*

$$W = (W_1, \ldots, W_n) = (\mathrm{sign}(Y_1), \ldots, \mathrm{sign}(Y_n)). \tag{2.43}$$

(a) *Determine the MAP decision rule based on the observable W. Give a simple sufficient statistic.*
(b) *Find the expression for the probability of error $\tilde{P}_e$ of the MAP decoder that observes W. You may assume that n is odd.*
(c) *Your answer to (b) contains a sum that cannot be expressed in closed form. Express the Bhattacharyya bound on $\tilde{P}_e$.*
(d) *For $n = 1, 3, 5, 7$, find the numerical values of P_e, $\tilde{P}_e$, and the Bhattacharyya bound on $\tilde{P}_e$.*

EXERCISE 2.29 (Tighter union Bhattacharyya bound: Binary case) *In this problem we derive a tighter version of the* union Bhattacharyya bound *for binary hypotheses. Let*

$$H = 0: \quad Y \sim f_{Y|H}(y|0)$$
$$H = 1: \quad Y \sim f_{Y|H}(y|1).$$

The MAP decision rule is

$$\hat{H}(y) = \arg\max_i P_H(i) f_{Y|H}(y|i),$$

and the resulting probability of error is

$$P_e = P_H(0) \int_{\mathcal{R}_1} f_{Y|H}(y|0) dy + P_H(1) \int_{\mathcal{R}_0} f_{Y|H}(y|1) dy.$$

(a) *Argue that*

$$P_e = \int_y \min\left\{P_H(0) f_{Y|H}(y|0), P_H(1) f_{Y|H}(y|1)\right\} dy.$$

(b) *Prove that for $a, b \geq 0$, $\min(a,b) \leq \sqrt{ab} \leq \frac{a+b}{2}$. Use this to prove the tighter version of the* Bhattacharyya bound, *i.e.*

$$P_e \leq \frac{1}{2} \int_y \sqrt{f_{Y|H}(y|0) f_{Y|H}(y|1)} dy.$$

(c) *Compare the above bound to (2.19) when there are two equiprobable hypotheses. How do you explain the improvement by a factor $\frac{1}{2}$?*

EXERCISE 2.30 (Tighter union Bhattacharyya bound: *M*-ary case) *In this problem we derive a tight version of the union bound for M-ary hypotheses. Let us analyze the following M-ary MAP detector:*

$$\hat{H}(y) = \text{smallest } i \text{ such that}$$
$$P_H(i) f_{Y|H}(y|i) = \max_j \{P_H(j) f_{Y|H}(y|j)\}.$$

Let

$$\mathcal{B}_{i,j} = \begin{cases} y : P_H(j) f_{Y|H}(y|j) \geq P_H(i) f_{Y|H}(y|i), & j < i \\ y : P_H(j) f_{Y|H}(y|j) > P_H(i) f_{Y|H}(y|i), & j > i. \end{cases}$$

(a) *Verify that $\mathcal{B}_{i,j} = \mathcal{B}_{j,i}^c$.*
(b) *Given $H = i$, the detector will make an error if and only if $y \in \bigcup_{j:j\neq i} \mathcal{B}_{i,j}$. The probability of error is $P_e = \sum_{i=0}^{M-1} P_e(i) P_H(i)$. Show that:*

$$P_e \leq \sum_{i=0}^{M-1} \sum_{j>i} [Pr\{Y \in \mathcal{B}_{i,j} | H = i\} P_H(i) + Pr\{Y \in \mathcal{B}_{j,i} | H = j\} P_H(j)]$$

$$= \sum_{i=0}^{M-1} \sum_{j>i} \left[\int_{\mathcal{B}_{i,j}} f_{Y|H}(y|i) P_H(i) dy + \int_{\mathcal{B}_{i,j}^c} f_{Y|H}(y|j) P_H(j) dy \right]$$

$$= \sum_{i=0}^{M-1} \sum_{j>i} \left[\int_y \min \{f_{Y|H}(y|i) P_H(i), f_{Y|H}(y|j) P_H(j)\} dy \right].$$

To prove the last part, go back to the definition of $\mathcal{B}_{i,j}$.
(c) *Hence show that:*

$$P_e \leq \sum_{i=0}^{M-1} \sum_{j>i} \left[\left(\frac{P_H(i) + P_H(j)}{2} \right) \int_y \sqrt{f_{Y|H}(y|i) f_{Y|H}(y|j)} dy \right].$$

Hint: *For $a, b \geq 0, \min(a,b) \leq \sqrt{ab} \leq \frac{a+b}{2}$.*

EXERCISE 2.31 (Applying the tight Bhattacharyya bound) *As an application of the tight Bhattacharyya bound (Exercise 2.29), consider the following binary hypothesis testing problem*

$$H = 0 : \quad Y \sim \mathcal{N}(-a, \sigma^2)$$
$$H = 1 : \quad Y \sim \mathcal{N}(+a, \sigma^2),$$

where the two hypotheses are equiprobable.

2.13. Exercises

(a) Use the tight Bhattacharyya bound to derive a bound on P_e.
(b) We know that the probability of error for this binary hypothesis testing problem is $Q(\frac{a}{\sigma}) \leq \frac{1}{2}\exp\left(-\frac{a^2}{2\sigma^2}\right)$, where we have used the result $Q(x) \leq \frac{1}{2}\exp\left(-\frac{x^2}{2}\right)$. How do the two bounds compare? Comment on the result.

EXERCISE 2.32 (Bhattacharyya bound for DMCs) *Consider a discrete memoryless channel (DMC). This is a channel model described by an input alphabet $\mathcal{X}$, an output alphabet $\mathcal{Y}$, and a transition probability[8] $P_{Y|X}(y|x)$. When we use this channel to transmit an n-tuple $x \in \mathcal{X}^n$, the transition probability is*

$$P_{Y|X}(y|x) = \prod_{i=1}^{n} P_{Y|X}(y_i|x_i).$$

So far, we have come across two DMCs, namely the BSC (binary symmetric channel) and the BEC (binary erasure channel). The purpose of this problem is to see that for DMCs, the Bhattacharyya bound takes a simple form, in particular when the channel input alphabet $\mathcal{X}$ contains only two letters.

(a) *Consider a transmitter that sends $c_0 \in \mathcal{X}^n$ and $c_1 \in \mathcal{X}^n$ with equal probability. Justify the following chain of (in)equalities.*

$$P_e \overset{(a)}{\leq} \sum_y \sqrt{P_{Y|X}(y|c_0)P_{Y|X}(y|c_1)}$$

$$\overset{(b)}{=} \sum_y \sqrt{\prod_{i=1}^{n} P_{Y|X}(y_i|c_{0,i})P_{Y|X}(y_i|c_{1,i})}$$

$$\overset{(c)}{=} \sum_{y_1,\ldots,y_n} \prod_{i=1}^{n} \sqrt{P_{Y|X}(y_i|c_{0,i})P_{Y|X}(y_i|c_{1,i})}$$

$$\overset{(d)}{=} \sum_{y_1} \sqrt{P_{Y|X}(y_1|c_{0,1})P_{Y|X}(y_1|c_{1,1})}$$

$$\cdots \sum_{y_n} \sqrt{P_{Y|X}(y_n|c_{0,n})P_{Y|X}(y_n|c_{1,n})}$$

$$\overset{(e)}{=} \prod_{i=1}^{n} \sum_y \sqrt{P_{Y|X}(y|c_{0,i})P_{Y|X}(y|c_{1,i})}$$

$$\overset{(f)}{=} \prod_{a \in \mathcal{X}, b \in \mathcal{X}, a \neq b} \left(\sum_y \sqrt{P_{Y|X}(y|a)P_{Y|X}(y|b)}\right)^{n(a,b)},$$

where $n(a,b)$ is the number of positions i in which $c_{0,i} = a$ and $c_{1,i} = b$.

---

[8] Here we are assuming that the output alphabet is discrete. Otherwise we use densities instead of probabilities.

(b) The Hamming distance $d_H(c_0, c_1)$ is defined as the number of positions in which c_0 and c_1 differ. Show that for a binary input channel, i.e. when $\mathcal{X} = \{a, b\}$, the Bhattacharyya bound becomes
$$P_e \leq z^{d_H(c_0, c_1)},$$
where
$$z = \sum_y \sqrt{P_{Y|X}(y|a) P_{Y|X}(y|b)}.$$
Notice that z depends only on the channel, whereas its exponent depends only on c_0 and c_1.

(c) Evaluate the channel parameter z for the following.
  (i) The binary input Gaussian channel described by the densities
$$f_{Y|X}(y|0) = \mathcal{N}(-\sqrt{E}, \sigma^2)$$
$$f_{Y|X}(y|1) = \mathcal{N}(\sqrt{E}, \sigma^2).$$
  (ii) The binary symmetric channel (BSC) with $\mathcal{X} = \mathcal{Y} = \{\pm 1\}$ and transition probabilities described by
$$P_{Y|X}(y|x) = \begin{cases} 1 - \delta, & \text{if } y = x, \\ \delta, & \text{otherwise.} \end{cases}$$
  (iii) The binary erasure channel (BEC) with $\mathcal{X} = \{\pm 1\}$, $\mathcal{Y} = \{-1, E, 1\}$, and transition probabilities given by
$$P_{Y|X}(y|x) = \begin{cases} 1 - \delta, & \text{if } y = x, \\ \delta, & \text{if } y = E, \\ 0, & \text{otherwise.} \end{cases}$$

EXERCISE 2.33 (Bhattacharyya bound and Laplacian noise) *Assuming two equiprobable hypotheses, evaluate the Bhattacharyya bound for the following (Laplacian noise) setting:*
$$H = 0: \quad Y = -a + Z$$
$$H = 1: \quad Y = a + Z,$$
where $a \in \mathbb{R}_+$ is a constant and Z is a random variable of probability density function $f_Z(z) = \frac{1}{2} \exp(-|z|)$, $z \in \mathbb{R}$.

EXERCISE 2.34 (Dice tossing) *You have two dice, one fair and one biased. A friend tells you that the biased die produces a 6 with probability $\frac{1}{4}$, and produces the other values with uniform probabilities. You do not know a priori which of the two is a fair die. You chose with uniform probabilities one of the two dice, and perform n consecutive tosses. Let $Y_i \in \{1, \ldots, 6\}$ be the random variable modeling the ith experiment and let $Y = (Y_1, \cdots, Y_n)$.*

(a) *Based on the observable Y, find the decision rule to determine whether the die you have chosen is biased. Your rule should maximize the probability that the decision is correct.*

2.13. Exercises

(b) *Identify a sufficient statistic* $S \in \mathbb{N}$.

(c) *Find the Bhattacharyya bound on the probability of error. You can either work with the observable* $(Y_1, \ldots, Y_n)$ *or with* $(Z_1, \ldots, Z_n)$, *where* Z_i *indicates whether the ith observation is a 6 or not. Yet another alternative is to work with* S. *Depending on the approach, the following may be useful:* $\sum_{i=0}^{n} \binom{n}{i} x^i = (1+x)^n$ *for* $n \in \mathbb{N}$.

EXERCISE 2.35 (ML receiver and union bound for orthogonal signaling) *Let* $H \in \{1, \ldots, m\}$ *be uniformly distributed and consider the communication problem described by:*

$$H = i : \qquad Y = c_i + Z, \quad Z \sim \mathcal{N}(0, \sigma^2 I_m),$$

where $c_1, \ldots, c_m$, $c_i \in \mathbb{R}^m$, *is a set of constant-energy orthogonal codewords. Without loss of generality we assume*

$$c_i = \sqrt{\mathcal{E}} e_i,$$

where e_i *is the ith unit vector in* $\mathbb{R}^m$, *i.e. the vector that contains* 1 *at position i and* 0 *elsewhere, and* $\mathcal{E}$ *is some positive constant.*

(a) *Describe the maximum-likelihood decision rule.*
(b) *Find the distances* $\|c_i - c_j\|$, $i \neq j$.
(c) *Using the union bound and the Q function, upper bound the probability* $P_e(i)$ *that the decision is incorrect when* $H = i$.

Exercises for Section 2.9

EXERCISE 2.36 (Uniform polar to Cartesian) *Let* R *and* Φ *be independent random variables.* R *is distributed uniformly over the unit interval,* Φ *is distributed uniformly over the interval* $[0, 2\pi)$.

(a) *Interpret* R *and* Φ *as the polar coordinates of a point in the plane. It is clear that the point lies inside (or on) the unit circle. Is the distribution of the point uniform over the unit disk? Take a guess!*
(b) *Define the random variables*

$$X = R \cos \Phi$$
$$Y = R \sin \Phi.$$

Find the joint distribution of the random variables X *and* Y *by using the Jacobian determinant.*

(c) *Does the result of part (b) support or contradict your guess from part (a)? Explain.*

Exercises for Section 2.10

EXERCISE 2.37 (Real-valued Gaussian random variables) *For the purpose of this exercise, two zero-mean real-valued Gaussian random variables* X *and* Y *are* called jointly *Gaussian* if and only if their joint density is

$$f_{XY}(x,y) = \frac{1}{2\pi\sqrt{\det \Sigma}} \exp\left(-\frac{1}{2}(x,y)\Sigma^{-1}(x,y)^\mathsf{T}\right), \qquad (2.44)$$

where (for zero-mean random vectors) the so-called covariance matrix Σ is

$$\Sigma = E\left[(X,Y)^\mathsf{T}(X,Y)\right] = \begin{pmatrix} \sigma_X^2 & \sigma_{XY} \\ \sigma_{XY} & \sigma_Y^2 \end{pmatrix}. \qquad (2.45)$$

(a) Show that if X and Y are zero-mean jointly Gaussian random variables, then X is a zero-mean Gaussian random variable, and so is Y.

(b) Show that if X and Y are independent zero-mean Gaussian random variables, then X and Y are zero-mean jointly Gaussian random variables.

(c) However, if X and Y are Gaussian random variables but not independent, then X and Y are not necessarily jointly Gaussian. Give an example where X and Y are Gaussian random variables, yet they are not jointly Gaussian.

(d) Let X and Y be independent Gaussian random variables with zero mean and variance σ_X^2 and σ_Y^2, respectively. Find the probability density function of $Z = X + Y$.

Observe that no computation is required if we use the definition of jointly Gaussian random variables given in Appendix 2.10.

EXERCISE 2.38 (Correlation vs. independence) *Let Z be a random variable with probability density function*

$$f_Z(z) = \begin{cases} 1/2, & -1 \leq z \leq 1 \\ 0, & \text{otherwise.} \end{cases}$$

Also, let $X = Z$ and $Y = Z^2$.

(a) Show that X and Y are uncorrelated.

(b) Are X and Y independent?

(c) Now let X and Y be jointly Gaussian, zero mean, uncorrelated with variances σ_X^2 and σ_Y^2, respectively. Are X and Y independent? Justify your answer.

Miscellaneous exercises

EXERCISE 2.39 (Data-storage channel) *The process of storing and retrieving binary data on a thin-film disk can be modeled as transmitting binary symbols across an additive white Gaussian noise channel where the noise Z has a variance that depends on the transmitted (stored) binary symbol X. The noise has the following input-dependent density:*

$$f_Z(z) = \begin{cases} \frac{1}{\sqrt{2\pi\sigma_1^2}} e^{-\frac{z^2}{2\sigma_1^2}} & \text{if } X = 1 \\ \frac{1}{\sqrt{2\pi\sigma_0^2}} e^{-\frac{z^2}{2\sigma_0^2}} & \text{if } X = 0, \end{cases}$$

where $\sigma_1 > \sigma_0$. The channel inputs are equally likely.

(a) On the same graph, plot the two possible output probability density functions. Indicate, qualitatively, the decision regions.
(b) Determine the optimal receiver in terms of σ_0 and σ_1.
(c) Write an expression for the error probability P_e as a function of σ_0 and σ_1.

EXERCISE 2.40 (A simple multiple-access scheme) Consider the following very simple model of a multiple-access scheme. There are two users. Each user has two hypotheses. Let $\mathcal{H}_1 = \mathcal{H}_2 = \{0, 1\}$ denote the respective set of hypotheses and assume that both users employ a uniform prior. Further, let X_1 and X_2 be the respective signals sent by user one and two. Assume that the transmissions of both users are independent and that $X_1 \in \{\pm 1\}$ and $X_2 \in \{\pm 2\}$ where X_1 and X_2 are positive if their respective hypothesis is zero and negative otherwise. Assume that the receiver observes the signal $Y = X_1 + X_2 + Z$, where Z is a zero-mean Gaussian random variable with variance σ^2 and is independent of the transmitted signal.

(a) Assume that the receiver observes Y and wants to estimate both H_1 and H_2. Let $\hat{H}_1$ and $\hat{H}_2$ be the estimates. What is the generic form of the optimal decision rule?
(b) For the specific set of signals given, what is the set of possible observations, assuming that $\sigma^2 = 0$? Label these signals by the corresponding (joint) hypotheses.
(c) Assuming now that $\sigma^2 > 0$, draw the optimal decision regions.
(d) What is the resulting probability of correct decision? That is, determine the probability $Pr\{\hat{H}_1 = H_1, \hat{H}_2 = H_2\}$.
(e) Finally, assume that we are interested in only the transmission of user two. Describe the receiver that minimizes the error probability and determine $Pr\{\hat{H}_2 = H_2\}$.

EXERCISE 2.41 (Data-dependent noise) Consider the following binary Gaussian hypothesis testing problem with data dependent noise. Under hypothesis $H = 0$ the transmitted signal is $c_0 = -1$ and the received signal is $Y = c_0 + Z_0$, where Z_0 is zero-mean Gaussian with variance one. Under hypothesis $H = 1$ the transmitted signal is $c_1 = 1$ and the received signal is $Y = c_1 + Z_1$, where Z_1 is zero-mean Gaussian with variance σ^2. Assume that the prior is uniform.

(a) Write the optimal decision rule as a function of the parameter σ^2 and the received signal Y.
(b) For the value $\sigma^2 = e^4$ compute the decision regions.
(c) Give expressions as simple as possible for the error probabilities $P_e(0)$ and $P_e(1)$.

EXERCISE 2.42 (Correlated noise) Consider the following communication problem. The message is represented by a uniformly distributed random variable H, that takes values in $\{0, 1, 2, 3\}$. When $H = i$ we send c_i, where $c_0 = (0, 1)^\mathsf{T}$, $c_1 = (1, 0)^\mathsf{T}$, $c_2 = (0, -1)^\mathsf{T}$, $c_3 = (-1, 0)^\mathsf{T}$ (see Figure 2.29). When $H = i$, the receiver observes the vector $Y = c_i + Z$, where Z is a zero-mean Gaussian random vector of covariance matrix $\Sigma = \begin{pmatrix} 4 & 2 \\ 2 & 5 \end{pmatrix}$.

```
                    x₂
                    ▲
                 1  •  c₀
          c₃        │      c₁
          •─────────┼──────•────▶ x₁
          -1        │      1
                 -1 •  c₂
```

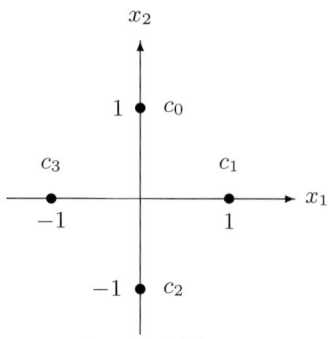

Figure 2.29.

(a) In order to simplify the decision problem, we transform Y into $\hat{Y} = BY = Bc_i + BZ$, where B is a 2-by-2 invertible matrix, and use $\hat{Y}$ as a sufficient statistic. Find a B such that BZ is a zero-mean Gaussian random vector with independent and identically distributed components. Hint: If $A = \frac{1}{4}\begin{pmatrix} 2 & 0 \\ -1 & 2 \end{pmatrix}$, then $A\Sigma A^\mathsf{T} = I$, with $I = \begin{pmatrix} 1 & 0 \\ 0 & 1 \end{pmatrix}$.

(b) Formulate the new hypothesis testing problem that has $\hat{Y}$ as the observable and depict the decision regions.

(c) Give an upper bound to the error probability in this decision problem.

3 Receiver design for the continuous-time AWGN channel: Second layer

3.1 Introduction

In Chapter 2 we focused on the receiver for the discrete-time AWGN (additive white Gaussian noise) channel. In this chapter, we address the same problem for a channel model closer to reality, namely the *continuous-time AWGN channel*. Apart from the channel model, the assumptions and the goal are the same: We assume that the source statistic, the transmitter, and the channel are given to us and we seek to understand what the receiver has to do to minimize the error probability. We are also interested in the resulting error probability, but this follows from Chapter 2 with no extra work. The setup is shown in Figure 3.1.

The channel of Figure 3.1 captures the most important aspect of all real-world channels, namely the presence of additive noise. Owing to the central limit theorem, the assumption that the noise is Gaussian is often a very good one. In Section 3.6 we discuss additional channel properties that also affect the design and performance of a communication system.

EXAMPLE 3.1 *A cable is a good example of a channel that can be modeled by the continuous-time AWGN channel. If the cable's frequency response cannot be considered as constant over the signal's bandwidth, then the cable's filtering effect also needs to be taken into consideration. We discuss this in Section 3.6. Another good example is the channel between the antenna of a geostationary satellite and the antenna of the corresponding Earth station. For the communication in either direction we can consider the model of Figure 3.1.*   □

Although our primary focus is on the receiver, in this chapter we also gain valuable insight into the transmitter structure. First we need to introduce the notion of signal's energy and specify two mild technical restrictions that we impose on the signal set $\mathcal{W} = \{w_0(t), \ldots, w_{m-1}(t)\}$.

EXAMPLE 3.2 *Suppose that $w_i(t)$ is the voltage feeding the antenna of a transmitter when $H = i$. An antenna has an internal impedance Z. A typical value for Z is 50 ohms. Assuming that Z is purely resistive, the current at the feeding point is $w_i(t)/Z$, the instantaneous power is $w_i^2(t)/Z$, and the energy transferred to the antenna is $\frac{1}{Z}\int w_i^2(t)dt$. Alternatively, if the $w_i(t)$ is the current feeding the antenna when $H = i$, the voltage at the feeding point is $w_i(t)Z$, the instantaneous power is*

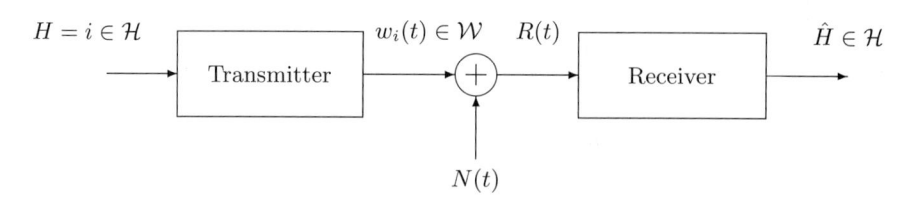

Figure 3.1. Communication across the continuous-time AWGN channel.

$w_i^2(t)Z$, and the energy is $Z \int w_i^2(t)dt$. In both cases the energy is proportional to $\|w_i\|^2 = \int |w_i(t)|^2 dt$. □

As in the above example, the squared norm of a signal $w_i(t)$ is generally associated with the signal's energy. It is quite natural to assume that we communicate via finite-energy signals. This is the first restriction on $\mathcal{W}$. A linear combination of a finite number of finite-energy signals is itself a finite-energy signal. Hence, every vector of the vector space $\mathcal{V}$ spanned by $\mathcal{W}$ is a square-integrable function. The second requirement is that if $v \in \mathcal{V}$ has a vanishing norm, then $v(t)$ vanishes for all t. Together, these requirements imply that $\mathcal{V}$ is an inner product space of square-integrable functions. (See Example 2.39.)

EXAMPLE 3.3 (Continuous functions) *Let $v : \mathbb{R} \to \mathbb{R}$ be a continuous function and suppose that $|v(t_0)| = a$ for some t_0 and some positive a. By continuity, there exists an $\epsilon > 0$ such that $|v(t)| > \frac{a}{2}$ for all $t \in \mathcal{I}$ where $\mathcal{I} = (t_0 - \frac{\epsilon}{2}, t_0 + \frac{\epsilon}{2})$. It follows that*

$$\|v(t)\|^2 \geq \|v(t)\mathbb{1}\{t \in \mathcal{I}\}\|^2 \geq \frac{a^2 \epsilon}{4} > 0.$$

We conclude that if a continuous function has a vanishing norm, then the function vanishes everywhere. □

All signals that represent real-world communication signals are finite-energy and continuous. Hence the vector space they span is always an inner product space.

This is a good place to mention the various reasons we are interested in the signal's energy or, somewhat equivalently, in the signal's power, which is the energy per second. First, for safety and for spectrum reusability, there are regulations that limit the power of a transmitted signal. Second, for mobile devices, the energy of the transmitted signal comes from the battery: a battery charge lasts longer if we decrease the signal's power. Third, with no limitation to the signal's power, we can transmit across a continuous-time AWGN channel at any desired rate, regardless of the available bandwidth and of the target error probability. Hence, it would be unfair to compare signaling methods that do not use the same power.

For now, we assume that $\mathcal{W}$ is given to us. The problem of choosing a suitable set $\mathcal{W}$ of signals will be studied in subsequent chapters.

The highlight of the chapter is the power of abstraction. The receiver design for the *discrete-time* AWGN channel relied on geometrical ideas that can be

3.2. White Gaussian noise

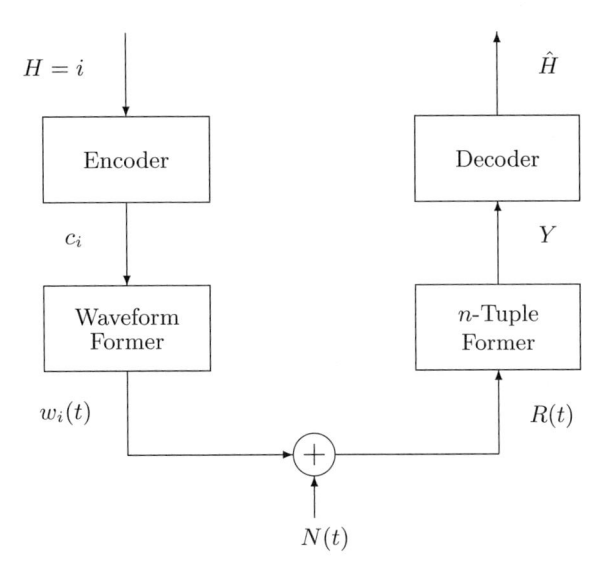

Figure 3.2. Waveform channel abstraction.

formulated whenever we are in an inner product space. We will use the same ideas for the *continuous-time* AWGN channel.

The main result is a decomposition of the sender and the receiver into the building blocks shown in Figure 3.2. We will see that, without loss of generality, we can (and should) think of the transmitter as consisting of an *encoder* that maps the message $i \in \mathcal{H}$ into an n-tuple c_i, as in the previous chapter, followed by a *waveform former* that maps c_i into a waveform $w_i(t)$. Similarly, we will see that the receiver can consist of an *n-tuple former* that takes the channel output and produces an n-tuple Y. The behavior from the waveform former input to the n-tuple former output is that of the discrete-time AWGN channel considered in the previous chapter. Hence we know already what the *decoder* of Figure 3.2 should do with the n-tuple former output.

In this chapter (like in the previous one) the vectors (functions) are real-valued. Hence, we could use the formalism that applies to real inner product spaces. Yet, in preparation of Chapter 7, we use the formalism for complex inner product spaces. This mainly concerns the standard inner product between functions, where we write $\langle a, b \rangle = \int a(t)b^*(t)dt$ instead of $\langle a, b \rangle = \int a(t)b(t)dt$. A similar comment applies to the definition of covariance, where for zero-mean random variables we use $\text{cov}(Z_i, Z_j) = \mathbb{E}\left[Z_i Z_j^*\right]$ instead of $\text{cov}(Z_i, Z_j) = \mathbb{E}\left[Z_i Z_j\right]$.

3.2 White Gaussian noise

The purpose of this section is to introduce the basics of *white Gaussian noise* $N(t)$. The standard approach is to give a mathematical description of $N(t)$, but this requires measure theory if done rigorously. The good news is that a mathematical model of $N(t)$ is not needed because $N(t)$ is not observable through physical experiments. (The reason will become clear shortly.) Our approach is to model what we can actually measure. We assume a working knowledge of Gaussian random vectors (reviewed in Appendix 2.10).

A receiver is an electrical instrument that connects to the channel output via a cable. For instance, in wireless communication, we might consider the channel output to be the output of the receiving antenna; in which case, the cable is the one that connects the antenna to the receiver. A cable is a linear time-invariant filter. Hence, we can assume that all the observations made by the receiver are through some linear time-invariant filter.

So if $N(t)$ represents the noise introduced by the channel, the receiver sees, at best, a filtered version $Z(t)$ of $N(t)$. We model $Z(t)$ as a stochastic process and, as such, it is described by the statistic of $Z(t_1), Z(t_2), \ldots, Z(t_k)$ for any positive integer k and any finite collection of sampling times $t_1, t_2, \ldots, t_k$.

If the filter impulse response is $h(t)$, then linear system theory suggests that

$$Z(t) = \int N(\alpha) h(t-\alpha) d\alpha$$

and

$$Z(t_i) = \int N(\alpha) h(t_i - \alpha) d\alpha, \qquad (3.1)$$

but the validity of these expressions needs to be justified, because $N(t)$ is not a deterministic signal. It is possible to define $N(t)$ as a stochastic process and prove that the (Lebesgue) integral in (3.1) is well defined; but we avoid this path which, as already mentioned, requires measure theory. In this text, equation (3.1) is shorthand for the statement "$Z(t_i)$ is the random variable that models the output at time t_i of a linear time-invariant filter of impulse response $h(t)$ fed with white Gaussian noise $N(t)$". Notice that $h(t_i - \alpha)$ is a function of α that we can rename as $g_i(\alpha)$. Now we are in the position to define white Gaussian noise.

DEFINITION 3.4 $N(t)$ *is white Gaussian noise of power spectral density* $\frac{N_0}{2}$ *if, for any finite collection of real-valued* $\mathcal{L}_2$ *functions* $g_1(\alpha), \ldots, g_k(\alpha)$,

$$Z_i = \int N(\alpha) g_i(\alpha) d\alpha, \quad i = 1, 2, \ldots, k \qquad (3.2)$$

is a collection of zero-mean jointly Gaussian random variables of covariance

$$\operatorname{cov}(Z_i, Z_j) = \mathbb{E}\left[Z_i Z_j^*\right] = \frac{N_0}{2} \int g_i(t) g_j^*(t) dt = \frac{N_0}{2} \langle g_i, g_j \rangle. \qquad (3.3)$$

□

If we are not evaluating the integral in (3.2), how do we know if $N(t)$ is white Gaussian noise? In this text, when applicable, we say that $N(t)$ is white Gaussian noise, in which case we can use (3.3) as we see fit. In the real world, often we know enough about the channel to know whether or not its noise can be modeled as

3.3. Observables and sufficient statistics

white and Gaussian. This knowledge could come from a mathematical model of the channel. Another possibility is that we perform measurements and verify that they behave according to Definition 3.4.

Owing to its importance and frequent use, we formulate the following special case as a lemma. It is the most important fact that should be remembered about white Gaussian noise.

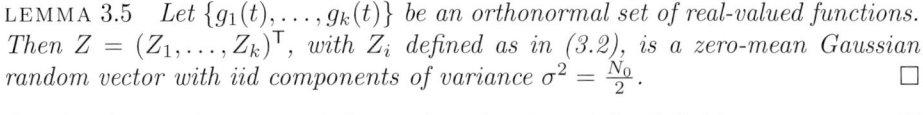

LEMMA 3.5 *Let $\{g_1(t), \ldots, g_k(t)\}$ be an orthonormal set of real-valued functions. Then $Z = (Z_1, \ldots, Z_k)^\mathsf{T}$, with Z_i defined as in (3.2), is a zero-mean Gaussian random vector with iid components of variance $\sigma^2 = \frac{N_0}{2}$.* □

Proof The proof is a straightforward application of the definitions. □

EXAMPLE 3.6 *Consider two bandpass filters that have non-overlapping frequency responses but are otherwise identical, i.e. if we frequency-translate the frequency response of one filter by the proper amount we obtain the frequency response of the other filter. By Parseval's relationship, the corresponding impulse responses are orthogonal to one another. If we feed the two filters with white Gaussian noise and sample their output (even at different times), we obtain two iid Gaussian random variables. We could extend the experiment (in the obvious way) to n filters of non-overlapping frequency responses, and would obtain n random variables that are iid – hence of identical variance. This explains why the noise is called white: like for white light, white Gaussian noise has its power equally distributed among all frequencies.* □

Are there other types of noise? Yes, there are. For instance, there are natural and man-made electromagnetic noises. The noise produced by electric motors and that produced by power lines are examples of man-made noise. Man-made noise is typically neither white nor Gaussian. The good news is that a careful design should be able to ensure that the receiver picks up a negligible amount of man-made noise (if any). Natural noise is unavoidable. Every conductor (resistor) produces thermal (Johnson) noise. (See Appendix 3.10.) The assumption that thermal noise is white and Gaussian is an excellent one. Other examples of natural noise are solar noise and cosmic noise. A receiving antenna picks up these noises, the intensity of which depends on the antenna's gain and pointing direction. A current in a conductor gives rise to shot noise. Shot noise originates from the discrete nature of the electric charges. Wikipedia is a good reference to learn more about various noise sources.

3.3 Observables and sufficient statistics

Recall that the setup is that of Figure 3.1, where $N(t)$ is white Gaussian noise. As discussed in Section 3.2, owing to the noise, the channel output $R(t)$ is not observable. What we can observe via physical experiments (measurements) are k-tuples $V = (V_1, \ldots, V_k)^\mathsf{T}$ such that

$$V_i = \int_{-\infty}^{\infty} R(\alpha) g_i^*(\alpha) d\alpha, \qquad i = 1, 2, \ldots, k, \tag{3.4}$$

where k is an arbitrary positive integer and $g_1(t), \ldots, g_k(t)$ are arbitrary finite-energy waveforms. The complex conjugate operator "*" on $g_i^*(\alpha)$ is superfluous for real-valued signals but, as we will see in Chapter 7, the baseband representation of a passband impulse response is complex-valued.

Notice that we assume that we can perform an arbitrarily large but *finite* number k of measurements. By disallowing infinite measurements we avoid distracting mathematical subtleties without losing anything of engineering relevance.

It is important to point out that the kind of measurements we consider is quite general. For instance, we can pass $R(t)$ through an ideal lowpass filter of cutoff frequency B for some huge B (say 10^{10} Hz) and collect an arbitrary large number of samples taken every $\frac{1}{2B}$ seconds so as to fulfill the sampling theorem (Theorem 5.2). In fact, by choosing $g_i(t) = h(\frac{i}{2B} - t)$, where $h(t)$ is the impulse response of the lowpass filter, V_i becomes the filter output sampled at time $t = \frac{i}{2B}$. As stated by the sampling theorem, from these samples we can reconstruct the filter output. If $R(t)$ consists of a signal plus noise, and the signal is bandlimited to less than B Hz, then from the samples we can reconstruct the signal plus the portion of the noise that has frequency components in $[-B, B]$.

Let $\mathcal{V}$ be the inner product space spanned by the elements of the signal set $\mathcal{W}$ and let $\{\psi_1(t), \ldots, \psi_n(t)\}$ be an arbitrary orthonormal basis for $\mathcal{V}$. We claim that the n-tuple $Y = (Y_1, \ldots, Y_n)^\mathsf{T}$ with ith component

$$Y_i = \int R(\alpha)\psi_i^*(\alpha)d\alpha$$

is a sufficient statistic (for the hypothesis H) among any collection of measurements that contains Y. To prove this claim, let $V = (V_1, \ldots, V_k)^\mathsf{T}$ be the collection of additional measurements made according to (3.4). Let $\mathcal{U}$ be the inner product space spanned by $\mathcal{V} \cup \{g_1(t), \ldots, g_k(t)\}$ and let $\{\psi_1(t), \ldots, \psi_n(t), \phi_1(t), \ldots, \phi_{\tilde{n}}(t)\}$ be an orthonormal basis for $\mathcal{U}$ obtained by extending the orthonormal basis $\{\psi_1(t), \ldots, \psi_n(t)\}$ for $\mathcal{V}$. Define

$$U_i = \int R(\alpha)\phi_i^*(\alpha)d\alpha, \qquad i = 1, \ldots, \tilde{n}.$$

It should be clear that we can recover V from Y and U. This is so because, from the projections onto a basis, we can obtain the projection onto any waveform in the span of the basis. Mathematically,

$$\begin{aligned}
V_i &= \int_{-\infty}^{\infty} R(\alpha)g_i^*(\alpha)d\alpha \\
&= \int_{-\infty}^{\infty} R(\alpha)\left[\sum_{j=1}^{n}\xi_{i,j}\psi_j(\alpha) + \sum_{j=1}^{\tilde{n}}\xi_{i,j+n}\phi_j(\alpha)\right]^* d\alpha \\
&= \sum_{j=1}^{n}\xi_{i,j}^*Y_j + \sum_{j=1}^{\tilde{n}}\xi_{i,j+n}^*U_j,
\end{aligned}$$

3.3. Observables and sufficient statistics

where $\xi_{i,1}, \ldots, \xi_{i,n+\tilde{n}}$ is the unique set of coefficients in the orthonormal expansion of $g_i(t)$ with respect to the basis $\{\psi_1(t), \ldots, \psi_n(t), \phi_1(t), \phi_2(t), \ldots, \phi_{\tilde{n}}(t)\}$.

Hence we can consider (Y, U) as *the observable* and it suffices to show that Y is a sufficient statistic. Note that when $H = i$,

$$Y_j = \int R(\alpha) \psi_j^*(\alpha) = \int \big(w_i(\alpha) + N(\alpha)\big) \psi_j^*(\alpha) d\alpha = c_{i,j} + Z_{|\mathcal{V},j},$$

where $c_{i,j}$ is the jth component of the n-tuple of coefficients c_i that represents the waveform $w_i(t)$ with respect to the chosen orthonormal basis, and $Z_{|\mathcal{V},j}$ is a zero-mean Gaussian random variable of variance $\frac{N_0}{2}$. The notation $Z_{|\mathcal{V},j}$ is meant to remind us that this random variable is obtained by "projecting" the noise onto the jth element of the chosen orthonormal basis for $\mathcal{V}$. Using n-tuple notation, we obtain the following statistic

$$H = i, \qquad Y = c_i + Z_{|\mathcal{V}},$$

where $Z_{|\mathcal{V}} \sim \mathcal{N}(0, \frac{N_0}{2} I_n)$. Similarly,

$$U_j = \int R(\alpha) \phi_j^*(\alpha) = \int \big(w_i(\alpha) + N(\alpha)\big) \phi_j^*(\alpha) d\alpha = \int N(\alpha) \phi_j^*(\alpha) d\alpha = Z_{\perp \mathcal{V},j},$$

where we used the fact that $w_i(t)$ is in the subspace spanned by $\{\psi_1(t), \ldots, \psi_n(t)\}$ and therefore it is orthogonal to $\phi_j(t)$ for each $j = 1, 2, \ldots, \tilde{n}$. The notation $Z_{\perp \mathcal{V},j}$ reminds us that this random variable is obtained by "projecting" the noise onto the jth element of an orthonormal basis that is orthogonal to $\mathcal{V}$. Using n-tuple notation, we obtain

$$H = i, \qquad U = Z_{\perp \mathcal{V}},$$

where $Z_{\perp \mathcal{V}} \sim \mathcal{N}(0, \frac{N_0}{2} I_{\tilde{n}})$. Furthermore, $Z_{|\mathcal{V}}$ and $Z_{\perp \mathcal{V}}$ are independent of each other and of H. The conditional density of Y, U given H is

$$f_{Y,U|H}(y, u|i) = f_{Y|H}(y|i) f_U(u).$$

From the Fisher–Neyman factorization theorem (Theorem 2.13, Chapter 2, with $h(y, u) = f_U(u)$, $T(y, u) = y$, and $g_i(T(y, u)) = f_{Y|H}(y|i)$), we see that Y is a sufficient statistic and U is irrelevant as claimed.

Figure 3.3 depicts what is going on, which we summarize as follows:

$Y = c_i + Z_{|\mathcal{V}}$ is a sufficient statistic: it is the projection of $R(t)$ onto the signal space $\mathcal{V}$;

$U = Z_{\perp \mathcal{V}}$ is irrelevant: it contains only independent noise.

Could we prove that a subset of the components of Y is *not* a sufficient statistic? Yes, we could. Here is the outline of a proof. Without loss of essential generality, let us think of Y as consisting of two parts, Y_a and Y_b. Similarly, we decompose every c_i into the corresponding parts c_{ia} and c_{ib}. The claim is that H followed by Y_a followed by (Y_a, Y_b) does *not* form a Markov chain in that order. In fact when $H = i$, Y_b consists of c_{ib} plus noise. Since c_{ib} cannot be deduced from Y_a in

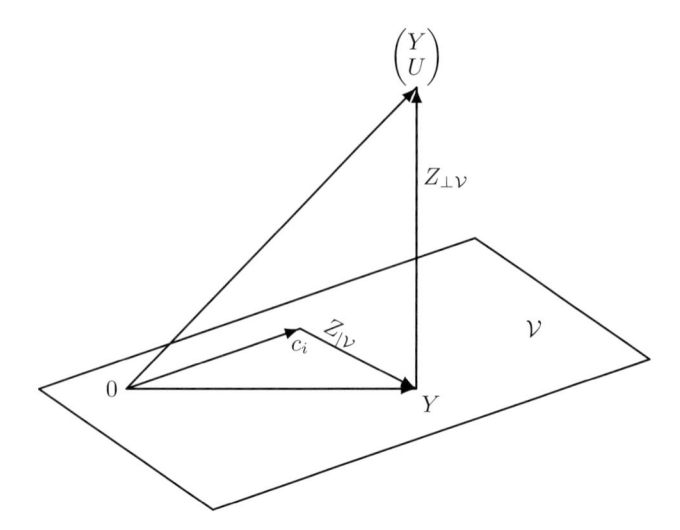

Figure 3.3. The vector of measurements $(Y^\mathsf{T}, U^\mathsf{T})^\mathsf{T}$ describes the projection of the received signal $R(t)$ onto $\mathcal{U}$. The vector Y describes the projection of $R(t)$ onto $\mathcal{V}$.

general (or else we would not bother sending c_{ib}), it follows that the statistic of Y_b depends on i even if we know the realization of Y_a.

3.4 Transmitter and receiver architecture

The results of the previous section tell us that a MAP receiver for the waveform AWGN channel can be structured as shown in Figure 3.4. We see that the receiver front end computes $Y \in \mathbb{R}^n$ from $R(t)$ in a block that we call *n-tuple former*. (The name is not standard.) Thus the n-tuple former performs a huge data reduction from the channel output $R(t)$ to the sufficient statistic Y. The hypothesis testing problem based on the observable Y is

$$H = i: \qquad Y = c_i + Z,$$

where $Z \sim \mathcal{N}(0, \frac{N_0}{2} I_n)$ is independent of H. This is precisely the hypothesis testing problem studied in Chapter 2 in conjunction with a transmitter that sends $c_i \in \mathbb{R}^n$ to signal message i across the *discrete-time* AWGN channel. As shown in the figure, we can also decompose the transmitter into a module that produces c_i, called *encoder*, and a module that produces $w_i(t)$, called *waveform former*. (Once again, the terminology is not standard.) Henceforth the n-tuple of coefficients c_i will be referred to as the *codeword* associated to $w_i(t)$. Figure 3.4 is the main result of the chapter. It implies that the decomposition of the transmitter and the receiver as depicted in Figure 3.2 is indeed completely general and it gives details about the waveform former and the n-tuple former.

Everything that we learned about a decoder for the discrete-time AWGN channel is applicable to the decoder of the continuous-time AWGN channel. Incidentally,

3.4. Transmitter and receiver architecture

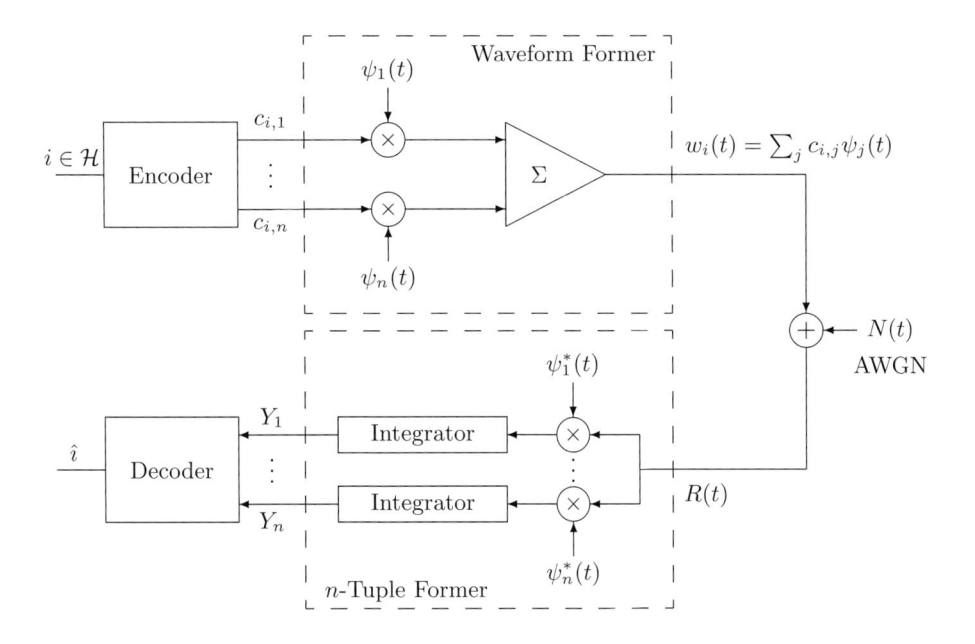

Figure 3.4. Decomposed transmitter and receiver.

the decomposition of Figure 3.4 is consistent with the layering philosophy of the OSI model (Section 1.1), in the sense that the encoder and decoder are designed as if they were talking to each other directly via a discrete-time AWGN channel. In reality, the channel seen by the encoder/decoder pair is the result of the "service" provided by the waveform former and the n-tuple former.

The above decomposition is useful for the system conception, for the performance analysis, as well as for the system implementation; but of course, we always have the option of implementing the transmitter as a straight map from the message set $\mathcal{H}$ to the waveform set $\mathcal{W}$ without passing through the codebook $\mathcal{C}$. Although such a straight map is a possibility and makes sense for relatively unsophisticated systems, the decomposition into an encoder and a waveform former is standard for modern designs. In fact, information theory, as well as coding theory, devote much attention to the study of encoder/decoder pairs.

The following example is meant to make two important points that apply when we communicate across the continuous-time AWGN channel and make an ML decision. First, sets of continuous-time signals may "look" very different yet they may share the same codebook, which is sufficient to guarantee that the error probability be the same; second, for binary constellations, what matters for the error probability is the distance between the two signals and nothing else.

EXAMPLE 3.7 (Orthogonal signals) *The following four choices of* $\mathcal{W} = \{w_0(t), w_1(t)\}$ *look very different yet, upon an appropriate choice of orthonormal basis, they share the same codebook* $\mathcal{C} = \{c_0, c_1\}$ *with* $c_0 = (\sqrt{\mathcal{E}}, 0)^\mathsf{T}$ *and* $c_1 = (0, \sqrt{\mathcal{E}})^\mathsf{T}$.

To see this, it suffices to verify that $\langle w_i, w_j \rangle$ equals $\mathcal{E}$ if $i = j$ and equals 0 otherwise. Hence the two signals are orthogonal to each other and they have squared norm $\mathcal{E}$. Figure 3.5 shows the signals and the associated codewords.

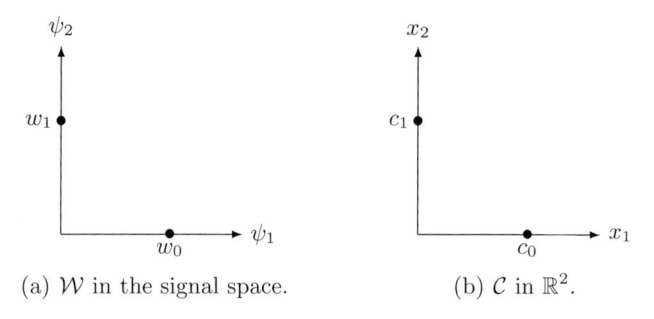

Figure 3.5. $\mathcal{W}$ and $\mathcal{C}$.

(a) $\mathcal{W}$ in the signal space.

(b) $\mathcal{C}$ in $\mathbb{R}^2$.

Choice 1 (Rectangular pulse position modulation):

$$w_0(t) = \sqrt{\frac{\mathcal{E}}{T}}\, \mathbb{1}\{t \in [0, T]\}$$

$$w_1(t) = \sqrt{\frac{\mathcal{E}}{T}}\, \mathbb{1}\{t \in [T, 2T]\},$$

where we have used the indicator function $\mathbb{1}\{t \in [a, b]\}$ to denote a rectangular pulse which is 1 in the interval $[a, b]$ and 0 elsewhere. Rectangular pulses can easily be generated, e.g. by a switch. They are used, for instance, to communicate a binary symbol within an electrical circuit. As we will see, in the frequency domain these pulses have side lobes that decay relatively slow, which is not desirable for high data rate over a channel for which bandwidth is at a premium.

Choice 2 (Frequency-shift keying):

$$w_0(t) = \sqrt{\frac{2\mathcal{E}}{T}} \sin\left(\pi k \frac{t}{T}\right) \mathbb{1}\{t \in [0, T]\}$$

$$w_1(t) = \sqrt{\frac{2\mathcal{E}}{T}} \sin\left(\pi l \frac{t}{T}\right) \mathbb{1}\{t \in [0, T]\},$$

where k and l are positive integers, $k \neq l$. With a large value of k and l, these signals could be used for wireless communication. To see that the two signals are orthogonal to each other we can use the trigonometric identity $\sin(\alpha)\sin(\beta) = 0.5[\cos(\alpha - \beta) - \cos(\alpha + \beta)]$.

Choice 3 (Sinc pulse position modulation):

$$w_0(t) = \sqrt{\frac{\mathcal{E}}{T}}\, \text{sinc}\left(\frac{t}{T}\right)$$

$$w_1(t) = \sqrt{\frac{\mathcal{E}}{T}}\, \text{sinc}\left(\frac{t - T}{T}\right).$$

3.4. Transmitter and receiver architecture

An advantage of sinc pulses is that they have a finite support in the frequency domain. By taking their Fourier transform, we quickly see that they are orthogonal to each other. See Appendix 5.10 for details.

Choice 4 (Spread spectrum):

$$w_0(t) = \sqrt{\mathcal{E}}\psi_1(t), \quad \text{with } \psi_1(t) = \sqrt{\frac{1}{T}}\sum_{j=1}^{n} s_{0,j}\mathbb{1}\left\{t - j\frac{T}{n} \in \left[0, \frac{T}{n}\right]\right\}$$

$$w_1(t) = \sqrt{\mathcal{E}}\psi_2(t), \quad \text{with } \psi_2(t) = \sqrt{\frac{1}{T}}\sum_{j=1}^{n} s_{1,j}\mathbb{1}\left\{t - j\frac{T}{n} \in \left[0, \frac{T}{n}\right]\right\},$$

where $(s_{0,1}, \ldots, s_{0,n}) \in \{\pm 1\}^n$ and $(s_{1,1}, \ldots, s_{1,n}) \in \{\pm 1\}^n$ are orthogonal. This signaling method is called spread spectrum. It is not hard to show that it uses much bandwidth but it has an inherent robustness with respect to interfering (non-white and possibly non-Gaussian) signals.

Now assume that we use one of the above choices to communicate across a continuous-time AWGN channel and that the receiver implements an ML decision rule. Since the codebook $\mathcal{C}$ is the same in all cases, the decoder and the error probability will be identical no matter which choice we make.

Computing the error probability is particularly easy when there are only two codewords. From the previous chapter we know that $P_e = Q\left(\frac{\|c_1 - c_0\|}{2\sigma}\right)$, where $\sigma^2 = \frac{N_0}{2}$. The distance

$$\|c_1 - c_0\| := \sqrt{\sum_{i=1}^{2}(c_{1,i} - c_{0,i})^2} = \sqrt{\mathcal{E} + \mathcal{E}} = \sqrt{2\mathcal{E}}$$

can also be computed as

$$\|w_1 - w_0\| := \sqrt{\int [w_1(t) - w_0(t)]^2 dt},$$

which requires neither an orthonormal basis nor the codebook. Yet another alternative is to use Pythagoras' theorem. As we know already that our signals have squared norm $\mathcal{E}$ and are orthogonal to each other, their distance is $\sqrt{\|w_0\|^2 + \|w_1\|^2} = \sqrt{2\mathcal{E}}$. Inserting, we obtain

$$P_e = Q\left(\sqrt{\frac{\mathcal{E}}{N_0}}\right).$$

□

EXAMPLE 3.8 (Single-shot PAM) Let $\psi(t)$ be a unit-energy pulse. We speak of single-shot pulse amplitude modulation when the transmitted signal is of the form

$$w_i(t) = c_i\psi(t),$$

where c_i takes a value in some discrete subset of $\mathbb{R}$ of the form $\{\pm a, \pm 3a, \pm 5a, \ldots, \pm(m-1)a\}$ for some positive number a. An example for $m = 6$ is shown in Figure 2.9, where $d = 2a$. □

EXAMPLE 3.9 (Single-shot PSK) *Let T and f_c be positive numbers and let m be a positive integer. We speak of single-shot phase-shift keying when the signal set consists of signals of the form*

$$w_i(t) = \sqrt{\frac{2\mathcal{E}}{T}} \cos\left(2\pi f_c t + \frac{2\pi}{m} i\right) \mathbb{1}\{t \in [0,T]\}, \quad i = 0, 1, \ldots, m-1. \quad (3.5)$$

For mathematical convenience, we assume that $2f_cT$ is an integer, so that $\|w_i\|^2 = \mathcal{E}$ for all i. (When $2f_cT$ is an integer, $w_i^2(t)$ has an integer number of periods in a length-T interval. This ensures that all $w_i(t)$ have the same norm, regardless of the initial phase. In practice, f_cT is very large, which implies that there are many periods in an interval of length T, in which case the energy difference due to an incomplete period is negligible.) The signal space representation can be obtained by using the trigonometric identity $\cos(\alpha + \beta) = \cos(\alpha)\cos(\beta) - \sin(\alpha)\sin(\beta)$ to rewrite (3.5) as

$$w_i(t) = c_{i,1}\psi_1(t) + c_{i,2}\psi_2(t),$$

where

$$c_{i,1} = \sqrt{\mathcal{E}} \cos\left(\frac{2\pi i}{m}\right), \quad \psi_1(t) = \sqrt{\frac{2}{T}} \cos(2\pi f_c t) \mathbb{1}\{t \in [0,T]\},$$

$$c_{i,2} = \sqrt{\mathcal{E}} \sin\left(\frac{2\pi i}{m}\right), \quad \psi_2(t) = -\sqrt{\frac{2}{T}} \sin(2\pi f_c t) \mathbb{1}\{t \in [0,T]\}.$$

The reader should verify that $\psi_1(t)$ and $\psi_2(t)$ are normalized functions and, because $2f_cT$ is an integer, they are orthogonal to each other. This can easily be verified using the trigonometric identity $\sin\alpha\cos\beta = \frac{1}{2}[\sin(\alpha+\beta) + \sin(\alpha-\beta)]$. Hence the codeword associated to $w_i(t)$ is

$$c_i = \sqrt{\mathcal{E}} \begin{pmatrix} \cos 2\pi i/m \\ \sin 2\pi i/m \end{pmatrix}.$$

In Example 2.15, we have already studied this constellation for the discrete-time AWGN channel. □

EXAMPLE 3.10 (Single-shot QAM) *Let T and f_c be positive numbers such that $2f_cT$ is an integer, let m be an even positive integer, and define*

$$\psi_1(t) = \sqrt{\frac{2}{T}} \cos(2\pi f_c t) \mathbb{1}\{t \in [0,T]\}$$

$$\psi_2(t) = \sqrt{\frac{2}{T}} \sin(2\pi f_c t) \mathbb{1}\{t \in [0,T]\}.$$

(We have already established in Example 3.9 that $\psi_1(t)$ and $\psi_2(t)$ are orthogonal to each other and have unit norm.) If the components of $c_i = (c_{i,1}, c_{i,2})^\mathsf{T}$, $i = 0, \ldots, m^2 - 1$, take values in some discrete subset of the form $\{\pm a, \pm 3a, \pm 5a, \ldots, \pm(m-1)a\}$ for some positive a, then

$$w_i(t) = c_{i,1}\psi_1(t) + c_{i,2}\psi_2(t),$$

3.5. Generalization and alternative receiver structures

is a single-shot quadrature amplitude modulation (QAM). The values of c_i for $m = 2$ and $m = 4$ are shown in Figures 2.10 and 2.22, respectively. $\square$

The signaling methods discussed in this section are the building blocks of many communication systems.

3.5 Generalization and alternative receiver structures

It is interesting to explore a refinement and a variation of the receiver structure shown in Figure 3.4. We also generalize to an arbitrary message distribution. We take the opportunity to review what we have so far.

The source produces $H = i$ with probability $P_H(i)$, $i \in \mathcal{H}$. When $H = i$, the channel output is $R(t) = w_i(t) + N(t)$, where $w_i(t) \in \mathcal{W} = \{w_0(t), w_1(t), \ldots, w_{m-1}(t)\}$ is the signal constellation composed of finite-energy signals (known to the receiver) and $N(t)$ is white Gaussian noise. Throughout this text, we make the natural assumption that the vector space $\mathcal{V}$ spanned by $\mathcal{W}$ forms an inner product space (with the standard inner product). This is guaranteed if the zero signal is the only signal that has vanishing norm, which is always the case in real-world situations. Let $\{\psi_1(t), \ldots, \psi_n(t)\}$ be an orthonormal basis for $\mathcal{V}$. We can use the Gram–Schmidt procedure to find an orthonormal basis, but sometimes we can pick a more convenient one "by hand". At the receiver, we obtain a sufficient statistic by taking the inner product of the received signal $R(t)$ with each element of the orthonormal basis. The result is

$$Y = (Y_1, Y_2, \ldots, Y_n)^\mathsf{T}, \text{ where}$$
$$Y_i = \langle R, \psi_i \rangle, \quad i = 1, \ldots, n.$$

We now face a hypothesis testing problem with prior $P_H(i)$, $i \in \mathcal{H}$, and observable Y distributed according to

$$f_{Y|H}(y|i) = \frac{1}{(2\pi\sigma^2)^{\frac{n}{2}}} \exp\left(-\frac{\|y - c_i\|^2}{2\sigma^2}\right),$$

where $\sigma^2 = \frac{N_0}{2}$. A MAP receiver that observes $Y = y$ decides $\hat{H} = i$ for one of the $i \in \mathcal{H}$ that maximize $P_H(i) f_{Y|H}(y|i)$ or any monotonic function thereof. Since $f_{Y|H}(y|i)$ is an exponential function of y, we simplify the test by taking the natural logarithm. We also remove terms that do not depend on i and rescale, keeping in mind that if we scale with a negative number, we have to change the maximization into minimization.

If we choose negative N_0 as the scaling factor we obtain the first of the following equivalent MAP tests.

(i) Choose $\hat{H}$ as one of the j that minimizes $\|y - c_j\|^2 - N_0 \ln P_H(j)$.
(ii) Choose $\hat{H}$ as one of the j that maximizes $\langle y, c_j \rangle - \frac{\|c_j\|^2}{2} + \frac{N_0}{2} \ln P_H(j)$.
(iii) Choose $\hat{H}$ as one of the j that maximizes $\int r(t) w_j^*(t) dt - \frac{\|w_j\|^2}{2} + \frac{N_0}{2} \ln P_H(j)$.

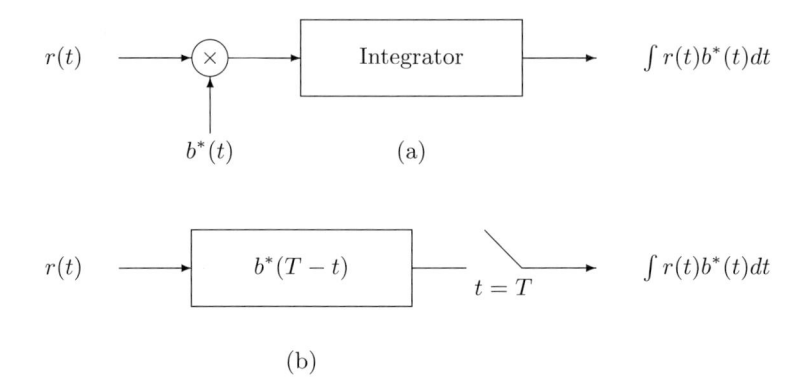

Figure 3.6. Two ways to implement $\int r(t)b^*(t)dt$, namely via a correlator (a) and via a matched filter (b) with the output sampled at time T.

The second is obtained from the first by using $\|y-c_i\|^2 = \|y\|^2 - 2\Re\{\langle y, c_i\rangle\} + \|c_i\|^2$. Once we drop the $\Re\{\cdot\}$ operator (the vectors are real-valued), remove the constant $\|y\|^2$, scale by $-1/2$, we obtain (ii).

Rules (ii) and (iii) are equivalent since $\int r(t)w_i^*(t)dt = \int r(t)\left(\sum_j c_{i,j}^*\psi_j^*(t)\right)dt = \sum_j y_j c_{i,j}^* = \langle y, c_i\rangle$.

The MAP rules (i)–(iii) require performing operations of the kind

$$\int r(t)b^*(t)dt, \qquad (3.6)$$

where $b(t)$ is some function ($\psi_j(t)$ or $w_j(t)$). There are two ways to implement (3.6). The obvious way, shown in Figure 3.6a is by means of a so-called *correlator*. A correlator is a device that multiplies and integrates two input signals. The other way to implement (3.6) is via a so-called *matched filter*. This is a filter that takes $r(t)$ as the input and has $h(t) = b^*(T-t)$ as impulse response (Figure 3.6b), where T is an arbitrary design parameter selected in such a way as to make $h(t)$ a causal impulse response. The matched filter output $y(t)$ is then

$$y(t) = \int r(\alpha)\, h(t-\alpha)\, d\alpha$$
$$= \int r(\alpha)\, b^*(T+\alpha-t)\, d\alpha,$$

and at $t = T$ it is

$$y(T) = \int r(\alpha)\, b^*(\alpha)\, d\alpha.$$

We see that the latter is indeed (3.6).

EXAMPLE 3.11 (Matched filter) If $b(t)$ is as in Figure 3.7, then $y = \langle r(t), b(t)\rangle$ is the output at $t = 0$ of a linear time-invariant filter that has input $r(t)$ and has $h_0(t) = b(-t)$ as the impulse response (see the figure). The impulse response $h_0(t)$ is non-causal. We obtain the same result with a causal filter by delaying the

3.5. Generalization and alternative receiver structures

impulse response by $3T$ and by sampling the output at $t = 3T$. The delayed impulse response is $h_{3T}(t)$, also shown in the figure. □

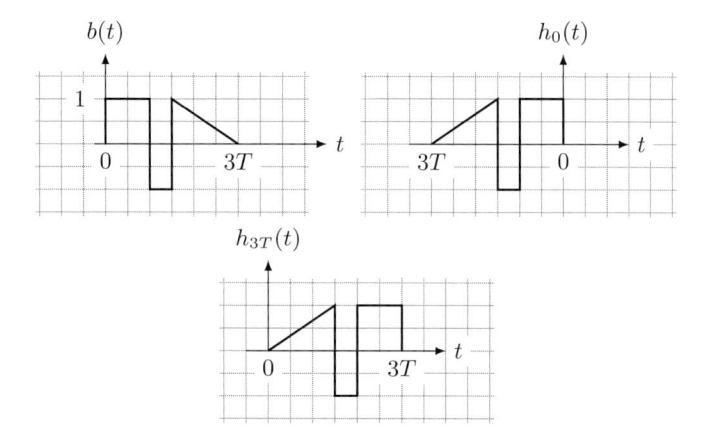

Figure 3.7.

It is instructive to plot the matched filter output as we do in the next example.

EXAMPLE 3.12 *Suppose that the signals are $w_0(t) = a\psi(t)$ and $w_1(t) = -a\psi(t)$, where a is some positive number and*

$$\psi(t) = \sqrt{\frac{1}{T}}\mathbb{1}\{0 \leq t \leq T\}.$$

The signals are plotted on the left of Figure 3.8. The n-tuple former consists of the matched filter of impulse response $h(t) = \psi^*(T - t) = \psi(t)$, with the output sampled at $t = T$. In the absence of noise, the matched filter output at the sampling time should be a when $w_0(t)$ is transmitted and $-a$ when $w_1(t)$ is transmitted. The

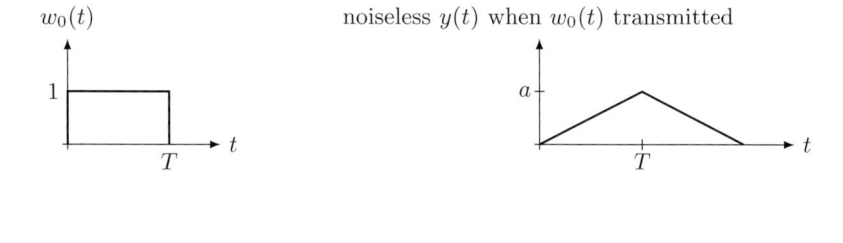

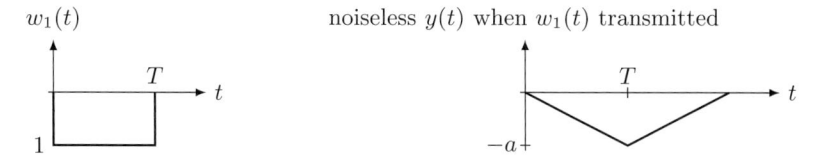

Figure 3.8. Matched filter response (right) to the input on the left.

plots on the right of the figure show the matched filter response $y(t)$ to the input on the left. Indeed, at $t = T$ we have a or $-a$. At any other time we have b or $-b$, for some b such that $0 \leq b \leq a$. This, and the fact that the noise variance does not depend on the sampling time, implies that $t = T$ is the sampling time at which the error probability is minimized. □

Figure 3.9 shows the block diagrams for the implementation of the three MAP rules (i)–(iii). In each case the front end has been implemented by using matched filters, but correlators could also be used, as in Figure 3.4.

Whether we use matched filters or correlators depends on the technology and on the waveforms. Implementing a correlator in analog technology is costly. But, if the processing is done by a microprocessor that has enough computational power, then a correlation can be done at no additional hardware cost. We would be inclined to use matched filters if there were easy-to-implement filters of the desired impulse response. In Exercise 3.10 of this chapter, we give an example where the matched filters can be implemented with passive components.

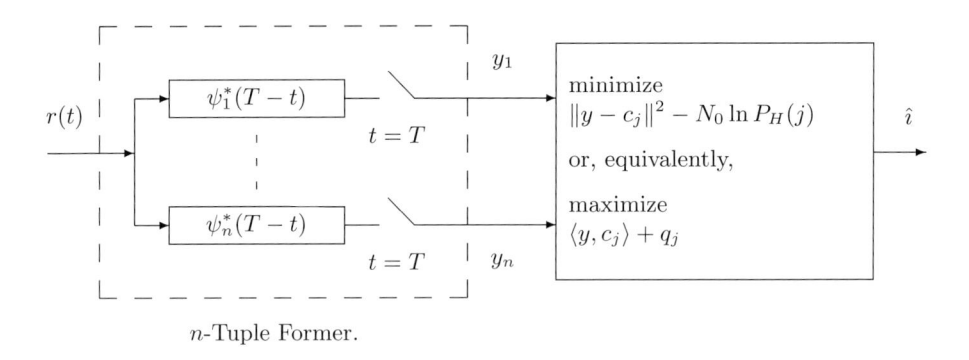

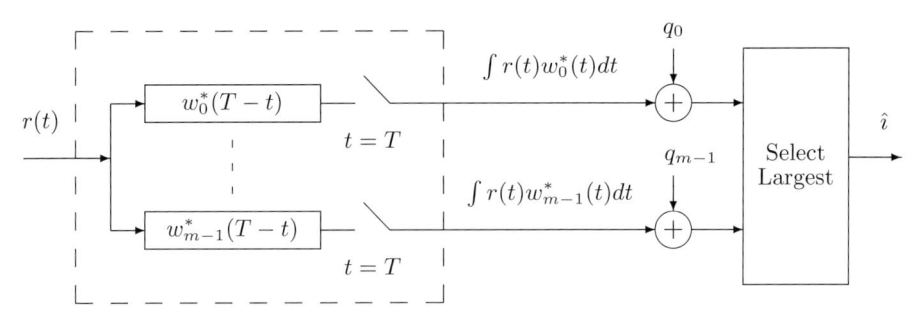

Figure 3.9. Block diagrams of a MAP receiver for the waveform AWGN channel, with $y = (y_1, \ldots, y_n)^\mathsf{T}$ and $q_j = -\|w_j\|^2/2 + (N_0/2) \ln P_H(j)$. The dashed boxes can alternatively be implemented via correlators.

Notice that the bottom implementation of Figure 3.9 requires neither an orthonormal basis nor knowledge of the codebook, but it does require m as opposed to n matched filters (or correlators). We know that $m \geq n$, and often m is much larger than n. Notice also that this implementation does *not* quite fit into the decomposition of Figure 3.2. In fact the receiver bypasses the need for the n-tuple Y. As a byproduct, this proves that the receiver performance is not affected by the choice of an orthonormal basis.

In a typical communication system, n and m are very large. So large that it is not realistic to have n or m matched filters (or correlators). Even if we disregard the cost of the matched filters, the number of operations required by a decoder that performs a "brute force" search to find the distance-minimizing index (or inner-product-maximizing index) is typically exorbitant. We will see that clever design choices can dramatically reduce the complexity of a MAP receiver.

The equivalence of the two operations of Figure 3.6 is very important. It should be known by heart.

3.6 Continuous-time channels revisited

Every channel adds noise and this is what makes the communication problem both challenging and interesting. In fact, noise is the only reason there is a fundamental limitation to the maximal rate at which we can communicate reliably through a cable, an optical fiber, and most other channels of practical interest. Without noise we could transmit reliably as many bits per second as we want, using as little energy as desired, even in the presence of the other channel imperfections that we describe next.

Attenuation and amplification Whether wireline or wireless, a passive channel always attenuates the signal. For a wireless channel, the attenuation can be of several orders of magnitude. Much of the attenuation is compensated for by a cascade of amplifiers in the first stage of the receiver, but an amplifier scales both the information-carrying signal and the noise, and adds some noise of its own.

The fact that the receiver front end incorporates a cascade of amplifiers needs some explanation. Why should the signal be amplified if the noise is amplified by the same factor? A first answer to this question is that electronic devices, such as an n-tuple former, are designed to process electrical signals that are in a certain range of amplitudes. For instance, the signal's amplitude should be large compared to the noise added by the circuit. This explains why the first amplification stage is done by a so-called *low-noise amplifier*. If the receiving antenna is connected to the receiver via a relatively long cable, as it would be the case for an outdoor antenna, then the low-noise amplifier is typically placed between the antenna and the cable.

The low-noise amplifier (or the stage that follows it) contains a noise-reduction filter that removes the out-of-band noise. With perfect electronic circuits, such a filter is superfluous because the out-of-band noise is removed by the n-tuple former.

But the out-of-band noise increases the chance that the electronic circuits – up to and including the n-tuple former – saturate, i.e. that the amplitude of the noise exceeds the range that can be tolerated by the circuits.

The typical next stage is the so-called *automatic gain control* (AGC) amplifier, designed to bring the signal's amplitude into the desired range. Hence the AGC amplifier introduces a scaling factor that depends on the strength of the input signal.

For the rest of this text, we ignore hardware imperfections. Therefore, we can also ignore the presence of the low-noise amplifier, of the noise-reduction filter, and of the automatic gain control amplifier. If the channel scales the signal by a factor α, the receiver front end can compensate by scaling the received signal by α^{-1}, but the noise is also scaled by the same factor. This explains why, in evaluating the error probability associated to a signaling scheme, we often consider channel models that only add noise. In such cases, the scaling factor α^{-1} is implicitly accounted for in the noise parameter $N_0/2$. An example of how to determine $N_0/2$ is given in Appendix 3.11, where we work out a case study based on satellite communication.

Propagation delay and clock misalignment Propagation delay refers to the time it takes a signal to reach a receiver. If the signal set is $\mathcal{W} = \{w_0(t), w_1(t), \ldots, w_{m-1}(t)\}$ and the propagation delay is τ, then for the receiver it is as if the signal set were $\tilde{\mathcal{W}} = \{w_0(t-\tau), w_1(t-\tau), \ldots, w_{m-1}(t-\tau)\}$. The common assumption is that the receiver does not know τ when the communication starts. For instance, in wireless communication, a receiver has no way to know that the propagation delay has changed because the transmitter has moved while it was turned off. It is the responsibility of the receiver to adapt to the propagation delay. We come to the same conclusion when we consider the fact that the clocks of different devices are often not synchronized. If the clock of the receiver reads $t - \tau$ when that of the transmitter reads t then, once again, for the receiver, the signal set is $\tilde{\mathcal{W}}$ for some unknown τ. Accounting for the unknown τ at the receiver goes under the general name of *clock synchronization*. For reasons that will become clear, the clock synchronization problem decomposes into the symbol synchronization and into the phase synchronization problems, discussed in Sections 5.7 and 7.5. Until then and unless otherwise specified, we assume that there is no propagation delay and that all clocks are synchronized.

Filtering In wireless communication, owing to reflections and diffractions on obstacles, the electromagnetic signal emitted by the transmitter reaches the receiver via multiple paths. Each path has its own delay and attenuation. If $w_i(t)$ is transmitted, the receiver antenna output has the form $R(t) = \sum_{l=1}^{L} w_i(t - \tau_l)h_l$ plus noise, where τ_l and h_l are the delay and the attenuation along the lth path. Unlike a mirror, the rough surface of certain objects creates a large number of small reflections that are best accounted for by the integral form $R(t) = \int w_i(t - \tau)h(\tau)d\tau$ plus noise. This is the same as saying that the channel contains a filter of impulse response $h(t)$. For a different reason, the same channel model applies to wireline communication. In fact, due to dispersion, the channel output to a unit-energy pulse applied to the input at time $t = 0$ is some impulse

3.6. Continuous-time channels revisited

response $h(t)$. Owing to the channel linearity, the output due to $w_i(t)$ at the input is, once again, $R(t) = \int w_i(t-\tau)h(\tau)d\tau$ plus noise.

The possibilities we have to cope with the channel filtering depend on whether the channel impulse response is known to the receiver alone, to both the transmitter and the receiver, or to neither. It is often realistic to assume that the receiver can measure the channel impulse response. The receiver can then communicate it to the transmitter via the reversed communication link (if it exists). Hence it is hardly the case that only the transmitter knows the channel impulse response.

If the transmitter uses the signal set $\mathcal{W} = \{w_0(t), w_1(t), \ldots, w_{m-1}(t)\}$ and the receiver knows $h(t)$, from the receiver's point of view, the signal set is $\tilde{\mathcal{W}}$ with the ith signal being $\tilde{w}_i(t) = (w_i \star h)(t)$ and the channel just adds white Gaussian noise. This is the familiar case. Realistically, the receiver knows at best an estimate $\tilde{h}(t)$ of $h(t)$ and uses it as the actual channel impulse response.

The most challenging situation occurs when the receiver does not know and cannot estimate $h(t)$. This is a realistic assumption in bursty communication, when a burst is too short for the receiver to estimate $h(t)$ and the impulse response changes from one burst to the next.

The most favorable situation occurs when both the receiver and the transmitter know $h(t)$ or an estimate thereof. Typically it is the receiver that estimates the channel impulse response and communicates it to the transmitter. This requires two-way communication, which is typically available. In this case, the transmitter can adapt the signal constellation to the channel characteristic. Arguably, the best strategy is the so-called water-filling (see e.g. [19]) that can be implemented via orthogonal frequency division multiplexing (OFDM).

We have assumed that the channel impulse response characterizes the channel filtering for the duration of the transmission. If the transmitter and/or the receiver move, which is often the case in mobile communication, then the channel is still linear but *time-varying*. Excellent graduate-level textbooks that discuss this kind of channel are [2] and [17].

Colored Gaussian noise We can think of colored noise as filtered white noise. It is safe to assume that, over the frequency range of interest, i.e. the frequency range occupied by the information-carrying signals, there is no positive-length interval over which there is no noise. (A frequency interval with no noise is physically unjustifiable and, if we insist on such a channel model, we no longer have an interesting communication problem because we can transmit infinitely many bits error-free by signaling where there is no noise.) For this reason, we assume that the frequency response of the noise-shaping filter cannot vanish over a positive-length interval in the frequency range of interest. In this case, we can modify the aforementioned noise-reduction filter in such a way that, in the frequency range of interest, it has the inverse frequency response of the noise-shaping filter. The noise at the output of the modified noise-reduction filter, called *whitening filter*, is zero-mean, Gaussian, and white (in the frequency range of interest). The minimum error probability with the whitening filter cannot be higher than without, because the filter is invertible in the frequency range of interest. What we gain with the noise-whitening filter is that we are back to the familiar situation where the noise is white and the signal set is $\tilde{\mathcal{W}} = \{\tilde{w}_0(t), \tilde{w}_1(t), \ldots, \tilde{w}_{m-1}(t)\}$, where $\tilde{w}_i(t) = (w_i \star h)(t)$ and $h(t)$ is the impulse response of the whitening filter.

3.7 Summary

In this chapter we have addressed the problem of communicating a message across a waveform AWGN channel. The importance of the continuous-time AWGN channel model comes from the fact that every conductor is a linear time-invariant system that smooths out and adds up the voltages created by the electron's motion. Owing to the central limit theorem, the result of adding up many contributions can be modeled as white Gaussian noise. No conductor can escape this phenomena, unless it is cooled to zero degrees kelvin. Hence every channel adds Gaussian noise. This does not imply that the continuous-time AWGN channel is the only channel model of interest. Depending on the situation, there can be other impairments such as fading, nonlinearities, and interference, that should be considered in the channel model, but they are outside the scope of this text.

As in the previous chapter, we have focused primarily on the receiver that minimizes the error probability assuming that the signal set is given to us. We were able to move forwards swiftly by identifying a sufficient statistic that reduces the receiver design problem to the one studied in Chapter 2. The receiver consists of an n-tuple former and a decoder. We have seen that the sender can also be decomposed into an encoder and a waveform former. This decomposition naturally fits the layering philosophy discussed in the introductory chapter: The waveform former at the sender and the n-tuple former at the receiver can be seen as providing a "service" to the encoder–decoder pair. The service consists in making the continuous-time AWGN channel look like a discrete-time AWGN channel.

Having established the link between the continuous-time and the discrete-time AWGN channel, we are in the position to evaluate the error probability of a communication system for the AWGN channel by means of simulation. An example is given in Appendix 3.8.

How do we proceed from here? First, we need to introduce the performance parameters we care mostly about, discuss how they relate to one another, and understand what options we have to control them. We start this discussion in the next chapter where we also develop some intuition about the kind of signals we want to use to transmit many bits.

Second, we need to start paying attention to cost and complexity because they can quickly get out of hand. For a brute-force implementation, the n-tuple former requires n correlators or matched filters and the decoder needs to compute and compare $\langle y, c_j \rangle + q_j$ for m codewords. With $k = 100$ (a very modest number of transmitted bits) and $n = 2k$ (a realistic relationship), the brute-force approach requires 200 matched filters or correlators and the decoder needs to evaluate roughly 10^{30} inner products. These are staggering numbers. In Chapter 5 we will learn how to choose the waveform former in such a way that the n-tuple former can be implemented with a single matched filter. In Chapter 6 we will see that

3.8 Appendix: A simple simulation

there are encoders for which the decoder needs to explore a number of possibilities that grows linearly rather than exponentially in k.

3.8 Appendix: A simple simulation

Here we give an example of a basic simulation. Instead of sending a continuous-time waveform $w(t)$, we send the corresponding codeword c; instead of adding a sample path of white Gaussian noise of power spectral density $N_0/2$, we add a realization z of a Gaussian random vector that consists of iid components that are zero mean and of variance $\sigma^2 = N_0/2$. The decoder observes $y = c+z$. MATLAB is a programming language that makes it possible to implement a simulation in a few lines of code. Here is how we can determine (by simulation) the error probability of m-PAM for $m = 6$, $d = 2$, and $\sigma^2 = 1$.

```
% define the parameters
m = 6 % alphabet size (positive even number)
d = 2 % distance between points
noiseVariance = 1
k = 1000 % number of transmitted symbols

% define the encoding function
encodingFunction = -(m-1)*d/2:d:(m-1)*d/2;

% generate the message
message = randi(m,k,1);

% encode
c = encodingFunction(message);

% generate the noise
z = normrnd(0,sqrt(noiseVariance),1,k);

% add the noise
y = c+z;

% decode
[distances,message_estimate] = min(abs(repmat(y',1,m)...
    -repmat(encodingFunction,k,1)),[],2);

% determine the symbol error probability and print
errorRate = symerr(message,message_estimate)/k
```

The above MATLAB code produces the following output (reformatted)

```
m = 6
d = 2
```

```
noiseVariance = 1
k = 1000
errorRate = 0.2660
```

3.9 Appendix: Dirac-delta-based definition of white Gaussian noise

It is common to define white Gaussian noise as a zero-mean WSS Gaussian random process $N(t)$ of autocovariance $K_N(\tau) = \frac{N_0}{2}\delta(\tau)$.

From the outset, the difference between this and the approach we chose (Section 3.2) lies on where we start with a mathematical model of the physical world. We chose to start with the measurements that the receiver can make about $N(t)$, whereas the standard approach starts with $N(t)$ itself.

To model and use $N(t)$ in a rigorous way requires familiarity with the notion of stochastic processes (typically not a problem), the ability to manipulate the Dirac delta (not a problem until something goes wrong), and measure theory to guarantee that integrals such as $\int N(\alpha)g(\alpha)d\alpha$ are well-defined. Most engineers are not familiar with measure theory. This results in situations that are undesirable for the instructor and for the student. Nevertheless it is important that the reader be aware of the standard procedure, which is the reason for this appendix.

As the following example shows, it is a simple exercise to derive (3.3) from the above definition of $N(t)$. (We take it for granted that the integrals exist.)

EXAMPLE 3.13 Let $g_1(t)$ and $g_2(t)$ be two finite-energy pulses and for $i = 1, 2$, define

$$Z_i = \int N(\alpha)g_i(\alpha)d\alpha, \qquad (3.7)$$

where $N(t)$ is white Gaussian noise as we just defined. We compute the covariance $\text{cov}(Z_i, Z_j)$ as follows:

$$\begin{aligned}
\text{cov}(Z_i, Z_j) &= \mathbb{E}\left[Z_i Z_j^*\right] \\
&= \mathbb{E}\left[\int N(\alpha)g_i(\alpha)d\alpha \int N^*(\beta)g_j^*(\beta)d\beta\right] \\
&= \int\int \mathbb{E}\left[N(\alpha)N^*(\beta)\right] g_i(\alpha)g_j^*(\beta)d\alpha d\beta \\
&= \int\int \frac{N_0}{2}\delta(\alpha-\beta)g_i(\alpha)g_j^*(\beta)d\alpha d\beta \\
&= \frac{N_0}{2}\int g_i(\beta)g_j^*(\beta)d\beta.
\end{aligned}$$

□

3.9. Appendix: Dirac-delta-based definition of white Gaussian noise 117

EXAMPLE 3.14 *Let $N(t)$ be white Gaussian noise at the input of a linear time-invariant circuit of impulse response $h(t)$ and let $Z(t)$ be the filter's output. Compute the autocovariance of the output $Z(t) = \int N(\alpha)h(t-\alpha)d\alpha$.*

Solution: The definition of autocovariance is $K_Z(\tau) := \mathbb{E}\left[Z(t+\tau)Z^*(t)\right]$. We proceed two ways. The computation using the definition of $N(t)$ given in this appendix mimics the derivation in Example 3.13. The result is $K_Z(\tau) = \frac{N_0}{2} \int h(t+\tau)h^*(t)dt$. If we use the definition of white Gaussian noise given in Section 3.2, we do not need to calculate (but we do need to know (3.3), which is part of the definition). In fact, the Z_i and Z_j defined in (3.2) and used in (3.3) become $Z(t+\tau)$ and $Z(t)$ if we set $g_i(\alpha) = h(t+\tau-\alpha)$ and $g_j(\alpha) = h(t-\alpha)$, respectively. Hence we can read the result directly out of 3.3, namely

$$K_Z(\tau) = \frac{N_0}{2} \int h(t+\tau-\alpha)h^*(t-\alpha)dt = \frac{N_0}{2} \int h(\beta+\tau)h^*(\beta)d\beta.$$

By defining the self-similarity function[1] *of $h(t)$*

$$R_h(\tau) = \int h(t+\tau)h^*(t)dt$$

we can summarize as follows

$$K_Z(\tau) = \frac{N_0}{2} R_h(\tau). \qquad \square$$

The definition of $N(t)$ given in this appendix is somewhat unsatisfactory also based on physical grounds. Recall that the Fourier transform of the autocovariance $K_N(\tau)$ is the *power spectral density* $S_N(f)$ (also called *power spectrum*). If $K_N(\tau) = \frac{N_0}{2}\delta(\tau)$ then $S_N(f) = \frac{N_0}{2}$, i.e. a constant. Integrating over the power spectral density yields the power, which in this case is infinite. The noise of a physical system cannot have infinite power. A related problem shows up from a different angle when we try to determine the variance of a sample $N(t_0)$ for an arbitrary time t_0. This is the autocovariance $K_N(\tau)$ evaluated at $\tau = 0$, but a Dirac delta is not defined as a stand-alone function.[2] Since we think of a Dirac delta as a very narrow and very tall function of unit area, we could argue that $\delta(0) = \infty$. This is also unsatisfactory because we would rather avoid having to define Gaussian random variables of infinite variance. More precisely, a stochastic process is characterized by specifying the joint distribution of each finite collection of samples, which implies that we would have to define the density of any collection of Gaussian random variables of *infinite* variance. Furthermore, we know that a random variable of infinite variance is not a good model for what we obtain when we sample noise. The reason the physically-unsustainable Dirac-delta-based model leads to physically meaningful results is that we use it only to describe filtered

---

[1] Also called the autocorrelation function. We reserve the term autocorrelation function for stochastic processes and use self-similarity function for deterministic pulses.
[2] Recall that a Dirac delta function is defined through what happens when we integrate it against a function, i.e. through the relationship $\int \delta(t)g(t) = g(0)$.

white Gaussian noise. (But then, why not bypass the mathematical description of $N(t)$ as we do in Section 3.2?)

As a final remark, note that defining an object indirectly through its behavior, as we have done in Section 3.2, is not new to us. We do something similar when we introduce the Dirac delta function by saying that it fulfills the relationship $\int f(t)\delta(t) = f(0)$. In both cases, we introduce the object of interest by saying how it behaves when integrated against a generic function.

3.10 Appendix: Thermal noise

Any conductor at non-zero temperature produces thermal (Johnson) noise. The motion of charges (electrons) that move inside a conductor yields many tiny electrical fields, the sum of which can be measured as a voltage at the conductor's terminals. Owing to the central limit theorem, the aggregate voltage can be modeled as white Gaussian noise. (It looks white, up to very high frequencies.)

Thermal noise was first measured by Johnson (Bell Labs, 1926) who made the following experiment. He took a number of different conducting substances, such as solutions of salt in water, copper sulfate, etc., and measured the intrinsic voltage fluctuations across these substances. He found that the thermal noise expresses itself as a voltage source $V_N(t)$ in series with the noise-free conductor (Figure 3.10). The mean square voltage of $V_N(t)$ per hertz (Hz) of bandwidth (accounting only for positive frequencies) equals $4Rk_BT$, where $k_B = 1.381 \times 10^{-23}$ is Bolzmann's constant in joules/kelvin, T is the absolute temperature of the substance in kelvin (290 K at room temperature), and R its resistance in ohms.

Johnson described his findings to Nyquist (also at Bell Labs) who was able to explain the results by using thermodynamics and statistical mechanics. (Nyquist's paper [25] is only four pages and very accessible. A recommended reading.) The expression for the mean of $V_N^2(t)$ per Hz of bandwidth derived by Nyquist is

$$\frac{4Rhf}{e^{\frac{hf}{k_BT}} - 1}, \qquad (3.8)$$

(a) Noisy conductor.   (b) Equivalent electrical circuit.

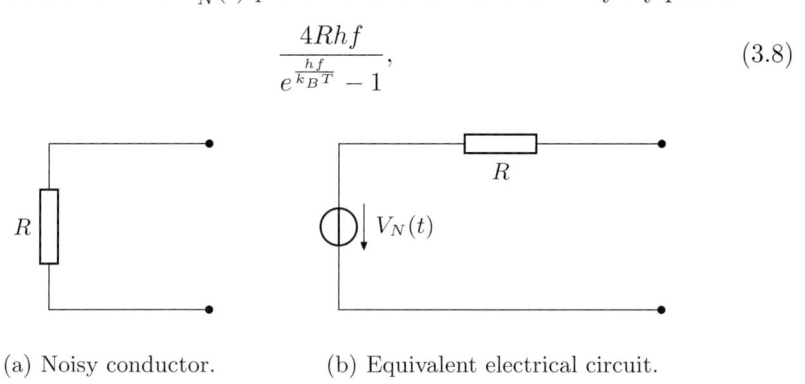

Figure 3.10. (a) Conductor of resistance R; (b) equivalent electrical circuit, where $V_N(t)$ is a voltage source modeled as white Gaussian noise of (single-sided) power spectral density $N_0 = 4k_BTR$ and R is an ideal (noise-free) resistor.

3.11. Appendix: Channel modeling, a case study

where $h = 6.626 \times 10^{-34}$ joules×seconds is Planck's constant. This expression also holds for the mean square voltage at the terminals of an impedance Z with $\Re\{Z\} = R$.

For small values of x, $e^x - 1$ is approximately x. Hence, as long as hf is much smaller than $k_B T$, the denominator of Nyquist's expression is approximately $\frac{hf}{k_B T}$, in which case (3.8) simplifies to

$$4Rk_B T,$$

in exact agreement with Johnson's measurements.

EXAMPLE 3.15   *At room temperature ($T = 290$ kelvin), $k_B T$ is about $4 \cdot 10^{-21}$. At 600 GHz, hf is about $4 \cdot 10^{-22}$. Hence, for applications in the frequency range from 0 to 600 GHz, we can pretend that $V_N(t)$ has a constant power spectral density.*   □

EXAMPLE 3.16   *Consider a resistor of 50 ohms at $T = 290$ kelvin. The mean square voltage per Hz of bandwidth due to thermal noise is $4k_B TR = 4 \times 1.381 \times 10^{-23} \times 290 \times 50 = 8 \times 10^{-19}$ volts^2/Hz.*   □

It is a straightforward exercise to check that the power per Hz of (single-sided) bandwidth that the voltage source of Figure 3.10b dissipates into a load of matching impedance is $k_B T$ watts. Because this is a very small number, it is convenient to describe it by means of its temperature T.

Even other noises, such as the noise produced by an amplifier or the one picked up by an antenna, are often characterized by a "noise temperature", defined as the number T that makes $k_B T$ equal to the spectral density of the noise injected into the receiver. (See Appendix 3.11.)

3.11 Appendix: Channel modeling, a case study

Once the signal set $\mathcal{W}$ is fixed (up to a scaling factor), the error probability of a MAP decoder for the AWGN channel depends only on the signal energy divided by the noise-power density (signal-to-noise ratio in short) $\mathcal{E}/N_0$ at the input of the n-tuple former. How do we determine $\mathcal{E}$ in terms of the design parameters that we can measure, such as the power P_T of the transmitter, the transmitting antenna gain G_T, the distance d between the transmitter and the receiver, the receiving antenna gain G_R? And how do we find the value for N_0? In this appendix, we work out a case study based on satellite communication.

Consider a transmitting antenna that radiates isotropically in free space at a power level of P_T watts. Imagine a sphere of radius d meters centered at the transmitting antenna. The surface of this sphere has an area of $4\pi d^2$, thus the power density at distance d is $\frac{P_T}{4\pi d^2}$ watts/m^2.

Satellites and the corresponding Earth stations use antennas that have directivity (typically a parabolic or a horn antenna for a satellite, and a parabolic antenna for an Earth station). Their directivity is specified by their gain G in the pointing direction. If the transmitting antenna has gain G_T, the power density in the pointing direction at distance d is $\frac{P_T G_T}{4\pi d^2}$ watts/m^2.

A receiving antenna at distance d gathers a portion of the transmitted power that is proportional to the antenna's effective area A_R. If we assume that the transmitting antenna is pointed in the direction of the receiving antenna, the received power is $P_R = \frac{P_T G_T A_R}{4\pi d^2}$.

Like the transmitting antenna, the receiving antenna can be described by its gain G_R. For a given effective area A_R, the gain is inversely proportional to λ^2, where λ is the wavelength. (As the bandwidth of the transmitted signal is small compared to the carrier frequency, we can use the carrier frequency wavelength.) Notice that this relationship between area, gain, and wavelength is rather intuitive. A familiar case is that of a flashlight. Owing to the small wavelength of light, a flashlight can create a focused beam even with a relatively small parabolic reflector. As we know from experience, the bigger the flashlight reflector, the narrower the beam. The precise relationship is $G_R = \frac{4\pi A_R}{\lambda^2}$. (Thanks to the ratio $\frac{A_R}{\lambda^2}$, the gain G_R is dimension-free.) Solving for A_R and plugging into P_R yields

$$P_R = \frac{P_T G_T G_R}{(4\pi d/\lambda)^2}. \tag{3.9}$$

The factor $L_S = (4\pi d/\lambda)^2$ is commonly called the *free-space path loss*, but this is a misnomer. In fact the free-space attenuation is independent of the wavelength. It is the relationship between the antenna's effective area and its gain that brings in the factor λ^2. Nevertheless, being able to write

$$P_R = P_T \frac{G_T G_R}{L_S} \tag{3.10}$$

has the advantage of underlining the "gains" and the "losses". Notice also that L_S is a factor on which the system designer has little control (for a geostationary satellite the distance is fixed and the carrier frequency is often dictated by regulations), whereas P_T, G_T, and G_R are parameters that a designer might be able to choose (within limits).

Now suppose that the receiving antenna is connected to the receiver via a lossless coaxial cable. The antenna and the receiver input have an impedance and the connecting cable has a characteristic impedance. For best power transfer, the three impedances should be resistive and have the same value, typically 50 ohms (see, e.g., Wikipedia, impedance matching). We assume that it is indeed the case and let R ohms be its value. Then, the impedance seen by the antenna looking into the cable is also R as if the receiver were connected directly to the antenna (see, e.g., Wikipedia, transmission line, or [14]). Figure 3.11 shows the electrical model for the receiving antenna and its load.[3] It shows the voltage source $W(t)$ that represents the intended signal, the voltage source $V_N(t)$ that represents all noise sources, the antenna impedance R and the antenna's load R.

---

[3] The circuit of Figure 3.11 is a suitable model for determining the voltage (and the current) at the receiver input (the load in the figure). There is a more complete model [26] that enables us to associate the power dissipated by the antenna's internal impedance with the power that the antenna radiates back to space.

3.11. Appendix: Channel modeling, a case study

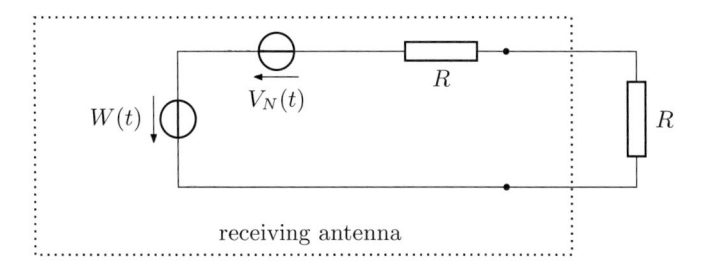

Figure 3.11. Electrical model for the receiving antenna and the load it sees looking into the first amplifier.

The advantage of having all the noise sources be represented by a single source which is co-located with the signal source $W(t)$ is that the signal-to-noise ratio at that point is the same as the signal-to-noise-ratio at the input of the n-tuple former. (Once all noise sources are accounted for at the input, the electronic circuits are considered as noise free). So, the $\mathcal{E}/N_0$ of interest to us is the signal energy absorbed by the load divided by the noise-power density absorbed by the same load.

The power harvested by the antenna is passed onto the load. This power is P_R, hence the energy is $P_R\tau$, where τ is the duration of the signals (assumed to be the same for all signals).

As mentioned in Appendix 3.10, it is customary to describe the noise-power density by the temperature T_N of a fictitious resistor that transfers the same noise-power density to the same load. This density is $k_B T_N$. If we know (for instance from measurements) the power density of each noise source, we can determine the equivalent density at the receiver input, sum all the densities, and divide by k_B to obtain the noise temperature T_N. Here we assume that this number is provided to us by the manufacturer of the receiver (see Example 3.17 for a numerical value). Putting things together, we obtain

$$\mathcal{E}/N_0 = \frac{P_R\tau}{k_B T_N} = \frac{P_T \tau G_T G_R}{L_S k_B T_N}. \tag{3.11}$$

To go one step further, we characterize the two voltage sources of Figure 3.11. This is a calculation that the hardware designer might want to do to determine the range of voltages and currents at the antenna output.

Recall that a voltage of v volts applied to a resistor of R ohms dissipates the power $P = v^2/R$ watts. When $H = i$, $W(t) = \alpha w_i(t)$ for some scaling factor α. We determine α by computing the resulting average power dissipated by the load and by equating to P_R. Thus $P_R = \frac{\alpha^2 \mathcal{E}}{4R\tau}$. Inserting the value of P_R and solving for α yields

$$\alpha = \sqrt{\frac{4R P_T G_T G_R}{L_S \mathcal{E}/\tau}}.$$

(a) Electrical circuit.

(b) System-engineering viewpoint.

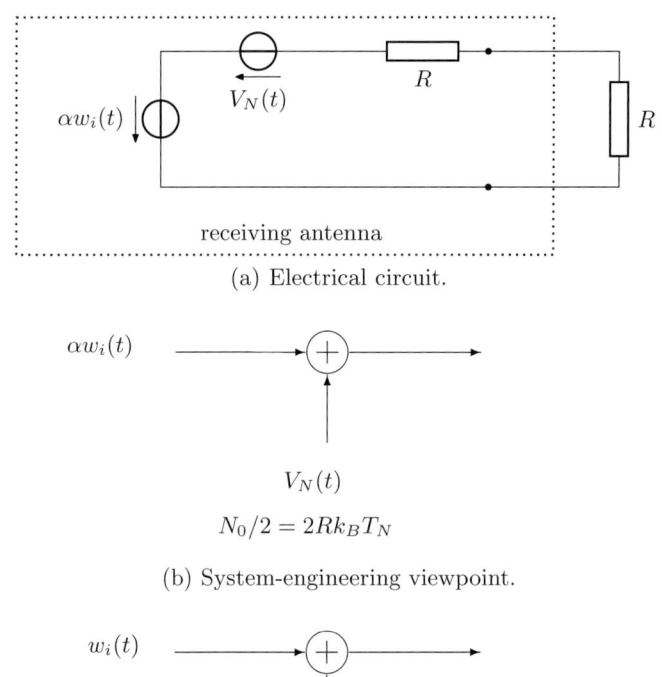

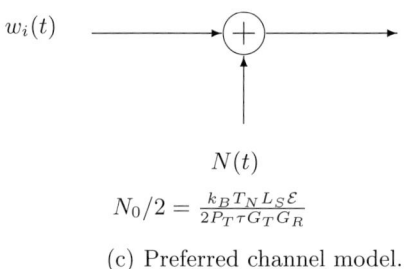

$$N_0/2 = 2Rk_BT_N$$

(c) Preferred channel model.

$$N_0/2 = \frac{k_BT_NL_S\mathcal{E}}{2P_T\tau G_TG_R}$$

Figure 3.12. Various viewpoints under hypothesis $H = i$.

Hence, when $H = i$, the received signal (before noise) is

$$W(t) = \alpha w_i(t) = \sqrt{\frac{4RP_TG_TG_R}{L_S\mathcal{E}/\tau}} w_i(t).$$

Figure 3.12a summarizes the equivalent electrical circuit under the hypothesis $H = i$. As determined in Appendix 3.10, the mean square voltage of the noise source $V_N(t)$ per Hz of (single-sided) bandwidth is $N_0 = 4Rk_BT_N$. Figure 3.12b is the equivalent representation from the point of view of a system designer. The usefulness of these models is that they give us actual voltages. As long as we are not concerned with hardware limitations, for the purpose of the channel model, we are allowed to scale the signal and the noise by the same factor. Specifically, if we divide the signal by α and divide the noise-power density by α^2, we obtain the channel model of Figure 3.12c. Observe that the impedance R has fallen out of the picture.

As a "sanity check", if we compute $\mathcal{E}/N_0$ using Figure 3.12c we obtain $\frac{\tau P_T G_T G_R}{L_S k_B T_N}$, which corresponds to (3.11). The following example gives numerical values.

EXAMPLE 3.17 *The following parameters pertain to Mariner-10, an American robotic space probe launched by NASA in 1973 to fly to the planets Mercury and Venus.*

$P_T = 16.8$ *watts (12.25 dBW).*
$\lambda = 0.13$ *m (carrier frequency at 2.3 GHz).*
$G_T = 575.44$ *(27.6 dB).*
$G_R = 1.38 \times 10^6$ *(61.4 dB).*
$d = 1.6 \times 10^{11}$ *meters.*
$T_N = 13.5$ *kelvin.*
$R_b = 117.6$ *kbps.*

The bit rate $R_b = 117.6$ kbps (kilobits per second) is the maximum data rate at which the space probe could transmit information. This can be achieved via antipodal signals of duration $\tau = 1/R_b = 8.5 \times 10^{-6}$ seconds. Under this assumption, plugging into (3.11) yields $\mathcal{E}/N_0 = 2.54$. The error rate for antipodal signaling is

$$P_e = Q\left(\sqrt{\frac{2\mathcal{E}_s}{N_0}}\right) = 0.0120.$$

We see that the error rate is fairly high, but by means of coding techniques (Chapter 6), it is possible to achieve reliable communication at the expense of some reduction in the bit rate. □

3.12 Exercises

Exercises for Section 3.1

EXERCISE 3.1 (Gram–Schmidt procedure on tuples) *By means of the Gram–Schmidt orthonormalization procedure, find an orthonormal basis for the subspace spanned by the four vectors $\beta_1 = (1, 0, 1, 1)^\mathsf{T}$, $\beta_2 = (2, 1, 0, 1)^\mathsf{T}$, $\beta_3 = (1, 0, 1, -2)^\mathsf{T}$, and $\beta_4 = (2, 0, 2, -1)^\mathsf{T}$.*

EXERCISE 3.2 (Gram–Schmidt procedure on two waveforms) *Use the Gram–Schmidt procedure to find an orthonormal basis for the vector space spanned by the functions shown in Figure 3.13.*

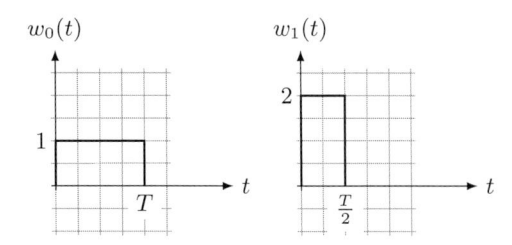

Figure 3.13.

EXERCISE 3.3 (Gram–Schmidt procedure on three waveforms)

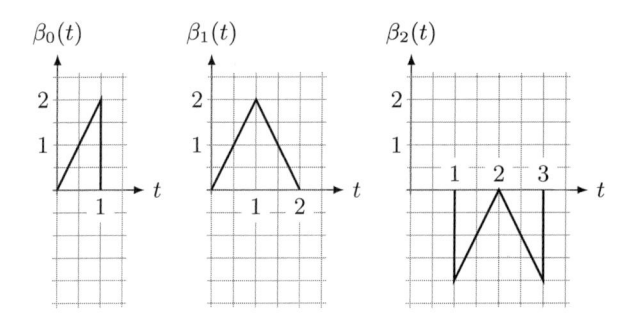

Figure 3.14.

(a) By means of the Gram–Schmidt procedure, find an orthonormal basis for the space spanned by the waveforms in Figure 3.14.
(b) In your chosen orthonormal basis, let $w_0(t)$ and $w_1(t)$ be represented by the codewords $c_0 = (3, -1, 1)^\mathsf{T}$ and $c_1 = (-1, 2, 3)^\mathsf{T}$, respectively. Plot $w_0(t)$ and $w_1(t)$.
(c) Compute the (standard) inner products $\langle c_0, c_1 \rangle$ and $\langle w_0, w_1 \rangle$ and compare them.
(d) Compute the norms $\|c_0\|$ and $\|w_0\|$ and compare them.

EXERCISE 3.4 (Orthonormal expansion) For the signal set of Figure 3.15, do the following.

(a) Find the orthonormal basis $\psi_1(t), \ldots, \psi_n(t)$ that you would find by following the Gram–Schmidt (GS) procedure. Note: No need to work out the intermediate steps of the GS procedure. The purpose of this exercise is to check, with hardly any calculation, your understanding of what the GS procedure does.
(b) Find the codeword $c_i \in \mathbb{R}^n$ that describes $w_i(t)$ with respect to your orthonormal basis. (No calculation needed.)

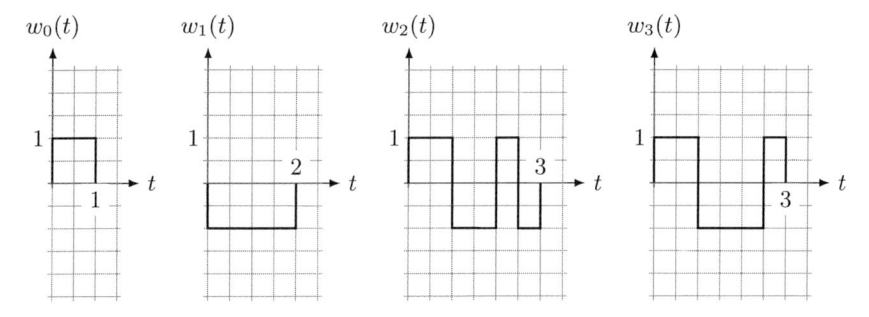

Figure 3.15.

3.12. Exercises

Exercises for Section 3.2

EXERCISE 3.5 (Noise in regions) *Let $N(t)$ be white Gaussian noise of power spectral density $\frac{N_0}{2}$. Let $g_1(t)$, $g_2(t)$, and $g_3(t)$ be waveforms as shown in Figure 3.16. For $i = 1, 2, 3$, let $Z_i = \int N(t)g_i^*(t)dt$, $Z = (Z_1, Z_2)^\mathsf{T}$, and $U = (Z_1, Z_3)^\mathsf{T}$.*

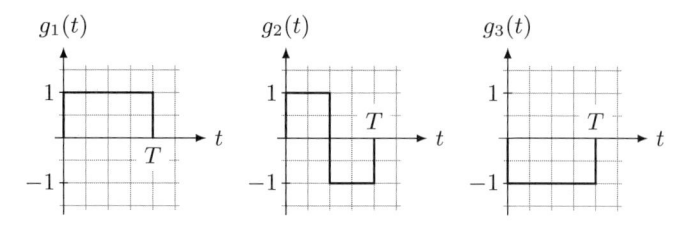

Figure 3.16.

(a) *Determine the norm $\|g_i\|$, $i = 1, 2, 3$.*
(b) *Are Z_1 and Z_2 independent? Justify your answer.*
(c) *Find the probability P_a that Z lies in the square of Figure 3.17a.*
(d) *Find the probability P_b that Z lies in the square of Figure 3.17b.*
(e) *Find the probability Q_a that U lies in the square of Figure 3.17a.*
(f) *Find the probability Q_c that U lies in the square of Figure 3.17c.*

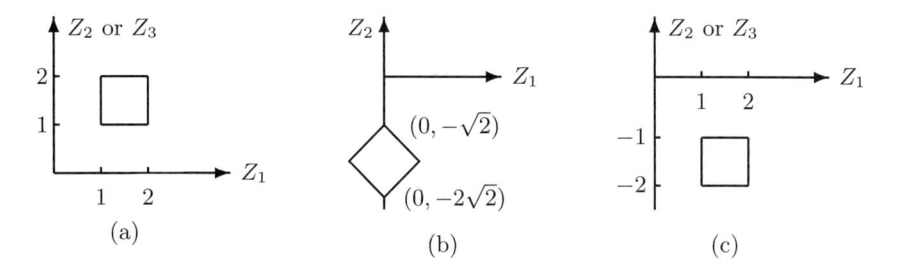

Figure 3.17.

Exercises for Sections 3.4 and 3.5

EXERCISE 3.6 (Two-signals error probability) *The two signals of Figure 3.18 are used to communicate one bit across the continuous-time AWGN channel of power spectral density $N_0/2 = 6$ W/Hz. Write an expression for the error probability of an ML receiver.*

EXERCISE 3.7 (On–off signaling) *Consider the binary hypothesis testing problem specified by:*

$$H = 0: \quad R(t) = w(t) + N(t)$$
$$H = 1: \quad R(t) = N(t),$$

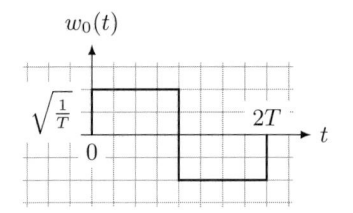

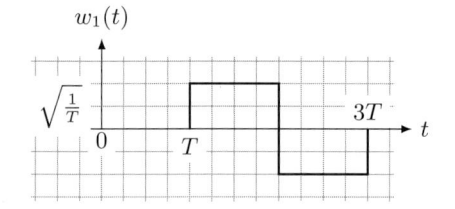

Figure 3.18.

where $N(t)$ is additive white Gaussian noise of power spectral density $N_0/2$ and $w(t)$ is the signal shown in the left of Figure 3.19

(a) Describe the maximum likelihood receiver for the received signal $R(t)$, $t \in \mathbb{R}$.
(b) Determine the error probability for the receiver you described in (a).
(c) Sketch a block diagram of your receiver of part (a) using a filter with impulse response $h(t)$ (or a scaled version thereof) shown in the right-hand part of Figure 3.19.

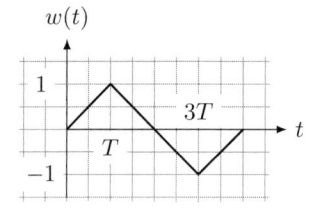

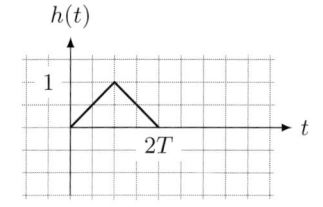

Figure 3.19.

EXERCISE 3.8 (QAM receiver) Let the channel output be

$$R(t) = W(t) + N(t),$$

where $W(t)$ has the form

$$W(t) = \begin{cases} X_1\sqrt{\frac{2}{T}}\cos 2\pi f_c t + X_2\sqrt{\frac{2}{T}}\sin 2\pi f_c t, & \text{for } 0 \leq t \leq T, \\ 0, & \text{otherwise,} \end{cases}$$

$2f_c T \in \mathbb{Z}$ is a constant known to the receiver, $X = (X_1, X_2)$ is a uniformly distributed random vector that takes values in

$$\{(\sqrt{\mathcal{E}}, \sqrt{\mathcal{E}}), (-\sqrt{\mathcal{E}}, \sqrt{\mathcal{E}}), (-\sqrt{\mathcal{E}}, -\sqrt{\mathcal{E}}), (\sqrt{\mathcal{E}}, -\sqrt{\mathcal{E}})\}$$

for some known constant $\mathcal{E}$, and $N(t)$ is white Gaussian noise of power spectral density $\frac{N_0}{2}$.

(a) Specify a receiver that, based on the channel output $R(t)$, decides on the value of the vector X with least probability of error.
(b) Find the error probability of the receiver you have specified.

3.12. Exercises

EXERCISE 3.9 (Signaling scheme example) *Let the message H be uniformly distributed over the message set $\mathcal{H} = \{0, 1, 2, \ldots, 2^k - 1\}$. When $H = i \in \mathcal{H}$, the transmitter sends $w_i(t) = w(t - \frac{iT}{2^k})$, where $w(t)$ is as shown in Figure 3.20. The channel output is $R(t) = w_i(t) + N(t)$, where $N(t)$ denotes white Gaussian noise of power spectral density $\frac{N_0}{2}$.*

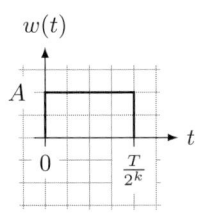

Figure 3.20.

Sketch a block diagram of a receiver that, based on $R(t)$, decides on the value of H with least probability of error. (See Example 4.6 for the probability of error.)

EXERCISE 3.10 (Matched filter implementation) *In this problem, we consider the implementation of matched filter receivers. In particular, we consider frequency-shift keying (FSK) with the following signals:*

$$w_j(t) = \begin{cases} \sqrt{\frac{2}{T}} \cos 2\pi \frac{n_j}{T} t, & \text{for } 0 \leq t \leq T, \\ 0, & \text{otherwise,} \end{cases} \quad (3.12)$$

where $n_j \in \mathbb{Z}$ and $0 \leq j \leq m - 1$. Thus, the communication scheme consists of m signals $w_j(t)$ of different frequencies $\frac{n_j}{T}$.

(a) *Determine the impulse response $h_j(t)$ of a causal matched filter for the signal $w_j(t)$. Plot $h_j(t)$ and specify the sampling time.*
(b) *Sketch the matched filter receiver. How many matched filters are needed?*
(c) *Sketch the output of the matched filter with impulse response $h_j(t)$ when the input is $w_j(t)$.*
(d) *Consider the ideal resonance circuit shown in Figure 3.21.*

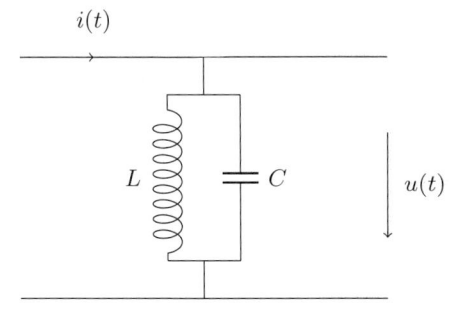

Figure 3.21.

For this circuit, the voltage response to the input current $i(t) = \delta(t)$ is

$$h(t) = \begin{cases} \frac{1}{C} \cos \frac{t}{\sqrt{LC}}, & t \geq 0 \\ 0 & \textit{otherwise.} \end{cases}$$

Show how this can be used to implement the matched filter for $w_j(t)$. Determine how L and C should be chosen. Hint: Suppose that $i(t) = w_j(t)$. In this case, what is $u(t)$?

EXERCISE 3.11 (Matched filter intuition) *In this problem, we develop further intuition about matched filters. You may assume that all waveforms are real-valued. Let $R(t) = \pm w(t) + N(t)$ be the channel output, where $N(t)$ is additive white Gaussian noise of power spectral density $N_0/2$ and $w(t)$ is an arbitrary but fixed pulse. Let $\phi(t)$ be a unit-norm but otherwise arbitrary pulse, and consider the receiver operation*

$$Y = \langle R, \phi \rangle = \langle w, \phi \rangle + \langle N, \phi \rangle. \tag{3.13}$$

The signal-to-noise ratio (SNR) is defined as

$$SNR = \frac{|\langle w, \phi \rangle|^2}{E\left[|\langle N, \phi \rangle|^2\right]}.$$

Notice that the SNR remains the same if we scale $\phi(t)$ by a constant factor. Notice also that

$$E\left[|\langle N, \phi \rangle|^2\right] = \frac{N_0}{2}. \tag{3.14}$$

(a) Use the Cauchy–Schwarz inequality to give an upper bound on the SNR. What is the condition for equality in the Cauchy–Schwarz inequality? Find the $\phi(t)$ that maximizes the SNR. What is the relationship between the maximizing $\phi(t)$ and the signal $w(t)$?

(b) Let us verify that we would get the same result using a pedestrian approach. Instead of waveforms we consider tuples. So let $c = (c_1, c_2)^\mathsf{T} \in \mathbb{R}^2$ and use calculus (instead of the Cauchy–Schwarz inequality) to find the $\phi = (\phi_1, \phi_2)^\mathsf{T} \in \mathbb{R}^2$ that maximizes $\langle c, \phi \rangle$ subject to the constraint that ϕ has unit norm.

(c) Verify with a picture (convolution) that the output at time T of a filter with input $w(t)$ and impulse response $h(t) = w(T - t)$ is indeed $\langle w, w \rangle = \int_{-\infty}^{\infty} w^2(t) dt$.

EXERCISE 3.12 (Two receive antennas) *Consider the following communication problem. The message is represented by a uniformly distributed random variable $X \in \{\pm 1\}$. The transmitter sends $Xw(t)$ where $w(t)$ is a unit-energy pulse known to the receiver. There are two channels with output $R_1(t)$ and $R_2(t)$, respectively, where*

$$R_1(t) = X\beta_1 w(t - \tau_1) + N_1(t)$$
$$R_2(t) = X\beta_2 w(t - \tau_2) + N_2(t),$$

3.12. Exercises

where $\beta_1, \beta_2, \tau_1, \tau_2$ are constants known to the receiver and $N_1(t)$ and $N_2(t)$ are white Gaussian noise of power spectral density $N_0/2$. We assume that $N_1(t)$ and $N_2(t)$ are independent of each other (in the obvious sense) and independent of X. We also assume that $\int w(t-\tau_1)w(t-\tau_2)dt = \gamma$, where $-1 \leq \gamma \leq 1$.

(a) Describe an ML receiver for X that observes both $R_1(t)$ and $R_2(t)$ and determine its probability of error in terms of the Q function, β_1, β_2, γ, and $N_0/2$.

(b) Repeat part (a) assuming that the receiver has access only to the sum-signal $R(t) = R_1(t) + R_2(t)$.

EXERCISE 3.13 (Receiver) The signal set

$$w_0(t) = \operatorname{sinc}^2(t)$$
$$w_1(t) = \sqrt{2}\operatorname{sinc}^2(t)\cos(4\pi t)$$

is used to communicate across the AWGN channel of noise power spectral density $\frac{N_0}{2}$.

(a) Sketch a block diagram of an ML receiver for the above signal set. (No need to worry about filter causality.)

(b) Determine the error probability of your receiver assuming that $w_0(t)$ and $w_1(t)$ are equally likely.

(c) If you keep the same receiver, but use $w_0(t)$ with probability $\frac{1}{3}$ and $w_1(t)$ with probability $\frac{2}{3}$, does the error probability increase, decrease, or remain the same?

EXERCISE 3.14 (ML receiver with single causal filter) Let $w_1(t)$ be as shown in Figure 3.22 and let $w_2(t) = w_1(t - T_d)$, where T_d is a fixed number known to the receiver. One of the two pulses is selected at random and transmitted across the AWGN channel of noise power spectral density $\frac{N_0}{2}$.

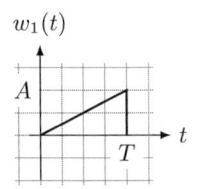

Figure 3.22.

(a) Describe an ML receiver that decides which pulse was transmitted. We ask that the n-tuple former contains a single matched filter. Make sure that the filter is causal and plot its impulse response.

(b) Express the probability of error in terms of T, A, T_d, N_0.

EXERCISE 3.15 (Delayed signals) One of the two signals shown in Figure 3.23 is selected at random and is transmitted over the additive white Gaussian noise channel of noise spectral density $\frac{N_0}{2}$. Draw a block diagram of a maximum likelihood receiver that uses a single matched filter and express its error probability.

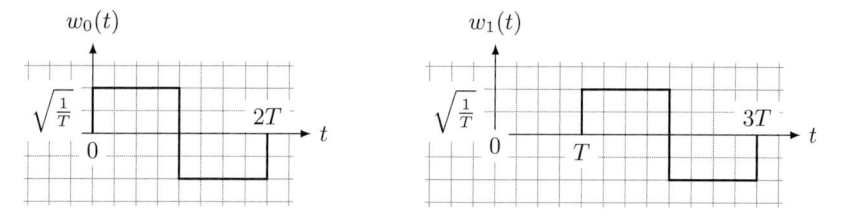

Figure 3.23.

EXERCISE 3.16 (ML decoder for AWGN channel) *The signal of Figure 3.24 is fed to an ML receiver designed for a transmitter that uses the four signals of Figure 3.15 to communicate across the AWGN channel. Determine the receiver output $\hat{H}$.*

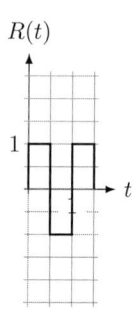

Figure 3.24.

Exercises for Section 3.6

EXERCISE 3.17 (AWGN channel and sufficient statistic) *Let $\mathcal{W} = \{w_0(t), w_1(t)\}$ be the signal constellation used to communicate an equiprobable bit across an additive Gaussian noise channel. In this exercise, we verify that the projection of the channel output onto the inner product space $\mathcal{V}$ spanned by $\mathcal{W}$ is not necessarily a sufficient statistic, unless the noise is white. Let $\psi_1(t), \psi_2(t)$ be an orthonormal basis for $\mathcal{V}$. We choose the additive noise to be $N(t) = Z_1\psi_1(t) + Z_2\psi_2(t) + Z_3\psi_3(t)$ for some normalized $\psi_3(t)$ that is orthogonal to $\psi_1(t)$ and $\psi_2(t)$ and choose Z_1, Z_2, and Z_3 to be zero-mean jointly Gaussian random variables of identical variance σ^2. Let $c_i = (c_{i,1}, c_{i,2}, 0)^\mathsf{T}$ be the codeword associated to $w_i(t)$ with respect to the extended orthonormal basis $\psi_1(t), \psi_2(t), \psi_3(t)$. There is a one-to-one correspondence between the channel output $R(t)$ and $Y = (Y_1, Y_2, Y_3)^\mathsf{T}$, where $Y_i = \langle R, \psi_i \rangle$. In terms of Y, the hypothesis testing problem is*

$$H = i \; : \; Y = c_i + Z \quad i = 0, 1,$$

where we have defined $Z = (Z_1, Z_2, Z_3)^\mathsf{T}$.

3.12. Exercises

(a) As a warm-up exercise, let us first assume that Z_1, Z_2, and Z_3 are independent. Use the Fisher–Neyman factorization theorem (Exercise 2.22 of Chapter 2) to show that Y_1, Y_2 is a sufficient statistic.

(b) Now assume that Z_1 and Z_2 are independent but $Z_3 = Z_2$. Prove that in this case Y_1, Y_2 is not a sufficient statistic.

(c) To check a specific case, consider $c_0 = (1, 0, 0)^\mathsf{T}$ and $c_1 = (0, 1, 0)^\mathsf{T}$. Determine the error probability of an ML receiver that observes $(Y_1, Y_2)^\mathsf{T}$ and that of another ML receiver that observes $(Y_1, Y_2, Y_3)^\mathsf{T}$.

EXERCISE 3.18 (Mismatched receiver) Let a channel output be

$$R(t) = c\, X\, w(t) + N(t), \tag{3.15}$$

where $c > 0$ is some deterministic constant, X is a uniformly distributed random variable that takes values in $\{3, 1, -1, -3\}$, $w(t)$ is the deterministic waveform

$$w(t) = \begin{cases} 1, & \text{if } 0 \leq t < 1 \\ 0, & \text{otherwise,} \end{cases} \tag{3.16}$$

and $N(t)$ is white Gaussian noise of power spectral density $\frac{N_0}{2}$.

(a) Describe the receiver that, based on the channel output $R(t)$, decides on the value of X with least probability of error.

(b) Find the error probability of the receiver you have described in part (a).

(c) Suppose now that you still use the receiver you have described in part (a), but that the received signal is actually

$$R(t) = \frac{3}{4} c\, X\, w(t) + N(t), \tag{3.17}$$

i.e. you were unaware that the channel was attenuating the signal. What is the probability of error now?

(d) Suppose now that you still use the receiver you have found in part (a) and that $R(t)$ is according to equation (3.15), but that the noise is colored. In fact, $N(t)$ is a zero-mean stationary Gaussian noise process of auto-covariance function

$$K_N(\tau) = \frac{1}{4\alpha} e^{-|\tau|/\alpha},$$

where $0 < \alpha < \infty$ is some deterministic real parameter. What is the probability of error now?

4 Signal design trade-offs

4.1 Introduction

In Chapters 2 and 3 we have focused on the receiver, assuming that the signal set was given to us. In this chapter we introduce the signal design.

The problem of choosing a convenient signal constellation is not as clean-cut as the receiver-design problem. The reason is that the receiver-design problem has a clear objective, to minimize the error probability, and one solution, namely the MAP rule. In contrast, when we choose a signal constellation we make trade-offs among conflicting objectives.

We have two main goals for this chapter: (i) to introduce the design parameters we care mostly about; and (ii) to sharpen our intuition about the role played by the dimensions of the signal space as we increase the number of bits to be transmitted. The continuous-time AWGN channel model is assumed.

4.2 Isometric transformations applied to the codebook

If the channel is AWGN and the receiver implements a MAP rule, the error probability is completely determined by the codebook $\mathcal{C} = \{c_0, \ldots, c_{m-1}\}$. The purpose of this section is to identify transformations to the codebook that do not affect the error probability. For the moment we assume that the codebook and the noise are real-valued. Generalization to complex-valued codebooks and complex-valued noise is straightforward but requires familiarity with the formalism of complex-valued random vectors (Appendix 7.9).

From the geometrical intuition gained in Chapter 2, it should be clear that the probability of error remains the same if a given codebook and the corresponding decoding regions are translated by the same n-tuple $b \in \mathbb{R}^n$.

A translation is a particular instance of an isometry. An isometry is a distance-preserving transformation. Formally, given an inner product space $\mathcal{V}$, $a : \mathcal{V} \to \mathcal{V}$ is an isometry if and only if for any $\alpha \in \mathcal{V}$ and $\beta \in \mathcal{V}$, the distance between α and β equals that between $a(\alpha)$ and $a(\beta)$. All isometries from $\mathbb{R}^n$ to $\mathbb{R}^n$ can be obtained from the composition of a reflection, a rotation, and a translation.

4.2. Isometric transformations applied to the codebook

EXAMPLE 4.1 *Figure 4.1 shows an original codebook $\mathcal{C} = \{c_0, c_1, c_2, c_3\}$ and three variations obtained by applying to $\mathcal{C}$ a reflection, a rotation, and a translation, respectively. In each case the isometry $a : \mathbb{R}^n \to \mathbb{R}^n$ sends c_i to $\tilde{c}_i = a(c_i)$.* □

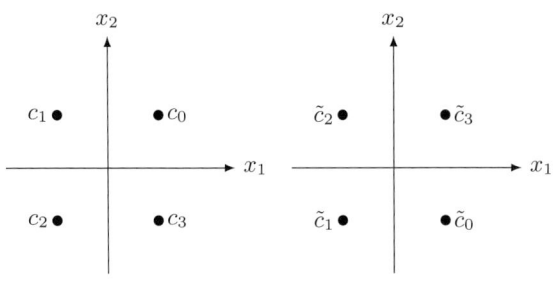

(a) Original codebook $\mathcal{C}$. (b) Reflected codebook.

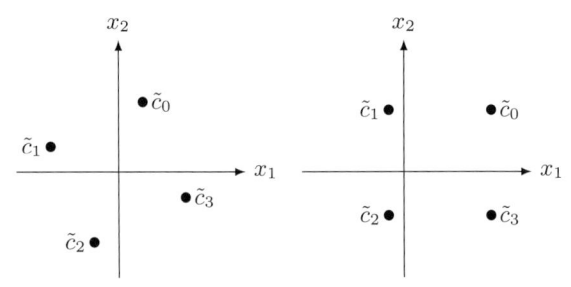

(c) Rotated codebook. (d) Translated codebook.

Figure 4.1. Isometries.

It should be intuitively clear that if we apply an isometry to a codebook and its decoding regions, then the error probability remains unchanged. A formal proof of this fact is given in Appendix 4.8.

If we apply a rotation or a reflection to an n-tuple, we do not change its norm. Hence reflections and rotations applied to a signal set do not change the average energy, but translations generally do. We determine the translation that minimizes the average energy.

Let $\tilde{Y}$ be a zero-mean random vector in $\mathbb{R}^n$. For any $b \in \mathbb{R}^n$,

$$\mathbb{E}\|\tilde{Y} + b\|^2 = \mathbb{E}\|\tilde{Y}\|^2 + \|b\|^2 + 2\mathbb{E}\langle \tilde{Y}, b \rangle = \mathbb{E}\|\tilde{Y}\|^2 + \|b\|^2 \geq \mathbb{E}\|\tilde{Y}\|^2$$

with equality if and only if $b = 0$. An arbitrary (not necessarily zero-mean) random vector $Y \in \mathbb{R}^n$ can be written as $Y = \tilde{Y} + m$, where $m = \mathbb{E}[Y]$ and $\tilde{Y} = Y - m$ is zero-mean. The above inequality can then be restated as

$$\mathbb{E}\|Y - b\|^2 = \mathbb{E}\|\tilde{Y} + m - b\|^2 \geq \mathbb{E}\|\tilde{Y}\|^2,$$

with equality if and only if $b = m$.

We apply the above to a codebook $\mathcal{C} = \{c_0, \ldots, c_{m-1}\}$. If we let Y be the random variable that takes value c_i with probability $P_H(i)$, then we see that the average energy $\mathcal{E} = \mathbb{E}\left[\|Y\|^2\right]$ can be decreased by a translation if and only if the mean $m = \mathbb{E}\left[Y\right] = \sum_i P_H(i)c_i$ is non-zero. If it is non-zero, then the translated constellation $\tilde{\mathcal{C}} = \{\tilde{c}_0, \ldots, \tilde{c}_{m-1}\}$, where $\tilde{c}_i = c_i - m$, will achieve the minimum energy among all possible translated versions of $\mathcal{C}$. The average energy associated to the translated constellation is $\tilde{\mathcal{E}} = \mathcal{E} - \|m\|^2$.

If $\mathcal{S} = \{w_0(t), \ldots, w_{m-1}(t)\}$ is the set of waveforms linked to $\mathcal{C}$ via some orthonormal basis, then through the same basis $\tilde{c}_i$ will be associated to $\tilde{w}_i(t) = w_i(t) - m(t)$ where $m(t) = \sum_i P_H(i)w_i(t)$. An example follows.

EXAMPLE 4.2 *Let $w_0(t)$ and $w_1(t)$ be rectangular pulses with support $[0,T]$ and $[T, 2T]$, respectively, as shown on the left of Figure 4.2a. Assuming that $P_H(0) = P_H(1) = \frac{1}{2}$, we calculate the average $m(t) = \frac{1}{2}w_0(t) + \frac{1}{2}w_1(t)$ and see that it is non-zero (center waveform). Hence we can save energy by using the new signal set defined by $\tilde{w}_i(t) = w_i(t) - m(t)$, $i = 0, 1$ (right). In Figure 4.2b we see the signals in the signal space, where $\psi_i(t) = \frac{w_{i-1}(t)}{\|w_{i-1}\|}$, $i = 1, 2$. As we see from the figures, $\tilde{w}_0(t)$ and $\tilde{w}_1(t)$ are antipodal signals. This is not a coincidence: After we remove the mean, any two signals become the negative of each other. As an alternative to representing the elements of $\mathcal{W}$ in the signal space, we could have represented the elements of the codebook $\mathcal{C}$ in $\mathbb{R}^2$, as we did in Figure 4.1. The two representations are equivalent.* □

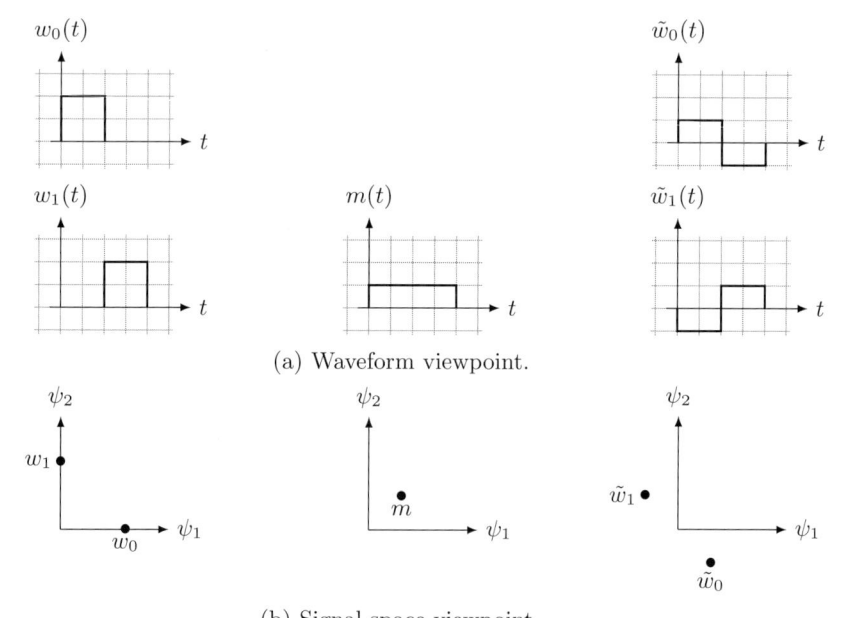

(a) Waveform viewpoint.

(b) Signal space viewpoint.

Figure 4.2. Energy minimization by translation.

4.3 Isometric transformations applied to the waveform set

The definition of isometry is based on the notion of distance, which is defined in every inner product space: the distance between $\alpha \in \mathcal{V}$ and $\beta \in \mathcal{V}$ is the norm $\|\beta - \alpha\|$.

Let $\mathcal{V}$ be the inner product space spanned by $\mathcal{W} = \{w_0(t), \ldots, w_{m-1}(t)\}$ and let $a : \mathcal{V} \to \mathcal{V}$ be an isometry. If we apply this isometry to $\mathcal{W}$, we obtain a new signal set $\tilde{\mathcal{W}} = \{\tilde{w}_0(t), \ldots, \tilde{w}_{m-1}(t)\} \subset \mathcal{V}$. Let $\mathcal{B} = \{\psi_1(t), \ldots, \psi_n(t)\}$ be an orthonormal basis for $\mathcal{V}$ and let $\mathcal{C} = \{c_0, \ldots, c_{m-1}\}$ be the codebook associated to $\mathcal{W}$ via $\mathcal{B}$. Could we have obtained $\tilde{\mathcal{W}}$ by applying some isometry to the codebook $\mathcal{C}$?

Yes, we could. Through $\mathcal{B}$, we obtain the codebook $\tilde{\mathcal{C}} = \{\tilde{c}_0, \ldots, \tilde{c}_{m-1}\}$ associated to $\tilde{\mathcal{W}}$. Through the composition that sends $c_i \to w_i(t) \to \tilde{w}_i(t) \to \tilde{c}_i$, we obtain a map from $\mathcal{C}$ to $\tilde{\mathcal{C}}$. It is easy to see that this map is an isometry of the kind considered in Section 4.2.

Are there other kinds of isometries applied to $\mathcal{W}$ that cannot be obtained simply by applying an isometry to $\mathcal{C}$? Yes, there are. The easiest way to see this is to keep the codebook the same and substitute the original orthonormal basis $\mathcal{B} = \{\psi_1(t), \ldots, \psi_n(t)\}$ with some other orthonormal basis $\tilde{\mathcal{B}} = \{\tilde{\psi}_1(t), \ldots, \tilde{\psi}_n(t)\}$. In so doing, we obtain an isometry from $\mathcal{V}$ to some other subspace $\tilde{\mathcal{V}}$ of the set of finite-energy signals.

The new signal set $\tilde{\mathcal{W}}$ might not bear any resemblance to $\mathcal{W}$, yet the resulting error probability will be identical since the codebook is unchanged. This sort of transformation is implicit in Example 3.7 of Section 3.4.

4.4 Building intuition about scalability: n versus k

The aim of this section is to sharpen our intuition by looking at a few examples of signal constellations that contain a large number m of signals. We are interested in exploring what happens to the probability of error when the number $k = \log_2 m$ of bits carried by one signal becomes large. In doing so, we will let the energy grow linearly with k so as to keep constant the energy per bit $\mathcal{E}_b$, which seems to be fair. The dimensionality of the signal space will be $n = 1$ for the first example (single-shot PAM) and $n = 2$ for the second (single-shot PSK). In the third example (bit-by-bit on a pulse train) n will be equal to k. In the final example (block-orthogonal signaling) we will let $n = 2^k$. These examples will provide us with useful insight on the asymptotic relationship between the number of transmitted bits and the dimensionality of the signal space.

What matters for all these examples is the choice of codebook. There is no need, in principle, to specify the waveform signal $w_i(t)$ associated to a codeword c_i. Nevertheless, we will specify $w_i(t)$ to make the examples more complete.

4.4.1 Keeping n fixed as k grows

EXAMPLE 4.3 (Single-shot PAM) *In this example, we fix $n = 1$ and consider PAM (see Example 3.8). We are interested in evaluating the error probability as* the number m of messages goes to infinity. Recall that the waveform associated to message i is

$$w_i(t) = c_i \psi(t),$$

where $\psi(t)$ is an arbitrary unit-energy waveform. (With $n = 1$ we do not have any choice other than modulating the amplitude of a pulse.) We are totally free to choose the pulse. For the sake of completeness we arbitrarily choose a rectangular pulse such as $\psi(t) = \frac{1}{\sqrt{T}}\mathbb{1}\{t \in [-\frac{T}{2}, \frac{T}{2}]\}$.

We have already computed the error probability of PAM in Example 2.5 of Section 2.4.3, namely

$$P_e = \left(2 - \frac{2}{m}\right) Q\left(\frac{a}{\sigma}\right),$$

where $\sigma^2 = N_0/2$. Following the instructions in Exercise 4.12, it is straightforward to prove that the average energy of the above constellation when signals are uniformly distributed is

$$\mathcal{E} = \frac{a^2(m^2 - 1)}{3}. \tag{4.1}$$

Equating to $\mathcal{E} = k\mathcal{E}_b$ and using the fact that $k = \log_2 m$ yields

$$a = \sqrt{\frac{3\mathcal{E}_b \log_2 m}{(m^2 - 1)}},$$

which goes to 0 as m goes to ∞. Hence P_e goes to 1 as m goes to ∞. □

EXAMPLE 4.4 *(Single-shot PSK)* In this example, we keep $n = 2$ and consider PSK (see Example 3.9). In Example 2.15, we have derived the following lower bound to the error probability of PSK

$$P_e \geq 2Q\left(\sqrt{\frac{\mathcal{E}}{\sigma^2}} \sin \frac{\pi}{m}\right) \frac{m-1}{m},$$

where $\sigma^2 = \frac{N_0}{2}$ is the variance of the noise in each coordinate. If we insert $\mathcal{E} = k\mathcal{E}_b$ and $m = 2^k$, we see that the lower bound goes to 1 as k goes to infinity. This happens because the circumference of the PSK constellation grows as $\sqrt{k}$ whereas the number of points grows as 2^k. Hence, the minimum distance between points goes to zero (indeed exponentially fast). □

As they are, the signal constellations used in the above two examples are not suitable to transmit a large amount k of bits by letting the constellation size $m = 2^k$ grow exponentially with k. The problem with the above two examples is that, as m grows, we are trying to pack an exponentially increasing number of points into a space that also grows in size but not fast enough. The space becomes "crowded" as m grows, meaning that the minimum distance becomes smaller and the probability of error increases.

4.4. Building intuition about scalability: n versus k

We should not conclude that PAM and PSK are not useful to send many bits. On the contrary, these signaling methods are widely used. In the next chapter we will see how. (See also the comment after the next example.)

4.4.2 Growing n linearly with k

EXAMPLE 4.5 *(Bit-by-bit on a pulse train)* The idea is to use a different dimension for each bit. Let $(b_{i,1}, b_{i,2}, \ldots, b_{i,k})$ be the binary sequence corresponding to message i. For mathematical convenience, we assume these bits to take value in $\{\pm 1\}$ rather than $\{0, 1\}$. We let the associated codeword $c_i = (c_{i,1}, c_{i,2}, \ldots, c_{i,k})^\mathsf{T}$ be defined by

$$c_{i,j} = b_{i,j}\sqrt{\mathcal{E}_b},$$

where $\mathcal{E}_b = \frac{\mathcal{E}}{k}$ is the energy per bit. The transmitted signal is

$$w_i(t) = \sum_{j=1}^{k} c_{i,j} \psi_j(t), \quad t \in \mathbb{R}. \tag{4.2}$$

As already mentioned, the choice of orthonormal basis is immaterial for the point we are making, but in practice some choices are more convenient than others. Specifically, if we choose $\psi_j(t) = \psi(t - jT_s)$ for some waveform $\psi(t)$ that fulfills $\langle \psi_i, \psi_j \rangle = \mathbb{1}\{i = j\}$, then the n-tuple former is drastically simplified because a single matched filter is sufficient to obtain all n projections (see Section 3.5). For instance, we can choose $\psi(t) = \frac{1}{\sqrt{T_s}} \mathbb{1}\{t \in [-T_s/2, T_s/2]\}$, which fulfills the mentioned constraints. We can now rewrite the waveform signal as

$$w_i(t) = \sum_{j=1}^{k} c_{i,j} \psi(t - jT_s), \quad t \in \mathbb{R}. \tag{4.3}$$

The above expression justifies the name *bit-by-bit on a pulse train* given to this signaling method (see Figure 4.3). As we will see in Chapter 5, there are many other possible choices for the pulse $\psi(t)$.

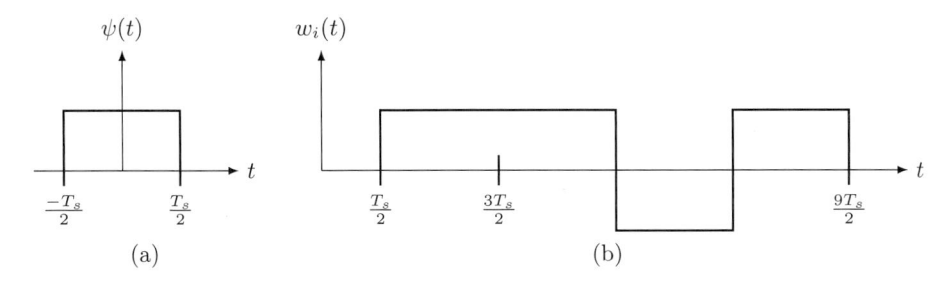

Figure 4.3. Example of (4.3) for $k = 4$ and $c_i = \sqrt{\mathcal{E}_b}(1, 1, -1, 1)^\mathsf{T}$.

The codewords $c_0, \ldots, c_{m-1}$ are the vertices of a k-dimensional hypercube as shown in Figure 4.4 for $k = 1, 2, 3$. For these values of k we immediately see from the figure what the decoding regions of an ML decoder are, but let us proceed analytically and find an ML decoding rule that works for any k. The ML receiver decides that the constellation point used by the sender is the $c_i \in \{\pm\sqrt{\mathcal{E}_b}\}^k$ that maximizes $\langle y, c_i \rangle - \frac{\|c_i\|^2}{2}$. Since $\|c_i\|^2$ is the same for all i, the previous expression is maximized by the c_i that maximizes $\langle y, c_i \rangle = \sum y_j c_{i,j}$. The maximum is achieved for the i for which $c_{i,j} = \text{sign}(y_j)\sqrt{\mathcal{E}_b}$, where

$$\text{sign}(y) = \begin{cases} 1, & y \geq 0 \\ -1, & y < 0. \end{cases}$$

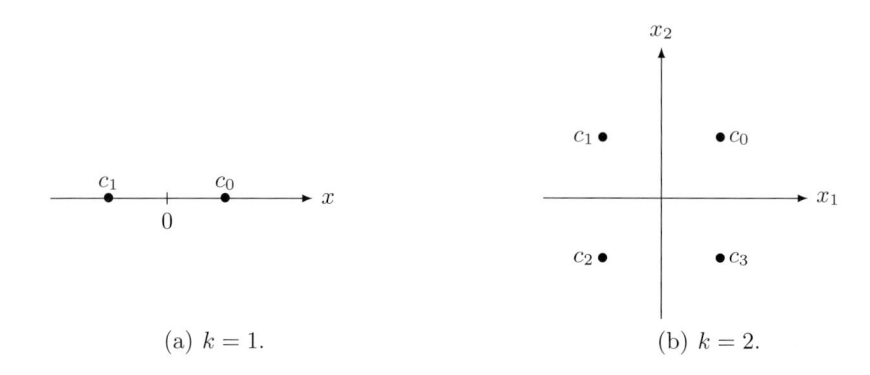

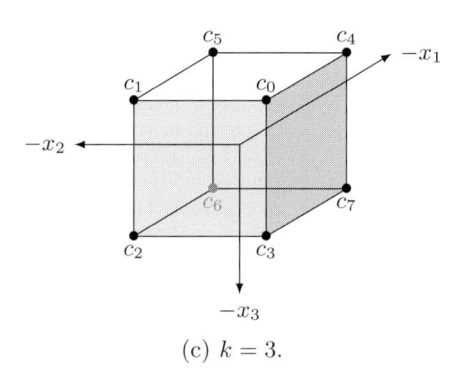

(a) $k = 1$.

(b) $k = 2$.

(c) $k = 3$.

Figure 4.4. Codebooks for bit-by-bit on a pulse train signaling.

We now compute the error probability. As usual, we first compute the error probability conditioned on a specific c_i. From the codebook symmetry, we expect that the error probability will not depend on i. If $c_{i,j}$ is positive, $Y_j = \sqrt{\mathcal{E}_b} + Z_j$ and a maximum likelihood decoder will make the correct decision if $Z_j > -\sqrt{\mathcal{E}_b}$. (The statement is an "if and only if" if we ignore the zero-probability event that $Z_j = -\sqrt{\mathcal{E}_b}$.) This happens with probability $1 - Q(\frac{\sqrt{\mathcal{E}_b}}{\sigma})$. Based on similar reasoning, it is

4.4. Building intuition about scalability: n versus k

straightforward to verify that the probability of error is the same if $c_{i,j}$ is negative. Now let C_j be the event that the decoder makes the correct decision about the jth bit. The probability of C_j depends only on Z_j. The independence of the noise components implies the independence of $C_1, C_2, \ldots, C_k$. Thus, the probability that all k bits are decoded correctly when $H = i$ is

$$P_c(i) = \left[1 - Q\left(\frac{\sqrt{\mathcal{E}_b}}{\sigma}\right)\right]^k,$$

which is the same for all i and, therefore, it is also the average P_c. Notice that $P_c \to 0$ as $k \to \infty$. However, the probability that any specific bit be decoded incorrectly is $P_b = Q(\frac{\sqrt{\mathcal{E}_b}}{\sigma})$, which does not depend on k. □

Although in this example we chose to transmit a single bit per dimension, we could have chosen to transmit $\log_2 m$ bits per dimension by letting the codeword components take value in an m-ary PAM constellation. In that case we call the signaling scheme *symbol-by-symbol on a pulse train*. Symbol-by-symbol on a pulse train and variations thereof is the basis for many digital communication systems. It will be studied in depth in Chapter 5.

The following question seems natural at this point: Is it possible to avoid that $P_c \to 0$ as $k \to \infty$? The next example gives us the answer.

4.4.3 Growing n exponentially with k

EXAMPLE 4.6 *(Block-orthogonal signaling)* Let $n = m = 2^k$, choose n orthonormal waveforms $\psi_1(t), \ldots, \psi_n(t)$, and define $w_1(t), \ldots, w_m(t)$ to be

$$w_i(t) = \sqrt{\mathcal{E}}\psi_i(t).$$

This is called block-orthogonal signaling. The name stems from the fact that in practice a block of k bits are collected and then mapped into one of m orthogonal waveforms (see Figure 4.5). Notice that $\|w_i\|^2 = \mathcal{E}$ for all i.

There are many ways to choose the 2^k waveforms $\psi_i(t)$. One way is to choose $\psi_i(t) = \psi(t - iT)$ for some normalized pulse $\psi(t)$ such that $\psi(t - iT)$ and $\psi(t - jT)$ are orthogonal when $i \neq j$. In this case the requirement for $\psi(t)$ is the same as that in bit-by-bit on a pulse train, but now we need 2^k rather than k shifted versions, and we send one pulse rather than a train of k weighted pulses. For obvious reasons this signaling scheme is called pulse position modulation.

Another example is to choose

$$w_i(t) = \sqrt{\frac{2\mathcal{E}}{T}} \cos(2\pi f_i t) \mathbb{1}\{t \in [0, T]\}. \tag{4.4}$$

This is called m-FSK (m-ary frequency-shift keying). If we choose $2f_iT = k_i$ for some integer k_i such that $k_i \neq k_j$ if $i \neq j$ then

$$\langle w_i, w_j \rangle = \frac{2\mathcal{E}}{T} \int_0^T \left[\frac{1}{2} \cos[2\pi(f_i + f_j)t] + \frac{1}{2} \cos[2\pi(f_i - f_j)t] \right] dt = \mathcal{E} \mathbb{1}\{i = j\},$$

as desired.

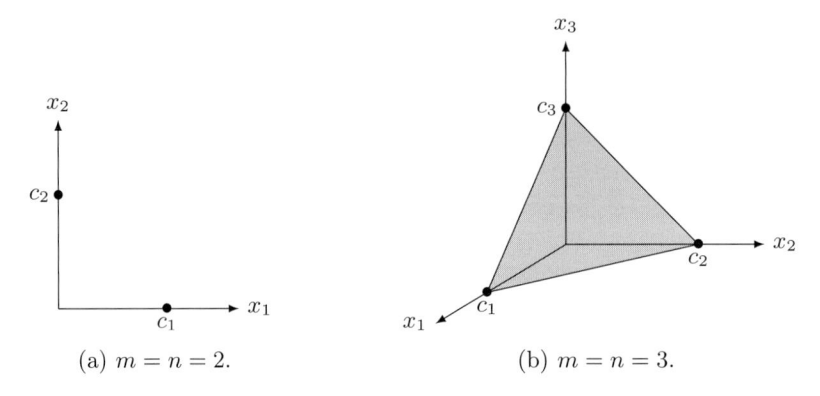

Figure 4.5. Codebooks for block-orthogonal signaling.

(a) $m = n = 2$. (b) $m = n = 3$.

When $m \geq 3$, it is not easy to visualize the decoding regions. However, we can proceed analytically, using the fact that all coordinates of c_i are 0 except for the ith, which has value $\sqrt{\mathcal{E}}$. Hence,

$$\hat{H}_{ML}(y) = \arg\max_i \langle y, c_i \rangle - \frac{\mathcal{E}}{2}$$
$$= \arg\max_i \langle y, c_i \rangle$$
$$= \arg\max_i y_i,$$

where y_i is the ith component of y. To compute (or bound) the error probability, we start as usual with a fixed c_i. We choose $i = 1$. When $H = 1$,

$$Y_j = \begin{cases} \sqrt{\mathcal{E}} + Z_j & \text{if } j = 1, \\ Z_j & \text{if } j \neq 1. \end{cases}$$

Then the probability of correct decoding is given by

$$P_c(1) = Pr\{Y_1 > Z_2, Y_1 > Z_3, \ldots, Y_1 > Z_m | H = 1\}.$$

To evaluate the right side, we first condition on $Y_1 = \alpha$, where $\alpha \in \mathbb{R}$ is an arbitrary number

$$Pr\{H = \hat{H} | H = 1, Y_1 = \alpha\} = Pr\{\alpha > Z_2, \ldots, \alpha > Z_m\} = \left[1 - Q\left(\frac{\alpha}{\sqrt{N_0/2}}\right)\right]^{m-1},$$

4.4. Building intuition about scalability: n versus k

and then remove the conditioning on Y_1,

$$P_c(1) = \int_{-\infty}^{\infty} f_{Y_1|H}(\alpha|1) \left[1 - Q\left(\frac{\alpha}{\sqrt{N_0/2}}\right)\right]^{m-1} d\alpha$$

$$= \int_{-\infty}^{\infty} \frac{1}{\sqrt{\pi N_0}} \exp\left\{-\frac{(\alpha - \sqrt{\mathcal{E}})^2}{N_0}\right\} \left[1 - Q\left(\frac{\alpha}{\sqrt{N_0/2}}\right)\right]^{m-1} d\alpha,$$

where we use the fact that when $H = 1$, $Y_1 \sim \mathcal{N}(\sqrt{\mathcal{E}}, \frac{N_0}{2})$. The above expression for $P_c(1)$ cannot be simplified further, but we can evaluate it numerically. By symmetry, $P_c(1) = P_c(i)$ for all i. Hence $P_c = P_c(1) = P_c(i)$.

The fact that the distance between any two distinct codewords is a constant simplifies the union bound considerably:

$$P_e = P_e(i) \leq (m-1)Q\left(\frac{d}{2\sigma}\right)$$

$$= (m-1)Q\left(\sqrt{\frac{\mathcal{E}}{N_0}}\right)$$

$$< 2^k \exp\left[-\frac{\mathcal{E}}{2N_0}\right]$$

$$= \exp\left[-k\left(\frac{\mathcal{E}/k}{2N_0} - \ln 2\right)\right],$$

where we used

$$\sigma^2 = \frac{N_0}{2},$$

$$d^2 = \|c_i - c_j\|^2$$

$$= \|c_i\|^2 + \|c_j\|^2 - 2\Re\{\langle c_i, c_j\rangle\}$$

$$= \|c_i\|^2 + \|c_j\|^2$$

$$= 2\mathcal{E},$$

$$Q(x) \leq \exp\left[-\frac{x^2}{2}\right], \quad x \geq 0,$$

$$m - 1 < 2^k.$$

By letting $\mathcal{E} = \mathcal{E}_b k$ we obtain

$$P_e < \exp\left\{-k\left(\frac{\mathcal{E}_b}{2N_0} - \ln 2\right)\right\}.$$

We see that $P_e \to 0$ as $k \to \infty$, provided that $\frac{\mathcal{E}_b}{N_0} > 2\ln 2$. (It is possible to prove that the weaker condition $\frac{\mathcal{E}_b}{N_0} > \ln 2$ is sufficient. See Exercise 4.3.) □

The result of the above example is quite surprising at first. The more bits we send, the larger is the probability P_c that they will *all* be decoded correctly. Yet what goes on is quite clear. In setting all but one component of each codeword to zero, we can make the non-zero component as large as $\sqrt{k\mathcal{E}_b}$. The decoder looks for the largest component. Because the variance of the noise is the same in all components and does not grow with k, when k is large it becomes almost impossible for the noise to alter the position of the largest component.

4.5 Duration, bandwidth, and dimensionality

Road traffic regulations restrict the length and width of vehicles that are allowed to circulate on highways. For similar and other reasons, the *duration* and *bandwidth* of signals that we use to communicate over a shared medium are often subject to limitations. Hence the question: What are the implications of having to use a given time and frequency interval? A precise answer to this question is known only in the limit of large time and frequency intervals, but this is good enough for our purpose.

First we need to define what it means for a signal to be time- and frequency-limited. We get a sense that the answer is not obvious by recalling that a signal that has finite support in the time domain must have infinite support in the frequency domain (and vice versa).

There are several meaningful options to define the duration and the bandwidth of a signal in such a way that both are finite for most signals of interest. Typically, people use the obvious definition for the duration of a signal, namely the length of the shortest interval that contains the signal's support and use a "softer" criterion for the bandwidth. The most common bandwidth definitions are listed in Appendix 4.9.

In this section we use an approach introduced by David Slepian in his Shannon Lecture [27].[1] In essence, Slepian's approach hinges on the idea that we should not make a distinction between signals that cannot be distinguished using a measuring instrument. Specifically, after fixing a small positive number $\eta < 1$ that accounts for the instrument's precision, we say that two signals are *indistinguishable at level* η if their difference has norm less than η.

We say that $s(t)$ is *time-limited* to (a, b) if it is indistinguishable at level η from $s(t)\mathbb{1}\{t \in (a, b)\}$. The length of the shortest such interval (a, b) is the signal's *duration* T (at level η).

EXAMPLE 4.7 *Consider the signal* $h(t) = e^{-|t|}$, $t \in \mathbb{R}$. *The norm of* $h(t) - h(t)\mathbb{1}\{t \in (-\frac{T}{2}, \frac{T}{2})\}$ *is* $\sqrt{e^{-T}}$. *Hence* $h(t)$ *has duration* $T = -2\ln\eta$ *at level* η. □

Similarly, we say that $s(t)$ is *frequency-limited* to (c, d) if $s_{\mathcal{F}}(f)$ and $s_{\mathcal{F}}(f)\mathbb{1}\{f \in (c, d)\}$ are indistinguishable at level η. The signal's *bandwidth* W is the width (length) of the shortest such interval.

---

[1] The Shannon Award is the most prestigious award bestowed by the Information Theory Society. Slepian was the first, after Shannon himself, to receive the award. The recipient presents the Shannon Lecture at the next IEEE International Symposium on Information Theory.

4.5. Duration, bandwidth, and dimensionality

A particularity of these definitions is that if we increase the strength of a signal, we could very well increase its duration and bandwidth. This makes good engineering sense. Yet it is in distinct contradiction with the usual (strict) definition of duration and the common definitions of bandwidth (Appendix 4.9). Another particularity is that all finite-energy signals have finite bandwidth and finite duration.

The dimensionality of a signal set is modified accordingly.[2] We say that a set $\mathcal{G}$ of signals has *approximate dimension n at level ϵ during the interval* $(-\frac{T}{2}, \frac{T}{2})$ if there is a fixed collection of $n = n(T, \epsilon)$ signals, say $\{\psi_1(t), \ldots, \psi_n(t)\}$, such that over the interval $(-\frac{T}{2}, \frac{T}{2})$ every signal in $\mathcal{G}$ is indistinguishable at level ϵ from some signal of the form $\sum_{i=1}^{n} a_i \psi_i(t)$. That is, we require for each $h(t) \in \mathcal{G}$ that there exists $a_1, \ldots, a_n$ such that $h(t)\mathbb{1}\{t \in (-\frac{T}{2}, \frac{T}{2})\}$ and $\sum_{i=1}^{n} a_i \psi_i(t)\mathbb{1}\{t \in (-\frac{T}{2}, \frac{T}{2})\}$ are indistinguishable at level ϵ. We further require that n be the smallest such number. We can now state the main result (without proof).

THEOREM 4.8 (Slepian) *Let $\mathcal{G}_\eta$ be the set of all signals frequency-limited to $(-\frac{W}{2}, \frac{W}{2})$ and time-limited to $(-\frac{T}{2}, \frac{T}{2})$ at level η. Let $n(W, T, \eta, \epsilon)$ be the approximate dimension of $\mathcal{G}_\eta$ at level ϵ during the interval $(-\frac{T}{2}, \frac{T}{2})$. Then, for every $\epsilon > \eta$,*

$$\lim_{T \to \infty} \frac{n(W, T, \eta, \epsilon)}{T} = W$$

$$\lim_{W \to \infty} \frac{n(W, T, \eta, \epsilon)}{W} = T.$$

□

In essence, this result says that for an arbitrary time interval (a, b) of length T and an arbitrary frequency interval (c, d) of width W, in the limit of large T and W, the set of finite-energy signals that are time-limited to (a, b) and frequency-limited to (c, d) is spanned by TW orthonormal functions. For later reference we summarize this by the expression

$$n \doteq TW, \qquad (4.5)$$

where the "$\cdot$" on top of the equal sign is meant to remind us that the relationship holds in the limit of large values for W and T.

Unlike Slepian's bandwidth definition, which applies also to complex-valued signals, the bandwidth definitions of Appendix 4.9 have been conceived with real-valued signals in mind. If $s(t)$ is real-valued, the conjugacy constraint implies that $|s_\mathcal{F}(f)|$ is an even function.[3] If, in addition, the signal is baseband, then it is frequency-limited to some interval of the form $(-B, B)$ and, according to a well-established practice, we say that the signal's bandwidth is B (not $2B$). To avoid confusions, we use the letter W for bandwidths that account for positive and

---

[2] We do not require that this signal set be closed under addition and under multiplication by scalars, i.e. we do not require that it forms a vector space.
[3] See Section 7.2 for a review of the conjugacy constraint.

negative frequencies and use B for so-called *single-sided* bandwidths. (We may call W a double-sided bandwidth.)

A result similar to (4.5) can be formulated for other meaningful definitions of time and frequency limitedness. The details depend on the definitions but the essence does not. What remains true for many meaningful definitions is that, asymptotically, there is a linear relationship between WT and n.

Two illustrative examples of this relationship follow. To avoid annoying calculations, for each example, we take the freedom to use the most convenient definition of duration and bandwidth.

EXAMPLE 4.9 Let $\psi(t) = \frac{1}{\sqrt{T_s}}\mathrm{sinc}(t/T_s)$ and

$$\psi_\mathcal{F}(f) = \sqrt{T_s}\mathbb{1}\{f \in [-1/(2T_s), 1/(2T_s)]\}$$

be a normalized pulse and its Fourier transform. Let $\psi_l(t) = \psi(t-lT_s)$, $l = 1,\ldots,n$. The collection $\mathcal{B} = \{\psi_1(t),\ldots,\psi_n(t)\}$ forms an orthonormal set. One way to see that $\psi_i(t)$ and $\psi_j(t)$ are orthogonal to each other when $i \neq j$ is to go to the Fourier domain and use Parseval's relationship. (Another way is to evoke Theorem 5.6 of Chapter 5.) Let $\mathcal{G}$ be the space spanned by the orthonormal basis $\mathcal{B}$. It has dimension n by construction. All signals of $\mathcal{G}$ are strictly frequency-limited to $(-W/2, W/2)$ for $W = 1/T_s$ and time-limited (for some η) to $(0,T)$ for $T = nT_s$. For this example $WT = n$. □

EXAMPLE 4.10 If we substitute an orthonormal basis $\{\psi_1(t),\ldots,\psi_n(t)\}$ with the related orthonormal basis $\{\varphi_1(t),\ldots,\varphi_n(t)\}$ obtained via the relationship $\varphi_i(t) = \sqrt{b}\psi_i(bt)$ for some $b \geq 1$, $i = 1,\ldots,n$, then all signals are time-compressed and frequency-expanded by the same factor b. This example shows that we can trade W for T without changing the dimensionality of the signal space, provided that WT is kept constant. □

Note that, in this section, n is the dimensionality of the signal space that may or may not be related to a codeword length (also denoted by n).

The relationship between n and WT establishes a fundamental relationship between the discrete-time and the continuous-time channel model. It says that if we are allowed to use a frequency interval of width W Hz during T seconds, then we can make approximately (asymptotically exactly) up to WT uses of the equivalent discrete-time channel model. In other words, we get to use the discrete-time channel at a rate of up to W channel uses per second.

The symmetry of (4.5) implies that time and frequency are on an equal footing in terms of providing the degrees of freedom exploited by the discrete-time channel. It is sometimes useful to think of T and W as the width and height of a rectangle in the time–frequency plane, as shown in Figure 4.6. We associate such a rectangle with the set of signals that have the corresponding time and frequency limitations. Like a piece of land, such a rectangle represents a natural resource and what matters for its exploitation is its area.

The fact that n can grow linearly with WT and not faster is bad news for block-orthogonal signaling. This means that n cannot grow exponentially in k unless WT does the same. In a typical system, W is fixed by regulatory constraints

4.6. Bit-by-bit versus block-orthogonal

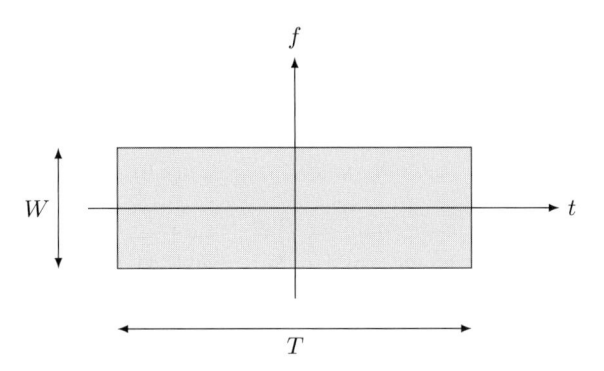

Figure 4.6. Time–frequency plane.

and T grows linearly with k. (T is essentially the time it takes to send k bits.) Hence WT cannot grow exponentially in k, which means that block-orthogonal is not scalable. Of the four examples studied in Section 4.4, only bit-by-bit on a pulse train seems to be a viable candidate for large values of k, provided that we can make it more robust to additive white Gaussian noise. The purpose of the next section is to gain valuable insight into what it takes to achieve this goal.

4.6 Bit-by-bit versus block-orthogonal

We have seen that the message error probability goes to 1 in bit-by-bit on a pulse train and goes to 0 (exponentially fast) in block-orthogonal signaling. The union bound is quite useful for understanding what goes on.

In computing the error probability when message i is transmitted, the union bound has one term for each $j \neq i$. The dominating terms correspond to the signals c_j that are closest to c_i. If we neglect the other terms, we obtain an expression of the form

$$P_e(i) \approx N_d \, Q\left(\frac{d_m}{2\sigma}\right),$$

where N_d is the number of dominant terms, i.e. the number of *nearest neighbors* to c_i, and d_m is the *minimum distance*, i.e. the distance to a nearest neighbor.

For bit-by-bit on a pulse train, there are k closest neighbors, each neighbor obtained by changing c_i in exactly one component, and each of them is at distance $2\sqrt{\mathcal{E}_b}$ from c_i. As k increases, N_d increases and $Q(\frac{d_m}{2\sigma})$ stays constant. The increase of N_d makes $P_e(i)$ increase.

Now consider block-orthogonal signaling. All signals are at the same distance from each other. Hence there are $N_d = 2^k - 1$ nearest neighbors to c_i, all at distance

$d_m = \sqrt{2\mathcal{E}} = \sqrt{2k\mathcal{E}_b}$. Hence

$$Q\left(\frac{d_m}{2\sigma}\right) \leq \frac{1}{2}\exp\left(-\frac{d_m^2}{8\sigma^2}\right) = \frac{1}{2}\exp\left(-\frac{k\mathcal{E}_b}{4\sigma^2}\right),$$
$$N_d = 2^k - 1 = \exp(k\ln 2) - 1.$$

We see that the probability that the noise carries a signal closer to a specific neighbor decreases as $\exp\left(-\frac{k\mathcal{E}_b}{4\sigma^2}\right)$, whereas the number of nearest neighbors increases as $\exp(k\ln 2)$. For $\frac{\mathcal{E}_b}{4\sigma^2} > \ln 2$ the product decreases, otherwise it increases.

In essence, to reduce the error probability we need to increase the minimum distance. If the number of dimensions remains constant, as in the first two examples of Section 4.4, the space occupied by the signals becomes crowded, the minimum distance decreases, and the error probability increases. For block-orthogonal signaling, the signal's norm increases as $\sqrt{k\mathcal{E}_b}$ and, by Pythagoras, the distance is a factor $\sqrt{2}$ larger than the norm – hence the distance grows as $\sqrt{2k\mathcal{E}_b}$. In bit-by-bit on a pulse train, the minimum distance remains constant. As we will see in Chapter 6, sophisticated coding techniques in conjunction with a generalized form of bit-by-bit on a pulse train can reduce the error probability by increasing the distance profile.

4.7 Summary

In this chapter we have introduced new design parameters and performance measures. The ones we are mostly concerned with are as follows.

- The cardinality m of the message set $\mathcal{H}$. Since in most cases the message consists of bits, typically we choose m to be a power of 2. Whether m is a power of 2 or not, we say that a message carries $k = \log_2 m$ bits of information (assuming that all messages are equiprobable).
- The *message* error probability P_e and the *bit* error rate P_b. The former, also called *block* error probability, is the error probability we have considered so far. The latter can be computed, in principle, once we specify the mapping between the set of k-bit sequences and the set of messages. Until then, the only statement we can make about P_b is that $\frac{P_e}{k} \leq P_b \leq P_e$. The right bound applies with equality if a message error always translates into 1-out-of-k bits being incorrectly reproduced. The left is an equality if all bits are incorrectly reproduced each time that there is a message error. Whether we care more about P_e or P_b depends on the application. If we send a file that contains a computer program, every single bit of the file has to be received correctly in order for the transmission to be successful. In this case we clearly want P_e to be small. However, there are sources that are more tolerant to occasional errors. This is the case of a digitized voice signal. For voice it is sufficient to have P_b small. To appreciate the difference between P_e and P_b, consider the hypothetical situation in which one message corresponds to $k = 10^3$ bits and 1 bit of every message is incorrectly reconstructed. Then the message error probability is 1 (every message is incorrectly reconstructed), whereas the bit-error probability is 10^{-3}.

4.7. Summary

- The average signal's energy $\mathcal{E}$ and the average energy per bit $\mathcal{E}_b$, where $\mathcal{E}_b = \frac{\mathcal{E}}{k}$. We are typically willing to double the energy to send twice as many bits. In this case we fix $\mathcal{E}_b$ and let $\mathcal{E}$ be a function of k.
- The transmission rate $R_b = \frac{k}{T} = \frac{\log_2(m)}{T}$ [bits/second].
- The single-sided bandwidth B and the two-sided bandwidth W. There are several meaningful criteria to determine the bandwidth.
- Scalability, in the sense that we ought to be able to communicate bit sequences of any length (provided we let WT scale in a sustainable way).
- The implementation cost and computational complexity. To keep the discussion as simple as possible, we assume that the cost is determined by the number of matched filters in the n-tuple former and the complexity is that of the decoder.

Clearly we desire scalability, high transmission rate, little energy spent per bit, small bandwidth, small error probability (message or bit, depending on the application), low cost and low complexity. As already mentioned, some of these goals conflict. For instance, starting from a given codebook we can trade energy for error probability by scaling down all the codewords by some factor. In so doing the average energy will decrease and so will the distance between codewords, which implies that the error probability will increase. Alternatively, once we have reduced the energy by scaling down the codewords we can add new codewords at the periphery of the codeword constellation, choosing their location in such a way that new codewords do not further increase the error probability. We keep doing this until the average energy has returned to the original value. In so doing we trade bit rate for error probability. By removing codewords at the periphery of the codeword constellation we can trade bit rate for energy. All these manipulations pertain to the encoder. By acting inside the waveform former, we can boost the bit rate at the expense of bandwidth. For instance, we can substitute $\psi_i(t)$ with $\phi_i(t) = \sqrt{b}\psi_i(bt)$ for some $b > 1$. This scales the duration of all signals by $1/b$ with two consequences. First, the bit rate is multiplied by b. (It takes a fraction b of time to send the same number of bits.) Second, the signal's bandwidth expands by b. (The scaling property of the Fourier transform asserts that the Fourier transform of $\psi(bt)$ is $\frac{1}{|b|}\psi_\mathcal{F}(\frac{f}{b})$.) These examples are meant to show that there is considerable margin for trading among bit rate, bandwidth, error probability, and average energy.

We have seen that, rather surprisingly, it is possible to transmit an increasing number k of bits at a fixed energy per bit $\mathcal{E}_b$ and to make the probability that even a single bit is decoded incorrectly go to zero as k increases. However, the scheme we used to prove this has the undesirable property of requiring an exponential growth of the time–bandwidth product. Such a growth would make us quickly run out of time and/or bandwidth even with moderate values of k. In real-world applications, we are given a fixed bandwidth and we let the duration grow linearly with k. It is not a coincidence that most signaling methods in use today can be seen one way or another as refinements of bit-by-bit on a pulse train. This line of signaling technique will be pursued in the next two chapters.

Information theory is a field that searches for the ultimate trade-offs, regardless of the signaling method. A main result from information theory is the famous formula

$$C = \frac{W}{2}\log_2\left(1 + \frac{2P}{N_0 W}\right) \tag{4.6}$$
$$= B\log_2\left(1 + \frac{P}{N_0 B}\right).$$

It gives a precise value to the ultimate rate C bps at which we can transmit reliably over a waveform AWGN channel of noise power spectral density $N_0/2$ watts/Hz if we are allowed to use signals of power not exceeding P watts and absolute (single-sided) bandwidth not exceeding B Hz.

This is a good time to clarify our non-standard use of the words *coding, encoder, codeword, and codebook*. We have seen that no matter which waveform signals we use to communicate, we can always break down the sender into a block that provides an n-tuple and one that maps the n-tuple into the corresponding waveform. This view is completely general and serves us well, whether we analyze or implement a system. Unfortunately there is no standard name for the first block. Calling it an encoder is a good name, but the reader should be aware that the current practice is to say that there is coding when the mapping from bits to codewords is non-trivial, and to say that there is no coding when the map is trivial as in bit-by-bit on a pulse train. Making such a distinction is not a satisfactory solution in our view. An example of a non-trivial encoder will be studied in depth in Chapter 6.

Calling the second block a *waveform former* is definitely non-standard, but we find this name to be more appropriate than calling it a modulator, which is the most common name used for it. The term modulator has been inherited from the old days of analog communication techniques such as amplitude modulation (AM) for which it was an appropriate name.

4.8 Appendix: Isometries and error probability

Here we give a formal proof that if we apply an isometry to a codebook and its decoding regions, then the error probability associated to the new codebook and the new regions is the same as that of the original codebook and original regions.

Let

$$g(\gamma) = \frac{1}{(2\pi\sigma^2)^{n/2}}\exp\left(-\frac{\gamma^2}{2\sigma^2}\right), \ \gamma \in \mathbb{R},$$

so that for $Z \sim \mathcal{N}(0, \sigma^2 I_n)$ we can write $f_Z(z) = g(\|z\|)$. Then for any codebook $\mathcal{C} = \{c_0, \ldots, c_{m-1}\}$, decoding regions $\mathcal{R}_0, \ldots, \mathcal{R}_{m-1}$, and isometry $a : \mathbb{R}^n \to \mathbb{R}^n$ we have

$$P_c(i) = Pr\{Y \in \mathcal{R}_i | \text{codeword } c_i \text{ is transmitted}\}$$
$$= \int_{y \in \mathcal{R}_i} g(\|y - c_i\|)dy$$

4.9. Appendix: Bandwidth definitions

$$\stackrel{(a)}{=} \int_{y \in \mathcal{R}_i} g(\|a(y) - a(c_i)\|)dy$$

$$\stackrel{(b)}{=} \int_{y:a(y) \in a(\mathcal{R}_i)} g(\|a(y) - a(c_i)\|)dy$$

$$\stackrel{(c)}{=} \int_{\alpha \in a(\mathcal{R}_i)} g(\|\alpha - a(c_i)\|)d\alpha$$

$$= Pr\{Y \in a(\mathcal{R}_i) | \text{codeword } a(c_i) \text{ is transmitted}\},$$

where in (a) we use the distance-preserving property of an isometry, in (b) we use the fact that $y \in \mathcal{R}_i$ if and only if $a(y) \in a(\mathcal{R}_i)$, and in (c) we make the change of variable $\alpha = a(y)$ and use the fact that the Jacobian of an isometry is ± 1. The last line is the probability of decoding correctly when the transmitter sends $a(c_i)$ and the corresponding decoding region is $a(\mathcal{R}_i)$.

4.9 Appendix: Bandwidth definitions

There are several widely accepted definitions of bandwidth, which we summarize in this appendix. They are all meant for real-valued baseband functions $h(t)$ that represent either a signal or an impulse response. Because $h(t)$ is real-valued, $|h_\mathcal{F}(f)|$ is an even function. This is a consequence of the conjugacy constraint (see Section 7.2). Baseband means that $|h_\mathcal{F}(f)|$ is negligible outside an interval of the form $(-B, B)$, where the meaning of "negligible" varies from one bandwidth definition to another. Extending to passband signals is straightforward. Here we limit ourselves to baseband signals because passband signals are treated in Chapter 7.

Since these bandwidth definitions were meant for functions $h(t)$ for which the essential support of $|h_\mathcal{F}(f)|$ has the form $(-B, B)$, the common practice is to say that their bandwidth is B rather than $2B$. An exception is the definition of bandwidth that we have given in Section 4.5, which applies also to complex-valued signals. As already mentioned, we avoid confusion by using B and W for single-sided and double-sided bandwidths, respectively.

For each bandwidth definition given below, the reader can find an example in Exercise 4.13.

- *Absolute bandwidth* This is the smallest positive number B such that $(-B, B)$ is the support of $h_\mathcal{F}(f)$. As mentioned in the previous paragraph, for signals that we use in practice the absolute bandwidth is infinite. The signals used in practice have finite (time-domain) support, which implies that their absolute bandwidth is ∞. However, in examples we sometimes use signals that do have a finite absolute bandwidth.
- *3-dB bandwidth* The 3-dB bandwidth, if it exists, is the positive number B such that $|h_\mathcal{F}(f)|^2 > \frac{|h_\mathcal{F}(0)|^2}{2}$ in the interval $\mathcal{I} = (-B, B)$ and $|h_\mathcal{F}(f)|^2 \leq \frac{|h_\mathcal{F}(0)|^2}{2}$ outside $\mathcal{I}$. In other words, outside $\mathcal{I}$ the value of $|h_\mathcal{F}(f)|$ is at least 3-dB smaller than at $f = 0$.

- *η-bandwidth* For any number $\eta \in (0,1)$, the η-bandwidth is the smallest positive number B such that

$$\int_{-B}^{B} |h_{\mathcal{F}}(f)|^2 df \geq (1-\eta) \int_{-\infty}^{\infty} |h_{\mathcal{F}}(f)|^2 df.$$

It defines the interval $(-B, B)$ that contains a fraction $(1-\eta)$ of the signal's energy. Reasonable values for η are $\eta = 0.1$ and $\eta = 0.01$. (Recall that, by Parseval's relationship, the integral on the right equals the squared norm $\|h\|^2$.)

- *First zero-crossing bandwidth* The first zero-crossing bandwidth, if it exists, is that B for which $|h_{\mathcal{F}}(f)|$ is positive in $\mathcal{I} = (-B, B)$ and vanishes at $\pm B$.

- *Equivalent noise bandwidth* This is B if

$$\int_{-\infty}^{\infty} |h_{\mathcal{F}}(f)|^2 df = 2B |h_{\mathcal{F}}(0)|^2.$$

The name comes from the fact that if we feed with white noise a filter of impulse response $h(t)$ and we feed with the same input an ideal lowpass filter of frequency response $|h_{\mathcal{F}}(0)|\mathbb{1}\{f \in [-B, B]\}$, then the output power is the same in both situations.

- *Root-mean-square (RMS) bandwidth* This is defined if $\int_{-\infty}^{\infty} |h_{\mathcal{F}}(f)|^2 df < \infty$, in which case it is

$$B = \left[\frac{\int_{-\infty}^{\infty} f^2 |h_{\mathcal{F}}(f)|^2 df}{\int_{-\infty}^{\infty} |h_{\mathcal{F}}(f)|^2 df} \right]^{\frac{1}{2}}.$$

To understand this definition, notice that the function $g(f) := \frac{|h_{\mathcal{F}}(f)|^2}{\int_{-\infty}^{\infty} |h_{\mathcal{F}}(f)|^2 df}$ is non-negative, even, and integrates to 1. Hence it is the density of some zero-mean random variable and $B = \sqrt{\int f^2 g(f) df}$ is the standard deviation of that random variable.

4.10 Exercises

Exercises for Section 4.3

EXERCISE 4.1 (Signal translation) *Consider the signals $w_0(t)$ and $w_1(t)$ shown in Figure 4.7, used to communicate 1 bit across the AWGN channel of power spectral density $N_0/2$.*

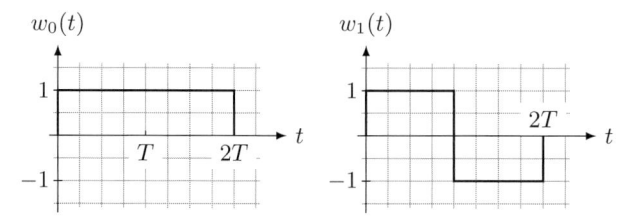

Figure 4.7.

4.10. Exercises

(a) Determine an orthonormal basis $\{\psi_0(t), \psi_1(t)\}$ for the space spanned by $\{w_0(t), w_1(t)\}$ and find the corresponding codewords c_0 and c_1. Work out two solutions, one obtained via Gram–Schmidt and one in which the second element of the orthonormal basis is a delayed version of the first. Which of the two solutions would you choose if you had to implement the system?

(b) Let X be a uniformly distributed binary random variable that takes values in $\{0,1\}$. We want to communicate the value of X over an additive white Gaussian noise channel. When $X = 0$, we send $w_0(t)$, and when $X = 1$, we send $w_1(t)$. Draw the block diagram of an ML receiver based on a single matched filter.

(c) Determine the error probability P_e of your receiver as a function of T and N_0.

(d) Find a suitable waveform $v(t)$, such that the new signals $\tilde{w}_0(t) = w_0(t) - v(t)$ and $\tilde{w}_1(t) = w_1(t) - v(t)$ have minimal energy and plot the resulting waveforms.

(e) What is the name of the kind of signaling scheme that uses $\tilde{w}_0(t)$ and $\tilde{w}_1(t)$? Argue that one obtains this kind of signaling scheme independently of the initial choice of $w_0(t)$ and $w_1(t)$.

EXERCISE 4.2 (Orthogonal signal sets) Consider a set $\mathcal{W} = \{w_0(t), \ldots, w_{m-1}(t)\}$ of mutually orthogonal signals with squared norm $\mathcal{E}$ each used with equal probability.

(a) Find the minimum-energy signal set $\tilde{\mathcal{W}} = \{\tilde{w}_0(t), \ldots, \tilde{w}_{m-1}(t)\}$ obtained by translating the original set.

(b) Let $\tilde{\mathcal{E}}$ be the average energy of a signal picked at random within $\tilde{\mathcal{W}}$. Determine $\tilde{\mathcal{E}}$ and the energy saving $\mathcal{E} - \tilde{\mathcal{E}}$.

(c) Determine the dimension of the inner product space spanned by $\tilde{\mathcal{W}}$.

Exercises for Section 4.4

EXERCISE 4.3 (Suboptimal receiver for orthogonal signaling) This exercise takes a different approach to the evaluation of the performance of block-orthogonal signaling (Example 4.6). Let the message $H \in \{1, \ldots, m\}$ be uniformly distributed and consider the communication problem described by

$$H = i: \quad Y = c_i + Z, \quad Z \sim \mathcal{N}(0, \sigma^2 I_m),$$

where $Y = (Y_1, \ldots, Y_m)^\mathsf{T} \in \mathbb{R}^m$ is the received vector and $\{c_1, \ldots, c_m\} \subset \mathbb{R}^m$ is the codebook consisting of constant-energy codewords that are orthogonal to each other. Without loss of essential generality, we can assume

$$c_i = \sqrt{\mathcal{E}} e_i,$$

where e_i is the ith unit vector in $\mathbb{R}^m$, i.e. the vector that contains 1 at position i and 0 elsewhere, and $\mathcal{E}$ is some positive constant.

(a) Describe the statistic of Y_j for $j = 1, \ldots, m$ given that $H = 1$.

(b) Consider a suboptimal receiver that uses a threshold $t = \alpha\sqrt{\mathcal{E}}$ where $0 < \alpha < 1$. The receiver declares $\hat{H} = i$ if i is the only integer such that $Y_i \geq t$.

If there is no such i or there is more than one index i for which $Y_i \geq t$, the receiver declares that it cannot decide. This will be viewed as an error. Let $E_i = \{Y_i \geq t\}$, $E_i^c = \{Y_i < t\}$, and describe, in words, the meaning of the event

$$E_1 \cap E_2^c \cap E_3^c \cap \cdots \cap E_m^c.$$

(c) Find an upper bound to the probability that the above event does not occur when $H = 1$. Express your result using the Q function.

(d) Now let $m = 2^k$ and let $\mathcal{E} = k\mathcal{E}_b$ for some fixed energy per bit $\mathcal{E}_b$. Prove that the error probability goes to 0 as $k \to \infty$, provided that $\frac{\mathcal{E}_b}{\sigma^2} > \frac{2\ln 2}{\alpha^2}$. (Notice that because we can choose α^2 as close to 1 as we wish, if we insert $\sigma^2 = \frac{N_0}{2}$, the condition becomes $\frac{\mathcal{E}_b}{N_0} > \ln 2$, which is a weaker condition than the one obtained in Example 4.6.) Hint: Use $m - 1 < m = \exp(\ln m)$ and $Q(x) < \frac{1}{2}\exp(-\frac{x^2}{2})$.

EXERCISE 4.4 (Receiver diagrams) For each signaling method discussed in Section 4.4, draw the block diagram of an ML receiver.

EXERCISE 4.5 (Bit-by-bit on a pulse train) A communication system uses bit-by-bit on a pulse train to communicate at 1 Mbps using a rectangular pulse. The transmitted signal is of the form

$$\sum_j B_j \mathbb{1}\{jT_s \leq t < (j+1)T_s\},$$

where $B_j \in \{\pm b\}$. Determine the value of b needed to achieve bit-error probability $P_b = 10^{-5}$ knowing that the channel corrupts the transmitted signal with additive white Gaussian noise of power spectral density $N_0/2 = 10^{-2}$ watts/Hz.

EXERCISE 4.6 (Bit-error probability) A discrete memoryless source produces bits at a rate of 10^6 bps. The bits, which are uniformly distributed and iid, are grouped into pairs and each pair is mapped into a distinct waveform and sent over the AWGN channel of noise power spectral density $N_0/2$. Specifically, the first two bits are mapped into one of the four waveforms shown in Figure 4.8 with $T_s = 2 \times 10^{-6}$ seconds, the next two bits are mapped onto the same set of waveforms delayed by T_s, etc.

(a) Describe an orthonormal basis for the inner product space $\mathcal{W}$ spanned by $w_i(t)$, $i = 0, \ldots, 3$ and plot the signal constellation in $\mathbb{R}^n$, where n is the dimensionality of $\mathcal{W}$.
(b) Determine an assignment between pairs of bits and waveforms such that the bit-error probability is minimized and derive an expression for P_b.
(c) Draw a block diagram of the receiver that achieves the above P_b using a single causal filter.
(d) Determine the energy per bit $\mathcal{E}_b$ and the power of the transmitted signal.

4.10. Exercises

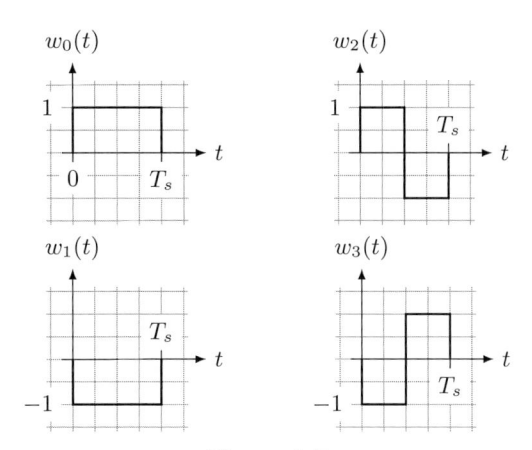

Figure 4.8.

EXERCISE 4.7 (*m-ary frequency-shift keying*) *m-ary frequency-shift keying (m-FSK) is a signaling method that uses signals of the form*

$$w_i(t) = \sqrt{\frac{2\mathcal{E}}{T}} \cos\left(2\pi(f_c + i\Delta f)t\right)\mathbb{1}\{t \in [0, T]\}, \quad i = 0, \ldots, m-1,$$

where $\mathcal{E}$, T, f_c, and Δf are fixed parameters, with $\Delta f \ll f_c$.

(a) *Determine the average energy $\mathcal{E}$. (You can assume that $f_c T$ is an integer.)*
(b) *Assuming that $f_c T$ is an integer, find the smallest value of Δf that makes $w_i(t)$ orthogonal to $w_j(t)$ when $i \neq j$.*
(c) *In practice the signals $w_i(t)$, $i = 0, 1, \ldots, m-1$ can be generated by changing the frequency of a single oscillator. In passing from one frequency to another a phase shift θ is introduced. Again, assuming that $f_c T$ is an integer, determine the smallest value Δf that ensures orthogonality between $\cos(2\pi(f_c + i\Delta f)t + \theta_i)$ and $\cos(2\pi(f_c + j\Delta f)t + \theta_j)$ whenever $i \neq j$ regardless of θ_i and θ_j.*
(d) *Sometimes we do not have complete control over f_c either, in which case it is not possible to set $f_c T$ to an integer. Argue that if we choose $f_c T \gg 1$ then for all practical purposes the signals will be orthogonal to one another if the condition found in part (c) is met.*
(e) *Give an approximate value for the bandwidth occupied by the signal constellation. How does the WT product behave as a function of $k = \log_2(m)$?*

Exercises for Section 4.5

EXERCISE 4.8 (*Packing rectangular pulses*) *This exercise is an interesting variation to Example 4.9. Let $\psi(t) = \mathbb{1}\{t \in [-T_s/2, T_s/2]\}/\sqrt{T_s}$ be a normalized rectangular pulse of duration T_s and let $\psi_{\mathcal{F}}(f) = \sqrt{T_s}\,\text{sinc}(T_s f)$ be its Fourier transform. The collection $\{\psi_1(t), \ldots, \psi_n(t)\}$, where $\psi_l(t) = \psi(t - lT_s)$, $l = 1, \ldots, n$, forms an orthonormal set. (This is obvious from the time domain.) It has dimension n by construction.*

(a) For the set $\mathcal{G}$ spanned by the above orthonormal basis, determine the relationship between n and WT.
(b) Compare with Example 4.9 and explain the difference.

EXERCISE 4.9 (Time- and frequency-limited orthonormal sets) *Complement Example 4.9 and Exercise 4.8 with similar examples in which the shifts occur in the frequency domain. The corresponding time-domain signals can be complex-valued.*

EXERCISE 4.10 (Root-mean-square bandwidth) *The root-mean-square bandwidth (abbreviated rms bandwidth) of a lowpass signal $g(t)$ of finite-energy is defined by*

$$B_{rms} = \left[\frac{\int_{-\infty}^{\infty} f^2 |g_{\mathcal{F}}(f)|^2 df}{\int_{-\infty}^{\infty} |g_{\mathcal{F}}(f)|^2 df}\right]^{1/2},$$

where $|g_{\mathcal{F}}(f)|^2$ is the energy spectral density of the signal. Correspondingly, the root-mean-square (rms) duration of the signal is defined by

$$T_{rms} = \left[\frac{\int_{-\infty}^{\infty} t^2 |g(t)|^2 dt}{\int_{-\infty}^{\infty} |g(t)|^2 dt}\right]^{1/2}.$$

We want to show that, with the above definitions and assuming that $|g(t)| \to 0$ faster than $1/\sqrt{|t|}$ as $|t| \to \infty$, the time–bandwidth product satisfies

$$T_{rms} B_{rms} \geq \frac{1}{4\pi}.$$

(a) Use the Schwarz inequality and the fact that for any $c \in \mathbb{C}$, $c + c^* = 2\Re\{c\} \leq 2|c|$ to prove that

$$\left\{\int_{-\infty}^{\infty} [g_1^*(t) g_2(t) + g_1(t) g_2^*(t)] dt\right\}^2 \leq 4 \int_{-\infty}^{\infty} |g_1(t)|^2 dt \int_{-\infty}^{\infty} |g_2(t)|^2 dt.$$

(b) In the above inequality insert $g_1(t) = tg(t)$ and $g_2(t) = \frac{dg(t)}{dt}$ and show that

$$\left[\int_{-\infty}^{\infty} t \frac{d}{dt}[g(t) g^*(t)] dt\right]^2 \leq 4 \int_{-\infty}^{\infty} t^2 |g(t)|^2 dt \int_{-\infty}^{\infty} \left|\frac{dg(t)}{dt}\right|^2 dt.$$

(c) Integrate the left-hand side by parts and use the fact that $|g(t)| \to 0$ faster than $1/\sqrt{|t|}$ as $|t| \to \infty$ to obtain

$$\left[\int_{-\infty}^{\infty} |g(t)|^2 dt\right]^2 \leq 4 \int_{-\infty}^{\infty} t^2 |g(t)|^2 dt \int_{-\infty}^{\infty} \left|\frac{dg(t)}{dt}\right|^2 dt.$$

(d) Argue that the above is equivalent to

$$\int_{-\infty}^{\infty} |g(t)|^2 dt \int_{-\infty}^{\infty} |g_{\mathcal{F}}(f)|^2 df \leq 4 \int_{-\infty}^{\infty} t^2 |g(t)|^2 dt \int_{-\infty}^{\infty} 4\pi^2 f^2 |g_{\mathcal{F}}(f)|^2 df.$$

4.10. Exercises

(e) Complete the proof to obtain $T_{rms}B_{rms} \geq \frac{1}{4\pi}$.
(f) As a special case, consider a Gaussian pulse defined by $g(t) = \exp(-\pi t^2)$. Show that for this signal $T_{rms}B_{rms} = \frac{1}{4\pi}$ i.e. the above inequality holds with equality. Hint: $\exp(-\pi t^2) \xleftrightarrow{\mathcal{F}} \exp(-\pi f^2)$.

EXERCISE 4.11 (Real basis for complex space) Let $\mathcal{G}$ be a complex inner product space of finite-energy waveforms with the property that $g(t) \in \mathcal{G}$ implies $g^*(t) \in \mathcal{G}$.

(a) Let $\mathcal{G}_R$ be the subset of $\mathcal{G}$ that contains only real-valued waveforms. Argue that $\mathcal{G}_R$ is a real inner product space.
(b) Prove that if $g(t) = a(t) + jb(t)$ is in $\mathcal{G}$, then both $a(t)$ and $b(t)$ are in $\mathcal{G}_R$.
(c) Prove that if $\{\psi_1(t), \ldots, \psi_n(t)\}$ is an orthonormal basis for the real inner product space $\mathcal{G}_R$ then it is also an orthonormal basis for the complex inner product space $\mathcal{G}$.

Comment: In this exercise we have shown that we can always find a real-valued orthonormal basis for an inner product space $\mathcal{G}$ such that $g(t) \in \mathcal{G}$ implies $g^*(t) \in \mathcal{G}$. An equivalent condition is that if $g(t) \in \mathcal{G}$ then also the inverse Fourier transform of $g_{\mathcal{F}}^*(-f)$ is in $\mathcal{G}$. The set $\mathcal{G}$ of complex-valued finite-energy waveforms that are strictly time-limited to $(-\frac{T}{2}, \frac{T}{2})$ and bandlimited to $(-B, B)$ (for any of the bandwidth definitions given in Appendix 4.9) fulfills the stated conjugacy condition.

EXERCISE 4.12 (Average energy of PAM) Let U be a random variable uniformly distributed in $[-a, a]$ and let S be a discrete random variable independent of U and uniformly distributed over the PAM constellation $\{\pm a, \pm 3a, \ldots, \pm(m-1)a\}$, where m is an even integer. Let $V = S + U$.

(a) Find the distribution of V.
(b) Find the variance of U and that of V.
(c) Use part (b) to determine the variance of S. Justify your steps. (Notice: by finding the variance of S, we have found the average energy of the PAM constellation used with uniform distribution.)

Exercises for Appendix 4.9

EXERCISE 4.13 (Bandwidth) Verify the following statements.

(a) The absolute bandwidth of $\text{sinc}(\frac{t}{T_s})$ is $B = \frac{1}{2T_s}$.
(b) The 3-dB bandwidth of an RC lowpass filter is $B = \frac{1}{2\pi RC}$. Hint: The impulse response of an RC lowpass filter is $h(t) = \frac{1}{RC} \exp\left(-\frac{t}{RC}\right)$, $t \geq 0$ and 0 otherwise. The squared magnitude of its Fourier transform is $|h_{\mathcal{F}}(f)|^2 = \frac{1}{1+(2\pi RC f)^2}$.
(c) The η-bandwidth of an RC lowpass filter is $B = \frac{1}{2\pi RC} \tan\left(\frac{\pi}{2}(1-\eta)\right)$. Hint: Same as in part (b).
(d) The zero-crossing bandwidth of $\mathbb{1}\{t \in [-\frac{T_s}{2}, \frac{T_s}{2}]\}$ is $B = \frac{2}{T_s}$.
(e) The equivalent noise bandwidth of an RC lowpass filter is $B = \frac{1}{4RC}$.

(f) The RMS bandwidth of $h(t) = \exp(-\pi t^2)$ is $B = \frac{1}{\sqrt{4\pi}}$. Hint: $h_\mathcal{F}(f) = \exp(-\pi f^2)$.

Miscellaneous exercises

EXERCISE 4.14 (Antipodal signaling and Rayleigh fading) *Consider using antipodal signaling, i.e. $w_0(t) = -w_1(t)$, to communicate 1 bit across a Rayleigh fading channel that we model as follows. When $w_i(t)$ is transmitted the channel output is*

$$R(t) = Aw_i(t) + N(t),$$

where $N(t)$ is white Gaussian noise of power spectral density $N_0/2$ and A is a random variable of probability density function

$$f_A(a) = \begin{cases} 2ae^{-a^2}, & \text{if } a \geq 0, \\ 0, & \text{otherwise.} \end{cases} \quad (4.7)$$

We assume that, unlike the transmitter, the receiver knows the realization of A. We also assume that the receiver implements a maximum likelihood decision, and that the signal's energy is $\mathcal{E}_b$.

(a) *Describe the receiver.*
(b) *Determine the error probability conditioned on the event $A = a$.*
(c) *Determine the unconditional error probability P_f. (The subscript stands for fading.)*
(d) *Compare P_f to the error probability P_e achieved by an ML receiver that observes $R(t) = mw_i(t) + N(t)$, where $m = \mathbb{E}[A]$. Comment on the different behavior of the two error probabilities. For each of them, find the $\mathcal{E}_b/N_0$ value necessary to obtain the probability of error 10^{-5}. (You may use $\frac{1}{2}\exp(-\frac{x^2}{2})$ as an approximation of $Q(x)$.)*

EXERCISE 4.15 (Non-white Gaussian noise) *Consider the following transmitter/receiver design problem for an additive non-white Gaussian noise channel.*

(a) *Let the hypothesis H be uniformly distributed in $\mathcal{H} = \{0, \ldots, m-1\}$ and when $H = i$, $i \in \mathcal{H}$, let $w_i(t)$ be the channel input. The channel output is then*

$$R(t) = w_i(t) + N(t),$$

where $N(t)$ is Gaussian noise of power spectral density $G(f)$, where we assume that $G(f) \neq 0$ for all f. Describe a receiver that, based on the channel output $R(t)$, decides on the value of H with least probability of error. Hint: Find a way to transform this problem into one that you can solve.

(b) *Consider the setting as in part (a) except that now you get to design the signal set with the restrictions that $m = 2$ and that the average energy cannot exceed $\mathcal{E}$. We also assume that $G^2(f)$ is constant in the interval $[a, b]$, $a < b$, where it also achieves its global minimum. Find two signals that achieve the smallest possible error probability under an ML decoding rule.*

4.10. Exercises

EXERCISE 4.16 (Continuous-time AWGN capacity) *To prove the formula for the capacity C of the continuous-time AWGN channel of noise power density $N_0/2$ when signals are power-limited to P and frequency-limited to $(-\frac{W}{2}, \frac{W}{2})$, we first derive the capacity C_d for the discrete-time AWGN channel of noise variance σ^2 and symbols constrained to average energy not exceeding $\mathcal{E}_s$. The two expressions are:*

$$C_d = \frac{1}{2}\log_2\left(1 + \frac{\mathcal{E}_s}{\sigma^2}\right) \quad \text{[bits per channel use]},$$

$$C = (W/2)\log_2\left(1 + \frac{P}{W(N_0/2)}\right) \quad \text{[bps]}.$$

To derive C_d we need tools from information theory. However, going from C_d to C using the relationship $n = WT$ is straightforward. To do so, let $\mathcal{G}_\eta$ be the set of all signals that are frequency-limited to $(-\frac{W}{2}, \frac{W}{2})$ and time-limited to $(-\frac{T}{2}, \frac{T}{2})$ at level η. We choose η small enough that for all practical purposes all signals of $\mathcal{G}_\eta$ are strictly frequency-limited to $(-\frac{W}{2}, \frac{W}{2})$ and strictly time-limited to $(-\frac{T}{2}, \frac{T}{2})$. Each waveform in $\mathcal{G}_\eta$ is represented by an n-tuple and as T goes to infinity n approaches WT. Complete the argument assuming $n = WT$ and without worrying about convergence issues.

EXERCISE 4.17 (Energy efficiency of single-shot PAM) *This exercise complements what we have learned in Example 4.3. Consider using the m-PAM constellation*

$$\{\pm a, \pm 3a, \pm 5a, \ldots, \pm(m-1)a\}$$

to communicate across the discrete-time AWGN channel of noise variance $\sigma^2 = 1$. Our goal is to communicate at some level of reliability, say with error probability $P_e = 10^{-5}$. We are interested in comparing the energy needed by PAM versus the energy needed by a system that operates at channel capacity, namely at $\frac{1}{2}\log_2\left(1 + \frac{\mathcal{E}_s}{\sigma^2}\right)$ bits per channel use.

(a) *Using the capacity formula, determine the energy per symbol $\mathcal{E}_s^C(k)$ needed to transmit k bits per channel use. (The superscript C stands for channel capacity.) At any rate below capacity it is possible to make the error probability arbitrarily small by increasing the codeword length. This implies that there is a way to achieve the desired error probability at energy per symbol $\mathcal{E}_s^C(k)$.*

(b) *Using single-shot m-PAM, we can achieve an arbitrary small error probability by making the parameter a sufficiently large. As the size m of the constellation increases, the edge effects become negligible, and the average error probability approaches $2Q(\frac{a}{\sigma})$, which is the probability of error conditioned on an interior point being transmitted. Find the numerical value of the parameter a for which $2Q(\frac{a}{\sigma}) = 10^{-5}$. (You may use $\frac{1}{2}\exp(-\frac{x^2}{2})$ as an approximation of $Q(x)$.)*

(c) *Having fixed the value of a, we can use equation (4.1) to determine the average energy $\mathcal{E}_s^P(k)$ needed by PAM to send k bits at the desired error probability.*

(The superscript P stands for PAM.) Find and compare the numerical values of $\mathcal{E}_s^P(k)$ and $\mathcal{E}_s^C(k)$ for $k = 1, 2, 4$.

(d) Find $\lim_{k \to \infty} \frac{\mathcal{E}_s^C(k+1)}{\mathcal{E}_s^C(k)}$ and $\lim_{k \to \infty} \frac{\mathcal{E}_s^P(k+1)}{\mathcal{E}_s^P(k)}$.

(e) Comment on PAM's efficiency in terms of energy per bit for small and large values of k. Comment also on the relationship between this exercise and Example 4.3.

5 Symbol-by-symbol on a pulse train: Second layer revisited

5.1 Introduction

In this and the following chapter, we focus on the signal design problem. This chapter is devoted to the waveform former and its receiver-side counterpart, the n-tuple former. In Chapter 6 we focus on the encoder/decoder pair.[1]

In principle, the results derived in this chapter can be applied to both baseband and passband communication. However, for reasons of flexibility, hardware costs, and robustness, we design the waveform former for baseband communication and assign to the up-converter, discussed in Chapter 7, the task of converting the waveform-former output into a signal suitable for passband communication.

Symbol-by-symbol on a pulse train will emerge as a natural signaling technique. To keep the notation to the minimum, we write

$$w(t) = \sum_{j=1}^{n} s_j \psi(t - jT) \tag{5.1}$$

instead of $w_i(t) = \sum_{j=1}^{n} c_{i,j} \psi(t - jT)$. We drop the message index i from $w_i(t)$ because we will be studying properties of the pulse $\psi(t)$, as well as properties of the stochastic process that models the transmitter output signal, neither of which depends on a particular message choice. Following common practice, we refer to s_j as a *symbol*.

EXAMPLE 5.1 (PAM signaling) *PAM signaling (PAM for short) is indeed symbol-by-symbol on a pulse train, with the symbols taking value in a PAM alphabet as described in Figure 2.9. It depends on the encoder whether or not all sequences with symbols taking value in the given PAM alphabet are allowed. As we will see in Chapter 6, we can decrease the error probability by allowing only a subset of the sequences.* □

We have seen the acronym PAM in three contexts that are related but should not be confused. Let us review them. (i) PAM alphabet as the constellation of

---

[1] The two chapters are essentially independent and could be studied in the reverse order, but the results of Section 5.3 (which is independent of the other sections) are needed for a few exercises in Chapter 6. The chosen order is preferable for continuity with the discussion in Chapter 4.

points of Figure 2.9. (ii) Single-shot PAM as in Example 3.8. We have seen that this signaling method is not appropriate for transmitting many bits. Therefore we will not discuss it further. (iii) PAM signaling as in Example 5.1. This is symbol-by-symbol on a pulse train with symbols taking value in a PAM alphabet. Similar comments apply to QAM and PSK, provided that we view their alphabets as subsets of $\mathbb{C}$ rather than of $\mathbb{R}^2$. The reason it is convenient to do so will become clear in Chapter 7.

As already mentioned, most modern communication systems rely on PAM, QAM, or PSK signaling. In this chapter we learn the main tool to design the pulse $\psi(t)$.

The chapter is organized as follows. In Section 5.2, we develop an instructive special case where the channel is strictly bandlimited and we rediscover symbol-by-symbol on a pulse train as a natural signaling technique for that situation. This also forms the basis for software-defined radio. In Section 5.3 we derive the expression for the power spectral density of the transmitted signal for an arbitrary pulse when the symbol sequence constitutes a discrete-time wide-sense stationary process. As a preview, we discover that when the symbols are uncorrelated, which is frequently the case, the spectrum is proportional to $|\psi_\mathcal{F}(f)|^2$. In Section 5.4, we derive the necessary and sufficient condition on $|\psi_\mathcal{F}(f)|^2$ in order for $\{\psi_j(t)\}_{j\in\mathbb{Z}}$ to be an orthonormal set when $\psi_j(t) = \psi(t-jT)$. (The condition is that $|\psi_\mathcal{F}(f)|^2$ fulfills the so-called *Nyquist criterion*.)

5.2 The ideal lowpass case

Suppose that the channel is as shown in Figure 5.1, where $N(t)$ is white Gaussian noise of spectral density $N_0/2$ and the filter has frequency response

$$h_\mathcal{F}(f) = \begin{cases} 1, & |f| \leq B \\ 0, & \text{otherwise.} \end{cases}$$

This is an idealized version of a lowpass channel.

Because the filter blocks all the signal's components that fall outside the frequency interval $[-B, B]$, without loss of optimality we consider signals that are strictly bandlimited to $[-B, B]$. The sampling theorem, stated below and proved in Appendix 5.12, tells us that such signals can be described by a sequence of

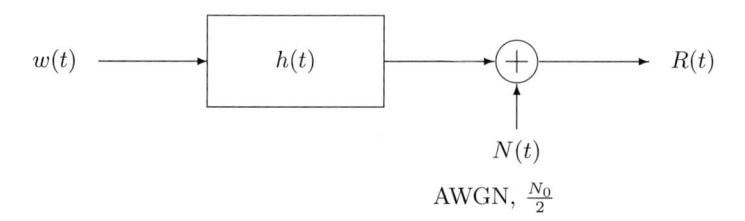

Figure 5.1. Lowpass channel model.

5.2. The ideal lowpass case

numbers. The idea is to let the encoder produce these numbers and let the waveform former do the "interpolation" that converts the samples into the desired $w(t)$.

THEOREM 5.2 (Sampling theorem) *Let $w(t)$ be a continuous $\mathcal{L}_2$ function (possibly complex-valued) and let its Fourier transform $w_{\mathcal{F}}(f)$ vanish for $f \notin [-B, B]$. Then $w(t)$ can be reconstructed from the sequence of T-spaced samples $w(nT)$, $n \in \mathbb{Z}$, provided that $T \leq \frac{1}{2B}$. Specifically,*

$$w(t) = \sum_{n=-\infty}^{\infty} w(nT) \operatorname{sinc}\left(\frac{t}{T} - n\right), \tag{5.2}$$

where $\operatorname{sinc}(t) = \frac{\sin(\pi t)}{\pi t}$. □

In the sampling theorem, we assume that the signal of interest is in $\mathcal{L}_2$. Essentially, $\mathcal{L}_2$ is the vector space of finite-energy functions, and this is all you need to know about $\mathcal{L}_2$ to get the most out of this text. However, it is recommended to read Appendix 5.9, which contains an informal introduction of $\mathcal{L}_2$, because we often read about $\mathcal{L}_2$ in the technical literature. The appendix is also necessary to understand some of the subtleties related to the Fourier transform (Appendix 5.10) and Fourier series (Appendix 5.11). Our reason for referring to $\mathcal{L}_2$ is that it is necessary for a rigorous proof of the sampling theorem. All finite-energy signals that we encounter in this text are $\mathcal{L}_2$ functions. It is safe to say that all finite-energy functions that model real-world signals are in $\mathcal{L}_2$.

In the statement of the sampling theorem, we require continuity. Continuity does not follow from the condition that $w(t)$ is bandlimited. In fact, if we take a nice (continuous and $\mathcal{L}_2$) function and modify it at a single point, say at $t = T/2$, then the original and the modified functions have the same Fourier transform. In particular, if the original is bandlimited, then so is the modified function. The sequence of samples is identical for both functions; and the reconstruction formula (5.2) will reconstruct the original (continuous) function. This eventuality is not a concern to practising engineers, because physical signals are continuous. Mathematically, when the difference between two $\mathcal{L}_2$ functions is a zero-norm function, we say that the functions are $\mathcal{L}_2$ equivalent (see Appendix 5.9).

To see what happens when we omit continuity in the sampling theorem, consider the situation where a continuous signal that fulfills the conditions of the sampling theorem is set to zero at all sampling points. Once again, the Fourier transform of the modified signal is identical to that of the original one. Yet, when we sample the modified signal and use the samples in the "reconstruction formula", we obtain the all-zero signal.

The sinc pulse used in the statement of the sampling theorem is not normalized to unit energy. If we normalize it, specifically define $\psi(t) = \frac{1}{\sqrt{T}} \operatorname{sinc}(\frac{t}{T})$, then $\{\psi(t - jT)\}_{j=-\infty}^{\infty}$ forms an orthonormal set. (This is implied by Theorem 5.6. The impatient reader can verify it by direct calculation, using Parseval's relationship. Useful facts about the sinc function and its Fourier transform are contained in Appendix 5.10.) Thus (5.2) can be rewritten as

$$w(t) = \sum_{j=-\infty}^{\infty} s_j \psi(t - jT), \qquad \psi(t) = \frac{1}{\sqrt{T}} \operatorname{sinc}(\frac{t}{T}), \tag{5.3}$$

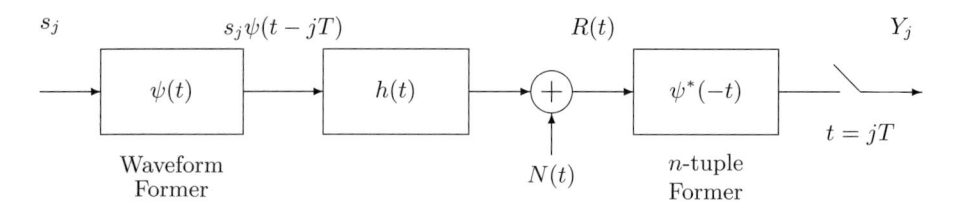

```
sⱼ           sⱼψ(t − jT)              R(t)                      Yⱼ
──→ [ ψ(t) ] ──────────→ [ h(t) ] ──→(+)──→ [ ψ*(−t) ] ──────/──→
                                      ↑                    t = jT
     Waveform                        N(t)            n-tuple
     Former                                          Former
```

Figure 5.2. Symbol-by-symbol on a pulse train obtained naturally from the sampling theorem.

where $s_j = w(jT)\sqrt{T}$. Hence a signal $w(t)$ that fulfills the conditions of the sampling theorem is one that lives in the inner product space spanned by $\{\psi(t - jT)\}_{j=-\infty}^{\infty}$. When we sample such a signal, we obtain (up to a scaling factor) the coefficients of its orthonormal expansion with respect to the orthonormal basis $\{\psi(t - jT)\}_{j=-\infty}^{\infty}$.

Now let us go back to our communication problem. We have just seen that any physical (continuous and $\mathcal{L}_2$) signal $w(t)$ that has no energy outside the frequency range $[-B, B]$ can be synthesized as $w(t) = \sum_j s_j \psi(t-jT)$. This signal has exactly the form of symbol-by-symbol on a pulse train. To implement this signaling method we let the jth encoder output be $s_j = w(jT)\sqrt{T}$, and let the waveform former be defined by the pulse $\psi(t) = \frac{1}{\sqrt{T}}\text{sinc}(\frac{t}{T})$. The waveform former, the channel, and the n-tuple former are shown in Figure 5.2.

It is interesting to observe that we use the sampling theorem somewhat backwards, in the following sense. In a typical application of the sampling theorem, the first step consists of sampling the source signal, then the samples are stored or transmitted, and finally the original signal is reconstructed from the samples. To the contrary, in the diagram of Figure 5.2, the transmitter does the (re)construction as the first step, the (re)constructed signal is transmitted, and finally the receiver does the sampling.

Notice also that $\psi^*(-t) = \psi(t)$ (the sinc function is even and real-valued) and its Fourier transform is

$$\psi_{\mathcal{F}}(f) = \begin{cases} \sqrt{T}, & |f| \leq \frac{1}{2T} \\ 0, & \text{otherwise} \end{cases}$$

(Appendix 5.10 explains an effortless method for relating the rectangle and the sinc as Fourier pairs). Therefore the matched filter at the receiver is a lowpass filter. It does exactly what seems to be the right thing to do – remove the out-of-band noise.

EXAMPLE 5.3 (*Software-defined radio*) *The sampling theorem is the theoretical underpinning of software-defined radio. No matter what the communications standard is (GSM, CDMA, EDGE, LTE, Bluetooth, 802.11, etc.), the transmitted signal can be described by a sequence of numbers. In a software-defined-radio implementation of a transmitter, the encoder that produces the samples is a computer program. Only the program is aware of the standard being implemented.*

5.3. Power spectral density

The hardware that converts the sequence of numbers into the transmitted signal (the waveform former of Figure 5.2) can be the same off-the-shelf device for all standards. Similarly, the receiver front end that converts the received signal into a sequence of numbers (the n-tuple former of Figure 5.2) can be the same for all standards. In a software-defined-radio receiver, the decoder is implemented in software. In principle, any past, present, and future standard can be implemented by changing the encoder/decoder program. The sampling theorem was brought to the engineering community by Shannon [24] in 1948, but only recently we have the technology and the tools needed to make software-defined radio a viable solution. In particular, computers are becoming fast enough, real-time operating systems such as RT Linux make it possible to schedule critical events with precision, and the prototyping is greatly facilitated by the availability of high-level programming-languages for signal-processing such as MATLAB. □

In the rest of the chapter we generalize symbol-by-symbol on a pulse train. The goal is to understand which pulses $\psi(t)$ are allowed and to determine their effect on the power spectral density of the communication signal they produce.

5.3 Power spectral density

A typical requirement imposed by regulators is that the power spectral density (PSD) (also called power spectrum) of the transmitter's output signal be below a given frequency-domain "mask". In this section we compute the PSD of the transmitter's output signal modeled as

$$X(t) = \sum_{i=-\infty}^{\infty} X_i \xi(t - iT - \Theta), \tag{5.4}$$

where $\{X_j\}_{j=-\infty}^{\infty}$ is a *zero-mean wide-sense stationary* (WSS) discrete-time process, $\xi(t)$ is an arbitrary $\mathcal{L}_2$ function (not necessarily normalized or orthogonal to its T-spaced time translates), and Θ is a random dither (or delay) independent of $\{X_j\}_{j=-\infty}^{\infty}$ and uniformly distributed in the interval $[0, T)$. (See Appendix 5.13 for a brief review on stochastic processes.)

The insertion of the random dither Θ, not considered so far in our signal's model, needs to be justified. It models the fact that a transmitter is switched on at a time unknown to an observer interested in measuring the signal's PSD. For this reason and because of the propagation delay, the observer has no information regarding the relative position of the signal with respect to his own time axis. The dither models this uncertainty. Thus far, we have not inserted the dither because we did not want to make the signal's model more complicated than necessary. After a sufficiently long observation time, the intended receiver can estimate the dither and compensate for it (see Section 5.7). From a mathematical point of view, the dither makes $X(t)$ a WSS process, thus greatly simplifying our derivation of the power spectral density.

In the derivation that follows we use the *autocovariance*

$$K_X[i] := \mathbb{E}[(X_{j+i} - \mathbb{E}[X_{j+i}])(X_j - \mathbb{E}[X_j])^*] = \mathbb{E}[X_{j+i}X_j^*],$$

which depends only on i since, by assumption, $\{X_j\}_{j=-\infty}^{\infty}$ is WSS. We use also the *self-similarity function* of the pulse $\xi(\tau)$, defined as

$$R_\xi(\tau) := \int_{-\infty}^{\infty} \xi(\alpha + \tau)\xi^*(\alpha)d\alpha. \tag{5.5}$$

(Think of the definition of an inner product if you tend to forget where to put the * in the above definition.)

The process $X(t)$ is zero-mean. Indeed, using the independence between X_i and Θ and the fact that $\mathbb{E}[X_i] = 0$, we obtain $\mathbb{E}[X(t)] = \sum_{i=-\infty}^{\infty} \mathbb{E}[X_i] \mathbb{E}[\xi(t - iT - \Theta)] = 0$.

The *autocovariance* of $X(t)$ is

$$K_X(t+\tau, t) = \mathbb{E}\big[\big(X(t+\tau) - \mathbb{E}[X(t+\tau)]\big)\big(X(t) - \mathbb{E}[X(t)]\big)^*\big]$$
$$= \mathbb{E}\big[X(t+\tau)X^*(t)\big]$$
$$= \mathbb{E}\Bigg[\sum_{i=-\infty}^{\infty} X_i\xi(t+\tau-iT-\Theta) \sum_{j=-\infty}^{\infty} X_j^*\xi^*(t-jT-\Theta)\Bigg]$$
$$= \mathbb{E}\Bigg[\sum_{i=-\infty}^{\infty}\sum_{j=-\infty}^{\infty} X_iX_j^*\xi(t+\tau-iT-\Theta)\xi^*(t-jT-\Theta)\Bigg]$$
$$\stackrel{(a)}{=} \sum_{i=-\infty}^{\infty}\sum_{j=-\infty}^{\infty} \mathbb{E}[X_iX_j^*]\mathbb{E}[\xi(t+\tau-iT-\Theta)\xi^*(t-jT-\Theta)]$$
$$= \sum_{i=-\infty}^{\infty}\sum_{j=-\infty}^{\infty} K_X[i-j]\mathbb{E}[\xi(t+\tau-iT-\Theta)\xi^*(t-jT-\Theta)]$$
$$\stackrel{(b)}{=} \sum_{k=-\infty}^{\infty} K_X[k] \sum_{i=-\infty}^{\infty} \frac{1}{T}\int_0^T \xi(t+\tau-iT-\theta)\xi^*(t-iT+kT-\theta)d\theta$$
$$\stackrel{(c)}{=} \sum_{k=-\infty}^{\infty} K_X[k] \frac{1}{T}\int_{-\infty}^{\infty} \xi(t+\tau-\theta)\xi^*(t+kT-\theta)d\theta$$
$$= \sum_k K_X[k]\frac{1}{T}R_\xi(\tau-kT),$$

where in (a) we use the fact that $X_iX_j^*$ and Θ are independent random variables, in (b) we make the change of variable $k = i - j$, and in (c) we use the fact that for an arbitrary function $u : \mathbb{R} \to \mathbb{R}$, an arbitrary number $a \in \mathbb{R}$, and a positive (interval length) b,

$$\sum_{i=-\infty}^{\infty} \int_a^{a+b} u(x+ib)dx = \int_{-\infty}^{\infty} u(x)dx. \tag{5.6}$$

5.3. Power spectral density

(If (5.6) is not clear to you, picture integrating from a to $a+2b$ by integrating first from a to $a+b$, then from $a+b$ to $a+2b$, and summing the results. This is the right-hand side. Now consider integrating both times from a to $a+b$, but before you perform the second integration you shift the function to the left by b. This is the left-hand side.)

We see that $K_X(t+\tau, t)$ depends only on τ. Hence we simplify notation and write $K_X(\tau)$ instead of $K_X(t+\tau, t)$. We summarize:

$$K_X(\tau) = \sum_k K_X[k] \frac{1}{T} R_\xi(\tau - kT). \tag{5.7}$$

The process $X(t)$ is WSS because neither its mean nor its autocovariance depend on t.

For the last step in the derivation of the power spectral density, we use the fact that the Fourier transform of $R_\xi(\tau)$ is $|\xi_\mathcal{F}(f)|^2$. This follows from Parseval's relationship,

$$R_\xi(\tau) = \int_{-\infty}^{\infty} \xi(\alpha+\tau)\xi^*(\alpha) d\alpha$$

$$= \int_{-\infty}^{\infty} \xi_\mathcal{F}(f) \xi_\mathcal{F}^*(f) \exp(j2\pi\tau f) df$$

$$= \int_{-\infty}^{\infty} |\xi_\mathcal{F}(f)|^2 \exp(j2\pi\tau f) df.$$

Now we can take the Fourier transform of $K_X(\tau)$ to obtain the power spectral density

$$S_X(f) = \frac{|\xi_\mathcal{F}(f)|^2}{T} \sum_k K_X[k] \exp(-j2\pi k fT). \tag{5.8}$$

The above expression is in a form that suits us. In many situations, the infinite sum has only a small number of non-zero terms. Note that the summation in (5.8) is the discrete-time Fourier transform of $\{K_X[k]\}_{k=-\infty}^{\infty}$, evaluated at fT. This is the power spectral density of the discrete-time process $\{X_i\}_{i=-\infty}^{\infty}$. If we think of $\frac{|\xi_\mathcal{F}(f)|^2}{T}$ as being the power spectral density of $\xi(t)$, we can interpret $S_X(f)$ as being the product of two PSDs, that of $\xi(t)$ and that of $\{X_i\}_{i=-\infty}^{\infty}$.

In many cases of interest, $K_X[k] = \mathcal{E}\mathbb{1}\{k=0\}$, where $\mathcal{E} = \mathbb{E}[|X_i|^2]$. In this case we say that the zero-mean WSS process $\{X_i\}_{i=-\infty}^{\infty}$ is *uncorrelated*. Then (5.7) and (5.8) simplify to

$$K_X(\tau) = \mathcal{E}\frac{R_\xi(\tau)}{T}, \tag{5.9}$$

$$S_X(f) = \mathcal{E}\frac{|\xi_\mathcal{F}(f)|^2}{T}. \tag{5.10}$$

EXAMPLE 5.4 *Suppose that $\{X_i\}_{i=-\infty}^{\infty}$ is an independent and uniformly distributed sequence taking values in $\{\pm\sqrt{\mathcal{E}}\}$, and $\xi(t) = \sqrt{1/T}\operatorname{sinc}(\frac{t}{T})$. Then*

$K_X[k] = \mathcal{E}\mathbb{1}\{k = 0\}$ and

$$S_X(f) = \mathcal{E}\mathbb{1}\{f \in [-B, B]\},$$

where $B = \frac{1}{2T}$. This is consistent with our intuition. When we use the pulse $\text{sinc}(\frac{t}{T})$, we expect a flat power spectrum over $[-B, B]$ and no power outside this interval. The energy per symbol is $\mathcal{E}$, hence the power is $\frac{\mathcal{E}}{T}$, and the power spectral density is $\frac{\mathcal{E}}{2BT} = \mathcal{E}$. □

In the next example, we work out a case where $K_X[k] \neq \mathcal{E}\mathbb{1}\{k = 0\}$. In this case, we say that the zero-mean WSS process $\{X_i\}_{i=-\infty}^{\infty}$ is *correlated*.

EXAMPLE 5.5 (Correlated symbol sequence) *Suppose that the pulse is as in Example 5.4, but the symbol sequence is now the output of an encoder described by $X_i = \sqrt{2\mathcal{E}}(B_i - B_{i-2})$, where $\{B_i\}_{i=-\infty}^{\infty}$ is a sequence of independent random variables that are uniformly distributed over the alphabet $\{0, 1\}$. After verifying that the symbol sequence is zero-mean, we insert $X_i = \sqrt{2\mathcal{E}}(B_i - B_{i-2})$ into $K_X[k] = \mathbb{E}[X_{j+k}X_j^*]$ and obtain*

$$K_X[k] = \begin{cases} \mathcal{E}, & k = 0, \\ -\mathcal{E}/2, & k = \pm 2, \\ 0, & \text{otherwise,} \end{cases}$$

$$S_X(f) = \frac{|\xi_\mathcal{F}(f)|^2}{T}\mathcal{E}\left(1 - \frac{1}{2}e^{j4\pi fT} - \frac{1}{2}e^{-j4\pi fT}\right) = 2\mathcal{E}\sin^2(2\pi fT)\mathbb{1}_{[-B,B]}(f). \tag{5.11}$$

When we compare this example to Example 5.4, we see that this encoder shapes the power spectral density from a rectangular shape to a squared sinusoid. Notice that the spectral density vanishes at $f = 0$. This is desirable if the channel blocks very low frequencies, which happens for instance for a cable that contains amplifiers. To avoid amplifying offset voltages and leaky currents, amplifiers are AC (alternating current) coupled. This means that amplifiers have a highpass filter at the input, often just a capacitor, that blocks DC (direct current) signals. Notice that the encoder is a linear time-invariant system (with respect to addition and multiplication in $\mathbb{R}$). Hence the cascade of the encoder and the pulse forms a linear time-invariant system. It is immediate to verify that its impulse response is $\tilde{\xi}(t) = \xi(t) - \xi(t - 2T)$. Hence in this case we can write

$$X(t) = \sum_l X_l \xi(t - lT) = \sum_l B_l \tilde{\xi}(t - lT).$$

The technique described in this example is called *correlative encoding* or *partial response signaling*. □

The encoder in Example 5.5 is linear with respect to (the field) $\mathbb{R}$ and this is the reason its effect can be incorporated into the pulse but this is not the case in general (see Exercise 5.14 and see Chapter 6).

5.4 Nyquist criterion for orthonormal bases

In the previous section we saw that symbol-by-symbol on a pulse train with uncorrelated symbols – a condition often met – generates a stochastic process of power spectral density $\mathcal{E}\frac{|\psi_\mathcal{F}(f)|^2}{T}$, where $\mathcal{E}$ is the symbol's average energy. This is progress, because it tells us how to choose $\psi(t)$ to obtain a desired power spectral density. But remember that $\psi(t)$ has another important constraint: it must be orthogonal to its T-spaced time translates. It would be nice to have a characterization of $|\psi_\mathcal{F}(f)|^2$ that is simple to work with and that guarantees orthogonality between $\psi(t)$ and $\psi(t - lT)$ for all $l \in \mathbb{Z}$. This seems to be asking too much, but it is exactly what the next result is all about.

So, we are looking for a frequency-domain equivalent of the condition

$$\int_{-\infty}^{\infty} \psi(t - nT)\psi^*(t)dt = \mathbb{1}\{n = 0\}. \tag{5.12}$$

The form of the left-hand side suggests using Parseval's relationship. Doing so yields

$$\mathbb{1}\{n = 0\} = \int_{-\infty}^{\infty} \psi(t - nT)\psi^*(t)dt = \int_{-\infty}^{\infty} \psi_\mathcal{F}(f)\psi_\mathcal{F}^*(f)e^{-\mathsf{j}2\pi nTf}df$$

$$= \int_{-\infty}^{\infty} |\psi_\mathcal{F}(f)|^2 e^{-\mathsf{j}2\pi nTf}df$$

$$\stackrel{(a)}{=} \int_{-\frac{1}{2T}}^{\frac{1}{2T}} \sum_{k \in \mathbb{Z}} |\psi_\mathcal{F}(f - \frac{k}{T})|^2 e^{-\mathsf{j}2\pi nT(f - \frac{k}{T})}df$$

$$\stackrel{(b)}{=} \int_{-\frac{1}{2T}}^{\frac{1}{2T}} \sum_{k \in \mathbb{Z}} |\psi_\mathcal{F}(f - \frac{k}{T})|^2 e^{-\mathsf{j}2\pi nTf}df$$

$$\stackrel{(c)}{=} \int_{-\frac{1}{2T}}^{\frac{1}{2T}} g(f)e^{-\mathsf{j}2\pi nTf}df,$$

where in (a) we use again (5.6) (but in the other direction), in (b) we use the fact that $e^{-\mathsf{j}2\pi nT(f - \frac{k}{T})} = e^{-\mathsf{j}2\pi nTf}$, and in (c) we introduce the function

$$g(f) = \sum_{k \in \mathbb{Z}} \left|\psi_\mathcal{F}\left(f - \frac{k}{T}\right)\right|^2.$$

Notice that $g(f)$ is a periodic function of period $1/T$ and the right-hand side of (c) is $1/T$ times the nth Fourier series coefficient A_n of the periodic function $g(f)$. (A review of Fourier series is given in Appendix 5.11.) Because $A_0 = T$ and $A_k = 0$ for $k \neq 0$, the Fourier series of $g(f)$ is the constant T. Up to a technicality discussed below, this proves the following result.

THEOREM 5.6 (Nyquist's criterion for orthonormal pulses) Let $\psi(t)$ be an $\mathcal{L}_2$ function. The set $\{\psi(t - jT)\}_{j=-\infty}^{\infty}$ consists of orthonormal functions if and only if

$$\text{l.i.m.} \sum_{k=-\infty}^{\infty} |\psi_{\mathcal{F}}(f - \frac{k}{T})|^2 = T, \quad f \in \mathbb{R}. \tag{5.13}$$

□

A frequency-domain function $a_{\mathcal{F}}(f)$ is said to satisfy *Nyquist's criterion* with parameter p if, for all $f \in \mathbb{R}$, l.i.m. $\sum_{k=-\infty}^{\infty} a_{\mathcal{F}}(f - \frac{k}{p}) = p$. Theorem 5.6 says that $\{\psi(t - jT)\}_{j=-\infty}^{\infty}$ is an orthonormal set if and only if $|\psi_{\mathcal{F}}(f)|^2$ satisfies Nyquist's criterion with parameter T.

The l.i.m. in (5.13) stands for *limit in $\mathcal{L}_2$ norm*. It means that as we add more and more terms to the sum on the left-hand side of (5.13) it becomes $\mathcal{L}_2$ equivalent to the constant on the right-hand side. The l.i.m. is a technicality due to the Fourier series. To see that the l.i.m. is needed, take a $\psi_{\mathcal{F}}(f)$ such that $|\psi_{\mathcal{F}}(f)|^2$ fulfills (5.13) without the l.i.m. An example of such a function is the rectangle $|\psi_{\mathcal{F}}(f)|^2 = T$ for $f \in [-\frac{1}{2T}, \frac{1}{2T})$ and zero elsewhere. Now, take a copy $|\tilde{\psi}_{\mathcal{F}}(f)|^2$ of $|\psi_{\mathcal{F}}(f)|^2$ and modify it at an arbitrary isolated point. For instance, we set $\tilde{\psi}_{\mathcal{F}}(0) = 0$. The inverse Fourier transform of $\tilde{\psi}_{\mathcal{F}}(f)$ is still $\psi(t)$. Hence $\psi(t)$ is orthogonal to its T-spaced time translates. Yet (5.13) is no longer fulfilled if we omit the l.i.m. For our specific example, the left and the right differ at exactly one point of each period. Equality still holds in the l.i.m. sense. In all practical applications, $\psi_{\mathcal{F}}(f)$ is a smooth function and we can ignore the l.i.m. in (5.13).

Notice that the left side of the equality in (5.13) is periodic with period $1/T$. Hence to verify that $|\psi_{\mathcal{F}}(f)|^2$ fulfills Nyquist's criterion with parameter T, it is sufficient to verify that (5.13) holds over an interval of length $1/T$.

EXAMPLE 5.7 *The following functions satisfy Nyquist's criterion with parameter T.*

(a) $|\psi_{\mathcal{F}}(f)|^2 = T\mathbb{1}\{-\frac{1}{2T} < f < \frac{1}{2T}\}(f)$.
(b) $|\psi_{\mathcal{F}}(f)|^2 = T\cos^2(\frac{\pi}{2}fT)\mathbb{1}\{-\frac{1}{T} < f < \frac{1}{T}\}(f)$.
(c) $|\psi_{\mathcal{F}}(f)|^2 = T(1 - T|f|)\mathbb{1}\{-\frac{1}{T} < f < \frac{1}{T}\}(f)$.

□

The following comments are in order.

(a) *(Constant but not T)* $\psi(t)$ is orthogonal to its T-spaced time translates even when the left-hand side of (5.13) is $\mathcal{L}_2$ equivalent to a constant other than T, but in this case $\|\psi(t)\|^2 \neq 1$. This is a minor issue, we just have to scale the pulse to make it unit-norm.
(b) *(Minimum bandwidth)* A function $|\psi_{\mathcal{F}}(f)|^2$ cannot fulfill Nyquist's criterion with parameter T if its support is contained in an interval of the form $[-B, B]$ with $0 < B < \frac{1}{2T}$. Hence, the minimum bandwidth to fulfill Nyquist's criterion is $\frac{1}{2T}$.
(c) *(Test for bandwidths between $\frac{1}{2T}$ and $\frac{1}{T}$)* If $|\psi_{\mathcal{F}}(f)|^2$ vanishes outside $[-\frac{1}{T}, \frac{1}{T}]$, the Nyquist criterion is satisfied if and only if $|\psi_{\mathcal{F}}(\frac{1}{2T} - \epsilon)|^2 + |\psi_{\mathcal{F}}(-\frac{1}{2T} - \epsilon)|^2 = T$ for $\epsilon \in [-\frac{1}{2T}, \frac{1}{2T}]$ (see Figure 5.3). If, in addition, $\psi(t)$ is real-valued, which is typically the case, then $|\psi_{\mathcal{F}}(-f)|^2 = |\psi_{\mathcal{F}}(f)|^2$. In this case, it is sufficient that we check the positive frequencies, i.e. Nyquist's criterion is met if

5.4. Nyquist criterion for orthonormal bases

$$\left|\psi_{\mathcal{F}}(\frac{1}{2T} - \epsilon)\right|^2 + \left|\psi_{\mathcal{F}}(\frac{1}{2T} + \epsilon)\right|^2 = T, \qquad \epsilon \in \left[0, \frac{1}{2T}\right].$$

This means that $|\psi_{\mathcal{F}}(\frac{1}{2T})|^2 = \frac{T}{2}$ and the amount by which the function $|\psi_{\mathcal{F}}(f)|^2$ varies when we go from $f = \frac{1}{2T}$ to $f = \frac{1}{2T} - \epsilon$ is compensated by the function's variation in going from $f = \frac{1}{2T}$ to $f = \frac{1}{2T} + \epsilon$. For examples of such a *band-edge symmetry* see Figure 5.3, Figure 5.6a, and the functions (b) and (c) in Example 5.7. The bandwidth $B_N = \frac{1}{2T}$ is sometimes called the *Nyquist bandwidth*.

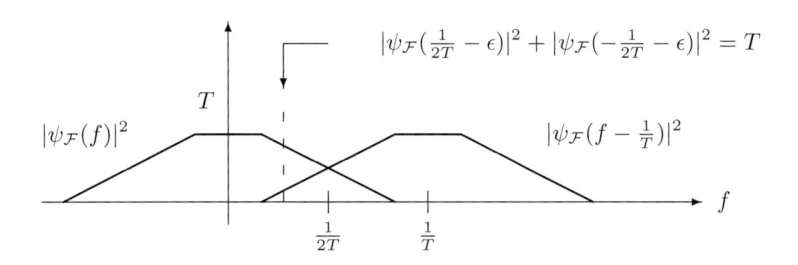

Figure 5.3. Band-edge symmetry for a pulse $|\psi_{\mathcal{F}}(f)|^2$ that vanishes outside $[-\frac{1}{T}, \frac{1}{T}]$ and fulfills Nyquist's criterion.

(d) *(Test for arbitrary finite bandwidths)* When the support of $|\psi_{\mathcal{F}}(f)|^2$ is wider than $\frac{1}{T}$, it is harder to see whether or not Nyquist's criterion is met with parameter T. A convenient way to organize the test goes as follows. Let $\mathcal{I}$ be the set of integers i for which the frequency interval of width $\frac{1}{T}$ centered at $f_i = \frac{1}{2T} + i\frac{1}{T}$ intersects with the support of $|\psi_{\mathcal{F}}(f)|^2$. For the example of Figure 5.4, $\mathcal{I} = \{-3, -2, 1, 2\}$, and the frequencies f_i, $i \in \mathcal{I}$ are marked with a "×". For each $i \in \mathcal{I}$, we consider the function $|\psi_{\mathcal{F}}(f_i + \epsilon)|^2$, $\epsilon \in [-\frac{1}{2T}, \frac{1}{2T}]$, as shown in Figure 5.5. Nyquist's criterion is met if and only if the sum of these functions,

$$g(\epsilon) = \sum_{i \in \mathcal{I}} |\psi_{\mathcal{F}}(f_i + \epsilon)|^2, \qquad \epsilon \in \left[-\frac{1}{2T}, \frac{1}{2T}\right],$$

is $\mathcal{L}_2$ equivalent to the constant T. From Figure 5.5, it is evident that the test is passed by the $|\psi_{\mathcal{F}}(f)|^2$ of Figure 5.4.

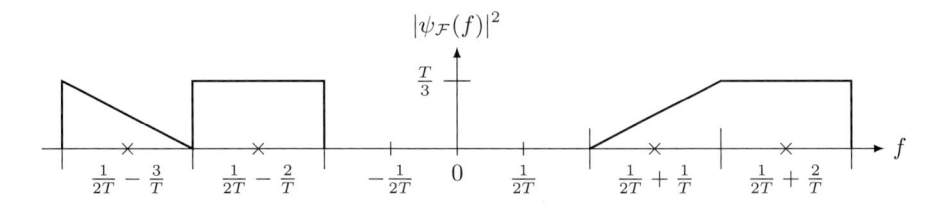

Figure 5.4. A $|\psi_{\mathcal{F}}(f)|^2$ of support wider than $\frac{1}{T}$ that fulfills Nyquist's criterion.

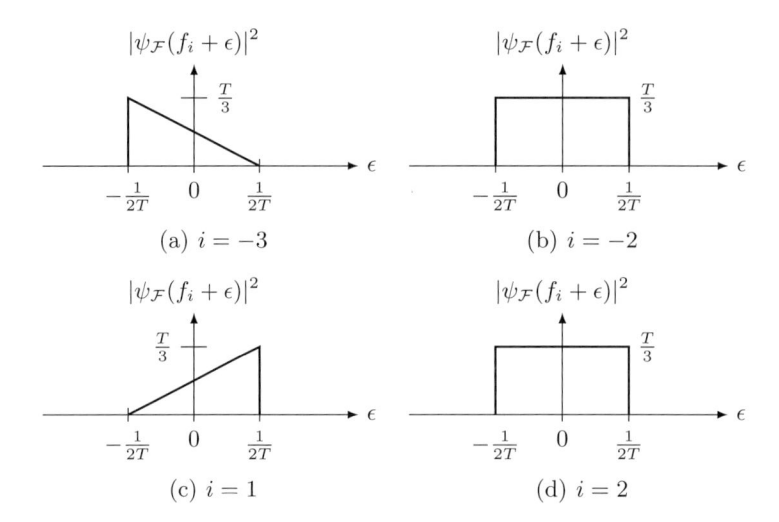

Figure 5.5. Functions of the form $|\psi_\mathcal{F}(f_i + \epsilon)|^2$ for $\epsilon \in [-\frac{1}{2T}, \frac{1}{2T}]$ and $i \in \mathcal{I} = \{-3, -2, 1, 2\}$. The sum-function $g(\epsilon)$ is the constant function T for $\epsilon \in [-\frac{1}{2T}, \frac{1}{2T}]$.

5.5 Root-raised-cosine family

For every $\beta \in (0, 1)$, and every $T > 0$, the *raised-cosine function*

$$|\psi_\mathcal{F}(f)|^2 = \begin{cases} T, & |f| \leq \frac{1-\beta}{2T} \\ \frac{T}{2}\left(1 + \cos\left[\frac{\pi T}{\beta}\left(|f| - \frac{1-\beta}{2T}\right)\right]\right), & \frac{1-\beta}{2T} < |f| < \frac{1+\beta}{2T} \\ 0, & \text{otherwise} \end{cases}$$

fulfills Nyquist's criterion with parameter T (see Figure 5.6a for a raised-cosine function with $\beta = \frac{1}{2}$). The expression might look complicated at first, but one can easily derive it by following the steps in Exercise 5.10.

By using the relationship $\frac{1}{2}(1 + \cos \alpha) = \cos^2 \frac{\alpha}{2}$, we can take the square root of $|\psi_\mathcal{F}(f)|^2$ and obtain the *root-raised-cosine function* (also called *square-root raised-cosine function*)

$$\psi_\mathcal{F}(f) = \begin{cases} \sqrt{T}, & |f| \leq \frac{1-\beta}{2T} \\ \sqrt{T}\cos\left[\frac{\pi T}{2\beta}(|f| - \frac{1-\beta}{2T})\right], & \frac{1-\beta}{2T} < |f| \leq \frac{1+\beta}{2T} \\ 0, & \text{otherwise.} \end{cases}$$

The inverse Fourier transform of $\psi_\mathcal{F}(f)$, derived in Appendix 5.14, is the *root-raised-cosine impulse response* (also called *root-raised-cosine pulse* or *impulse response of a root-raised cosine filter*)[2]

---

[2] A root-raised-cosine impulse response should not be confused with a root-raised-cosine function.

5.5. Root-raised-cosine family

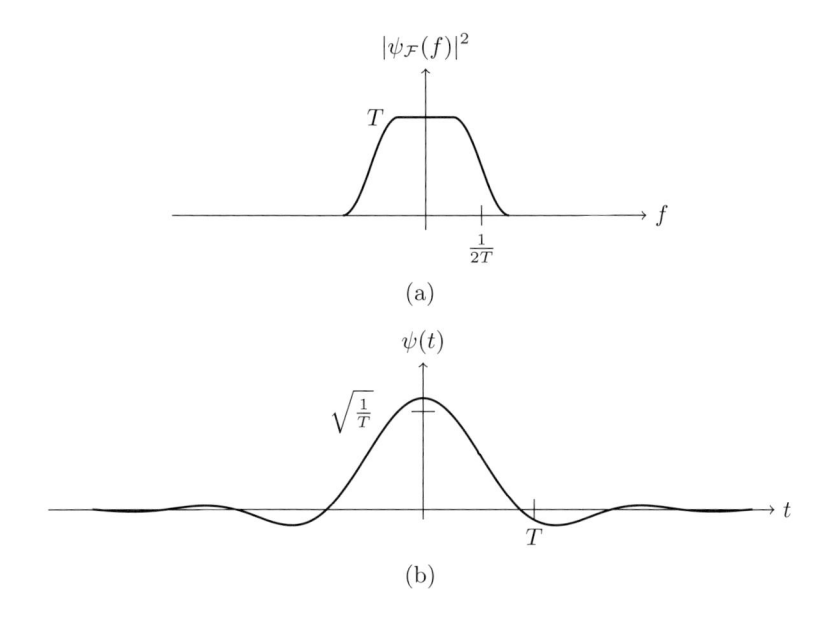

Figure 5.6. (a) Raised-cosine function $|\psi_{\mathcal{F}}(f)|^2$ with $\beta = \frac{1}{2}$; (b) corresponding pulse $\psi(t)$.

$$\psi(t) = \frac{4\beta}{\pi\sqrt{T}} \frac{\cos\left[(1+\beta)\pi\frac{t}{T}\right] + \frac{(1-\beta)\pi}{4\beta}\operatorname{sinc}\left[(1-\beta)\frac{t}{T}\right]}{1 - \left(4\beta\frac{t}{T}\right)^2}, \qquad (5.14)$$

where $\operatorname{sinc}(x) = \frac{\sin(\pi x)}{\pi x}$. The pulse $\psi(t)$ is plotted in Figure 5.6b for $\beta = \frac{1}{2}$. At $t = \pm\frac{T}{4\beta}$, both the numerator and the denominator of (5.14) vanish. Using L'Hospital's rule we determine that

$$\lim_{t \to \pm\frac{T}{4\beta}} \psi(t) = \frac{\beta}{\pi\sqrt{2T}}\left((\pi+2)\sin\left(\frac{\pi}{4\beta}\right) + (\pi-2)\cos\left(\frac{\pi}{4\beta}\right)\right).$$

The root-raised-cosine method is the most popular way of constructing pulses $\psi(t)$ that are orthogonal to their T-spaced time translates. When $\beta = 0$, the pulse becomes $\psi(t) = \sqrt{1/T}\operatorname{sinc}(\frac{t}{T})$.

Figure 5.7a shows a train of root-raised-cosine impulse responses, with each pulse scaled by a symbol taking value in $\{\pm 1\}$. Figure 5.7b shows the corresponding sum-signal.

A root-raised-cosine impulse response $\psi(t)$ is real-valued, even, and of infinite support (in the time domain). In practice, such a pulse has to be truncated to finite length and to make it causal, it has to be delayed. In general, as β increases from 0 to 1, the pulse $\psi(t)$ decays faster. The faster a pulse decays, the shorter we can truncate it without noticeable difference in its main property, which is to be orthogonal to its shifts by integer multiples of T. The eye diagram, described next, is a good way to visualize what goes on as we vary the roll-off factor β. The drawback of increasing β is that the bandwidth increases as well.

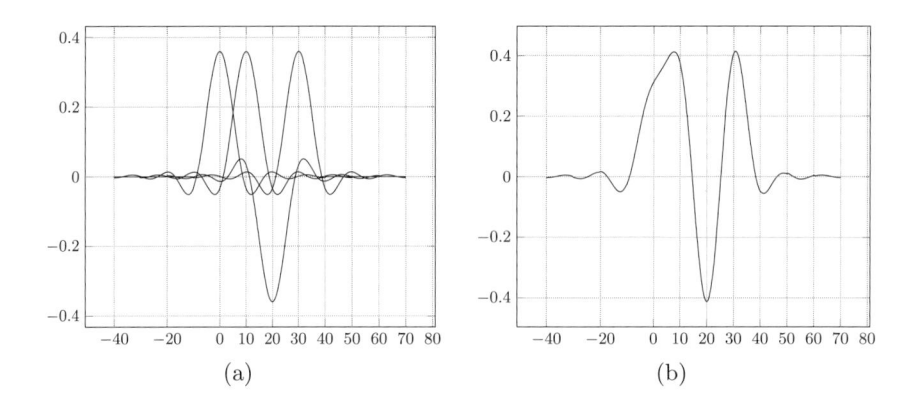

Figure 5.7. (a) Superimposed sequence of scaled root-raised-cosine impulse responses of the form $s_i\psi(t-iT)$, $i = 0, \ldots, 3$, with symbols s_i taking value in $\{\pm 1\}$, and (b) corresponding sum-signal. The design parameters are $\beta = 0.5$ and $T = 10$. The abscissa is the time.

5.6 Eye diagrams

The fact that $\psi(t)$ has unit-norm and is orthogonal to $\psi(t-iT)$ for all non-zero integers i, implies that the self-similarity function $R_\psi(\tau)$ satisfies

$$R_\psi(iT) = \begin{cases} 1, & i = 0 \\ 0, & i \text{ non-zero integer.} \end{cases} \quad (5.15)$$

This is important for the following reason. The noiseless signal $w(t) = \sum_i s_i \psi(t - iT)$ applied at the input of the matched filter of impulse response $\psi^*(-t)$ produces the noiseless output

$$y(t) = \sum_i s_i R_\psi(t - iT) \quad (5.16)$$

that, when sampled at $t = jT$, yields

$$y(jT) = \sum_i s_i R_\psi((j-i)T) = s_j.$$

Figure 5.8 shows the matched filter outputs obtained when the functions of Figure 5.7 are applied at the filter's input. Specifically, Figure 5.8a shows the train of symbol-scaled self-similarity functions. From the figure we see that (5.15) is satisfied. (When a pulse achieves its maximum, which has value 1, the other pulses vanish.) We see it also from Figure 5.8b, in that the signal $y(t)$ takes values in the symbol alphabet $\{\pm 1\}$ at the sampling times $t = 0, 10, 20, 30$.

If $\psi(t)$ is not orthogonal to $\psi(t-iT)$, which can happen for instance if a truncated pulse is made too short, then $R_\psi(iT)$ will be non-zero for several integers i. If we define $l_i = R_\psi(iT)$, then we can write

5.6. Eye diagrams

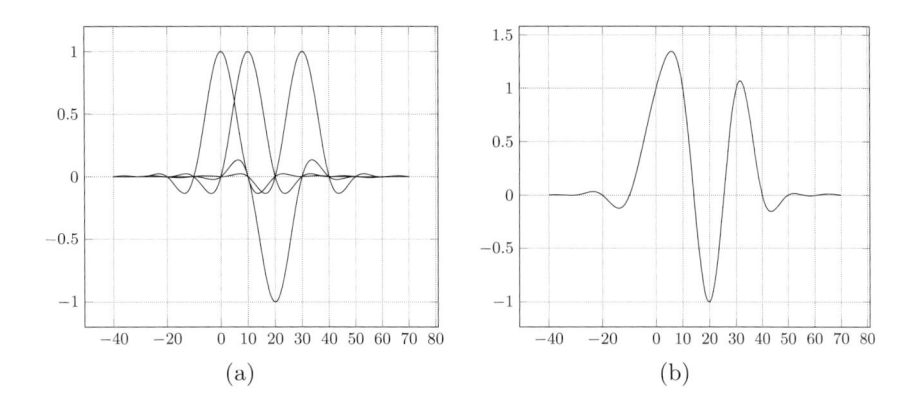

Figure 5.8. Each pulse in (a) has the form $s_i R_\psi(t - iT)$, $i = 0, \ldots, 3$. It is the response of the matched filter of impulse response $\psi^*(-t)$ to the input $s_i \psi(t - iT)$. Part (b) shows $y(t)$. The parameters are as in Figure 5.7.

$$y(jT) = \sum_i s_i l_{j-i}.$$

The fact that the noiseless $y(jT)$ depends on multiple symbols is referred to as *inter-symbol interference* (ISI).

There are two main causes to ISI. We have already mentioned one, specifically when $R_\psi(\tau)$ is non-zero for more than one sample. ISI happens also if the matched-filter output is not sampled at the correct times. In this case, we obtain $y(jT+\Delta) = \sum_i s_i R_\psi\big((j-i)T + \Delta\big)$, which is again of the form $\sum_i s_i l_{j-i}$ for $l_i = R_\psi(iT + \Delta)$.

The *eye diagram* is a technique that allows us to visualize if there is ISI and to see how critical it is that the sampling time be precise. The eye diagram is obtained from the matched-filter output before sampling. Let $y(t) = \sum_i s_i R_\psi(t - iT)$ be the noiseless matched filter output, with symbols taking value in some discrete set $\mathcal{S}$. For the example that follows, $\mathcal{S} = \{\pm 1\}$. To obtain the eye diagram, we plot the superposition of traces of the form $y(t - iT)$, $t \in [-T, T]$, for various integers i. Figure 5.9 gives examples of eye diagrams for various roll-off factors and pulse truncation lengths. Parts (a), (c), and (d) show no sign of ISI. Indeed, all traces go through ± 1 at $t = 0$, which implies that $y(iT) \in \mathcal{S}$. We see that truncating the pulse to length $20T$ does not lead to ISI for either roll-off factor. However, ISI is present when $\beta = 0.25$ and the pulse is truncated to $4T$ (part (b)). We see its presence from the fact that the traces go through various values at $t = 0$. This means that $y(iT)$ takes on values outside $\mathcal{S}$. These examples are meant to illustrate the point made in the last paragraph of the previous section.

Note also that the eye, the untraced space in the middle of the eye diagram, is wider in (c) than it is in (a). The advantage of a wider eye is that the system is more tolerant to small variations (jitter) in the sampling time. This is characteristic of a larger β and it is a consequence of the fact that as β increases, the pulse decays faster as a function of $|t|$. For the same reason, a pulse with larger beta can be truncated to a shorter length, at the price of a larger bandwidth.

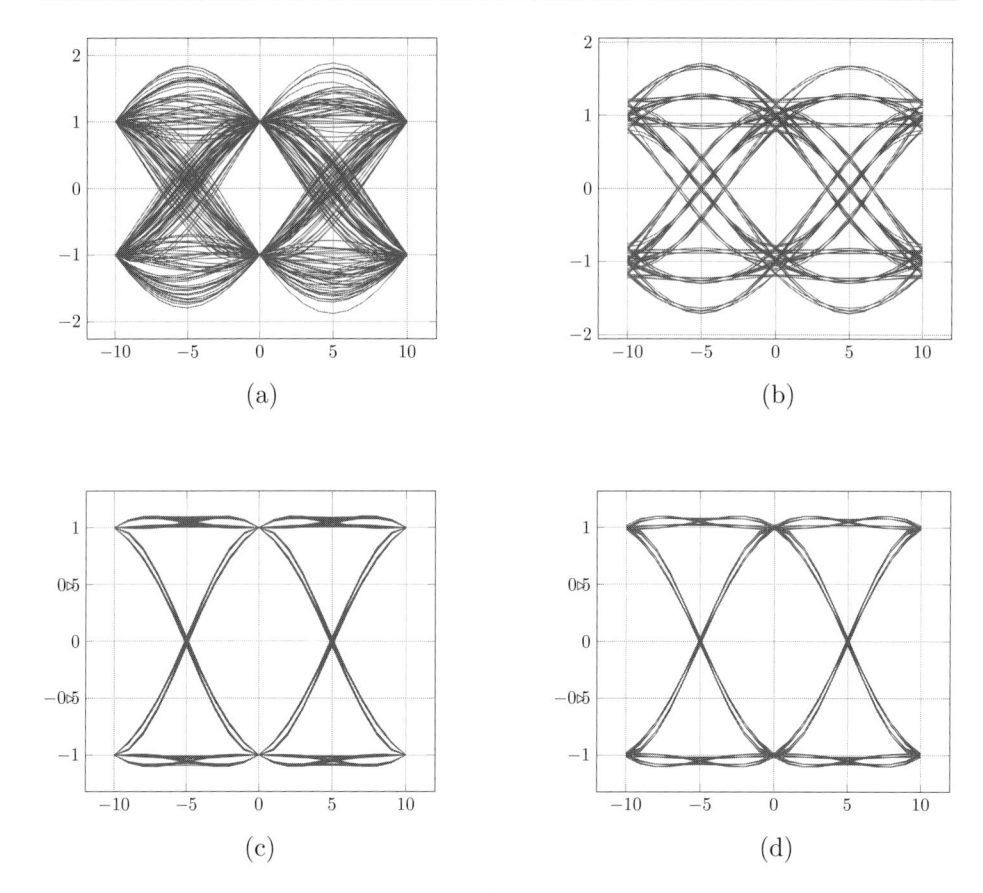

Figure 5.9. Eye diagrams of $\sum_i s_i R_\psi(t - iT)$ for $s_i \in \{\pm 1\}$ and pulse of the root-raised-cosine family with $T = 10$. The abscissa is the time. The roll-off factor is $\beta = 0.25$ for the top figures and $\beta = 0.9$ for the bottom ones. The pulse is truncated to length $20T$ for the figures on the left, and to length $4T$ for those on the right. The eye diagram of part (b) shows the presence of ISI.

The `MATLAB` program used to produce the above plots can be downloaded from the book web page.

The popularity of the eye diagram lies in the fact that it can be obtained quite easily by looking at the matched-filter output with an oscilloscope triggered by the clock that produces the sampling time. The eye diagram is very informative even if the channel has attenuated the signal and/or has added noise.

5.7 Symbol synchronization

The n-tuple former for symbol-by-symbol on a pulse train contains a matched filter with the output sampled at $\theta + lT$, for some θ and all integers l in some interval.

5.7. Symbol synchronization

Without loss of essential generality, we can assume that $\theta \in [0,T)$. In general, θ is not known to the receiver when the communication starts. This is obviously the case for a cell phone after a flight during which the phone was switched off (or in "airplane mode"). The n-tuple former is unable to produce the sufficient statistics until it has a sufficiently accurate estimate of θ. (The eye diagram gives us an indication of how accurate the estimate needs to be.) Finding the correct sampling times at the matched-filter output goes under the topic of *symbol synchronization*.

Estimating a parameter θ is a *parameter estimation* problem. Parameter estimation, like detection, is a well-established field. The difference between the two is that – in detection – the hypothesis takes values in a discrete set and – in estimation – it takes values in a continuous set. In detection, typically we are interested in the error probability under various hypotheses or in the average error probability. In estimation, the decision is almost always wrong, so it does not make sense to minimize the error probability, yet a maximum likelihood (ML) approach is a sensible choice. We discuss the ML approach in Section 5.7.1 and, in Section 5.7.2, we discuss a more pragmatic approach, the *delay locked loop* (DLL). In both cases, we assume that the transmitter starts with a *training sequence*.

5.7.1 Maximum likelihood approach

Suppose that the transmission starts with a training signal $s(t)$ known to the receiver. For the moment, we assume that $s(t)$ is real-valued. Extension to a complex-valued $s(t)$ is done in Section 7.5.

The channel output signal is

$$R(t) = \alpha s(t - \theta) + N(t),$$

where $N(t)$ is white Gaussian noise of power spectral density $\frac{N_0}{2}$, α is an unknown scaling factor (channel attenuation and receiver front-end amplification), and θ is the unknown parameter to be estimated. We can assume that the receiver knows that θ is in some interval, say $[0, \theta_{max}]$, for some possibly large constant θ_{max}.

To describe the ML estimate of θ, we need a statistical description of the received signal as a function of θ and α. Towards this goal, suppose that we have an orthonormal basis $\phi_1(t), \phi_2(t), \ldots, \phi_n(t)$ that spans the set $\{s(t-\hat{\theta}) : \hat{\theta} \in [0, \theta_{max}]\}$. To simplify the notation, we assume that the orthonormal basis is finite, but an infinite basis is also a possibility. For instance, if $s(t)$ is continuous, has finite duration, and is essentially bandlimited, then the sampling theorem tells us that we can use sinc functions for such a basis.

For $i = 1, \ldots, n$, let $Y_i = \langle R(t), \phi_i(t) \rangle$ and let y_i be the observed sample value of Y_i. The random vector $Y = (Y_1, \ldots, Y_n)^\mathsf{T}$ consists of independent random variables with $Y_i \sim \mathcal{N}(\alpha m_i(\theta), \sigma^2)$, where $m_i(\theta) = \langle s(t-\theta), \phi_i(t) \rangle$ and $\sigma^2 = \frac{N_0}{2}$. Hence, the density of Y parameterized by θ and α is

$$f(y; \theta, \alpha) = \frac{1}{(2\pi\sigma^2)^{\frac{n}{2}}} e^{-\frac{\sum_{i=1}^n (y_i - \alpha m_i(\theta))^2}{2\sigma^2}}. \tag{5.17}$$

A maximum likelihood (ML) estimate $\hat{\theta}_{ML}$ is a value of $\hat{\theta}$ that maximizes the *likelihood function* $f(y;\hat{\theta},\alpha)$. This is the same as maximizing $\sum_{i=1}^{n}(y_i\alpha m_i(\hat{\theta}) - \frac{\alpha^2 m_i(\hat{\theta})^2}{2})$, obtained by taking the natural log of (5.17), removing terms that do not depend on $\hat{\theta}$, and multiplying by σ^2.

Because the collection $\phi_1(t),\phi_2(t),\ldots,\phi_n(t)$ spans the set $\{s(t-\hat{\theta}) : \hat{\theta} \in [0,\theta_{max}]\}$, additional projections of $R(t)$ onto normalized functions that are orthogonal to $\phi_1(t),\ldots,\phi_n(t)$ lead to zero-mean Gaussian random variables of variance $N_0/2$. Their inclusion in the likelihood function has no effect on the maximizer θ_{ML}. Furthermore, $\sum_i m_i^2(\hat{\theta})$ equals $\int s^2(t-\hat{\theta})dt = \|s\|^2$ (see (2.32)). Finally, $\int R(t)s(t-\hat{\theta})dt$ equals $\sum_i Y_i m_i(\hat{\theta})$. (If this last step is not clear to you, substitute $s(t-\hat{\theta})$ with $\sum_i m_i(\hat{\theta})\phi_i(t)$ and swap the sum and the integral.) Putting everything together, we obtain that the maximum likelihood estimate $\hat{\theta}_{ML}$ is the $\hat{\theta}$ that maximizes

$$\int r(t)s(t-\hat{\theta})dt, \qquad (5.18)$$

where $r(t)$ is the observed sample path of $R(t)$.

This result is intuitive. If we write $r(t) = \alpha s(t-\theta) + n(t)$, where $n(t)$ is the sample path of $N(t)$, we see that we are maximizing $\alpha R_s(\hat{\theta}-\theta) + \int n(t)s(t-\hat{\theta})dt$, where $R_s(\cdot)$ is the self-similarity function of $s(t)$ and $\int n(t)s(t-\hat{\theta})dt$ is a sample of a zero-mean Gaussian random variable of variance that does not depend on $\hat{\theta}$. The self-similarity function of a real-valued function achieves its maximum at the origin, hence $R_s(\hat{\theta}-\theta)$ achieves its maximum at $\hat{\theta}=\theta$.

Notice that we have introduced an orthonormal basis for an intermediate step in our derivations, but it is not needed to maximize (5.18).

The ML approach to parameter estimation is a well-established method but it is not the only one. If we model θ as the sample of a random variable Θ and we have a cost function $c(\theta,\hat{\theta})$ that quantifies the "cost" of deciding that the parameter is $\hat{\theta}$ when it is θ, then we can aim for the estimate that minimizes the expected cost. This is the Bayesian approach to parameter estimation. Another approach is the least-squares estimation (LSE). It seeks the $\hat{\theta}$ for which $\sum_i(y_i - \alpha m_i(\hat{\theta}))^2$ is minimized. In words, the LSE approach chooses the parameter that provides the most accurate description of the measurements, where accuracy is measured in terms of squared distance. This is a different objective than that of the ML approach, which chooses the parameter for which the measurements are the most likely. When the noise is additive and Gaussian, as in our case, the ML approach and the LSE approach lead to the same estimate. This is due to the fact that the likelihood function $f(y;\hat{\theta},a)$ depends on the squared distance $\sum_i(y_i - \alpha m_i(\hat{\theta}))^2$.

5.7.2 Delay locked loop approach

The shape of the training signal did not matter for the derivation of the ML estimate, but it does matter for the delay locked loop approach. We assume that the training signal takes the same form as the communication signal, namely

5.7. Symbol synchronization

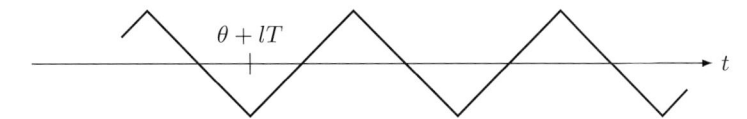

Figure 5.10. Shape of $M(t)$.

$$s(t) = \sum_{l=0}^{L-1} c_l \psi(t - lT),$$

where $c_0, \ldots, c_{L-1}$ are *training symbols* known to the receiver. The easiest way to see how the delay locked loop works is to assume that $\psi(t)$ is a rectangular pulse

$$\psi(t) = \sqrt{\frac{1}{T}} \mathbb{1}\{0 \leq t \leq T\}$$

and let the training symbol sequence $c_0, \ldots, c_{L-1}$ be an alternating sequence of $\sqrt{\mathcal{E}_s}$ and $-\sqrt{\mathcal{E}_s}$. The corresponding received signal $R(t)$ is $\alpha \sum_{l=0}^{L-1} c_l \psi(t - lT - \theta)$ plus white Gaussian noise, where α is the unknown scaling factor. If we neglect the noise for the moment, the matched filter output (before sampling) is the convolution of $R(t)$ with $\psi^*(-t)$, which can be written as

$$M(t) = \alpha \sum_{l=0}^{L-1} c_l R_\psi(t - lT - \theta),$$

where

$$R_\psi(\tau) = \left(1 - \frac{|\tau|}{T}\right) \mathbb{1}\{-T \leq \tau \leq T\}$$

is the self-similarity function of $\psi(t)$. Figure 5.10 plots a piece of $M(t)$.

The desired sampling times of the form $\theta + lT$, l integer, correspond to the maxima and minima of $M(t)$. Let t_k be the kth sampling point. Until symbol synchronization is achieved, $M(t_k)$ is not necessarily near a maximum or a minimum of $M(t)$. For every sample point t_k, we also collect an early sample at $t_k^E = t_k - \Delta$ and a late sample at $t_k^L = t_k + \Delta$, where Δ is some small positive value (smaller than $T/2$). The dots in Figure 5.11 are examples of sample values. Consider the cases when $M(t_k)$ is positive (parts (a) and (b)). We see that $M(t_k^L) - M(t_k^E)$ is negative when t_k is late with respect to the target, and it is positive when t_k is early. The opposite is true when $M(t_k)$ is negative (parts (c) and (d)). Hence, in general, $\bigl(M(t_k^L) - M(t_k^E)\bigr) M(t_k)$ can be used as a feedback signal to the clock that determines the sampling times. A positive feedback signal is a sign for the clock to speed up, and a negative value is a sign to slow down. This can be implemented via a *voltage-controlled oscillator* (VCO), with the feedback signal as the controlling voltage.

Now consider the effect of noise. The noise added to $M(t)$ is zero-mean. Intuitively, if the VCO does not react too quickly to the feedback signal, or equivalently if the feedback signal is lowpass filtered, then we expect the sampling point to settle

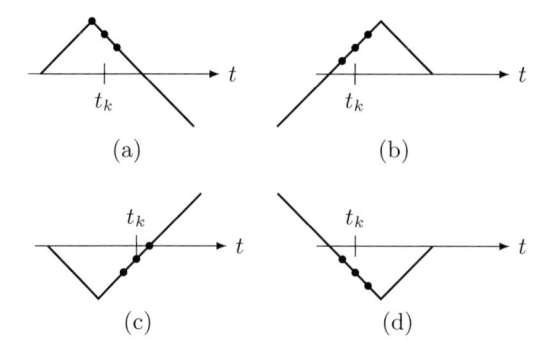

Figure 5.11. DLL sampling points. The three consecutive dots of each part are examples of $M(t_k^E)$, $M(t_k)$, and $M(t_k^L)$, respectively.

at the correct position even when the feedback signal is noisy. A rigorous analysis is possible but it is outside the scope of this text. For a more detailed introduction on synchronization we recommend [14, Chapters 14–16], [15, Chapter 4], and the references therein.

Notice the similarity between the ML and the DLL solution. Ultimately they both make use of the fact that a self-similarity function achieves its maximum at the origin. It is also useful to think of a correlation such as

$$\frac{\int r(t)s(t-\hat{\theta})dt}{\|r(t)\|\|s(t)\|}$$

as a measure for the degree of similarity between two functions, where the denominator serves to make the result invariant to a scaling of either function. In this case the two functions are $r(t) = \alpha s(t-\theta) + n(t)$ and $s(t-\hat{\theta})$ and the maximum is achieved when $s(t-\hat{\theta})$ lines up with $s(t-\theta)$. The solution to the ML approach correlates with the entire training signal $s(t)$, whereas the DLL correlates with the pulse $\psi(t)$; but it does so repeatedly and averages the results. The DLL is designed to work with a VCO, and together they provide a complete and easy-to-implement solution that tracks the sampling times.[3] It is easy to see that the DLL provides valuable feedback even after the transition from the training symbols to the regular symbols, provided that the symbols change polarity sufficiently often. To implement the ML approach, we still need a good way to find the maximum of (5.18) and to offset the clock accordingly. This can easily be done if the receiver is implemented in a digital signal processor (DSP) but could be costly in terms of additional hardware if the receiver is implemented with analog technology.

---

[3] The DLL can be interpreted as a stochastic gradient descent method that seeks the ML estimate of θ.

5.8 Summary

The signal design consists of choosing the finite-energy signals $w_0(t), \ldots, w_{m-1}(t)$ that represent the messages. Rather than choosing the signal set directly, we choose an orthonormal basis $\{\psi_1(t), \ldots, \psi_n(t)\}$ and a codebook $\{c_0, \ldots, c_{m-1}\} \subset \mathbb{R}^n$ and, for $i = 0, \ldots, m-1$, we define

$$w_i(t) = \sum_{j=1}^{n} c_{i,j} \psi_j(t).$$

In doing so, we separate the signal design problem into two subproblems: finding an appropriate orthonormal basis and codebook. In this chapter, we have focused on the choice of the orthonormal basis. The choice of the codebook will be the topic of the next chapter.

Particularly interesting are those orthonormal bases that consist of an appropriately chosen unit-norm function $\psi(t)$ and its T-spaced translates. Then the generic form of a signal is

$$w(t) = \sum_{j=1}^{n} s_j \psi(t - jT).$$

In this case, the n inner products performed by the n-tuple former can be obtained by means of a single matched filter with the output sampled at n time instants.

We call s_j the jth symbol. In theory, a symbol can take values in $\mathbb{R}$ or in $\mathbb{C}$; in practice, the symbol alphabet is some *discrete* subset $\mathcal{S}$ of $\mathbb{R}$ or of $\mathbb{C}$. For instance, PAM symbols are in $\mathbb{R}$, and QAM or PSK symbols, viewed as complex-valued numbers, are standard examples of symbol in $\mathbb{C}$ (see Chapter 7).

If the symbol sequence is a realization of an uncorrelated WSS process, then the power spectral density of the transmitted signal is $S_X(f) = \mathcal{E}|\psi_\mathcal{F}(f)|^2/T$, where $\mathcal{E}$ is the average energy per symbol. The pulse $\psi(t)$ has a unit norm and is orthogonal to its T-spaced translates if and only if $|\psi_\mathcal{F}(f)|^2$ fulfills Nyquist's criterion with parameter T (Theorem 5.6).

Typically $\psi(t)$ is real-valued, in which case $|\psi_\mathcal{F}(f)|^2$ is an even function. To save bandwidth, we often choose $|\psi_\mathcal{F}(f)|^2$ in such a way that it vanishes for $f \notin [-\frac{1}{T}, \frac{1}{T}]$. When these two conditions are satisfied, Nyquist's criterion is fulfilled if and only if $|\psi_\mathcal{F}(f)|^2$ has the so-called band-edge symmetry.

It is instructive to compare the sampling theorem to the Nyquist criterion. Both are meant for signals of the form $\sum_i s_i \psi(t - iT)$, where $\psi(t)$ is orthogonal to its T-spaced time translates. Without loss of generality, we can assume that $\psi(t)$ has a unit norm. In this case, $|\psi_\mathcal{F}(f)|^2$ fulfills the Nyquist criterion with parameter T. Typically we use the Nyquist criterion to select a pulse that leads to symbol-by-symbol on a pulse train of a desired power spectral density. In the sampling theorem, we choose $\psi(t) = \mathrm{sinc}(\frac{t}{T})/\sqrt{T}$ because the orthonormal basis $\mathcal{B} = \{\mathrm{sinc}[(t - iT)/T]/\sqrt{T}\}_{i \in \mathbb{Z}}$ spans the inner product space of continuous $\mathcal{L}_2$ functions that have a vanishing Fourier transform outside $[-\frac{1}{2T}, \frac{1}{2T}]$. Any signal in this space can be represented by the coefficients of its orthonormal expansion with respect to $\mathcal{B}$.

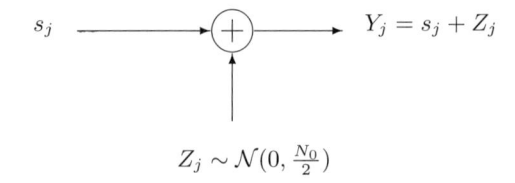

$$s_j \longrightarrow \boxed{+} \longrightarrow Y_j = s_j + Z_j$$

$$Z_j \sim \mathcal{N}(0, \tfrac{N_0}{2})$$

Figure 5.12. Equivalent discrete-time channel used at the rate of one channel use every T seconds. The noise is iid.

The eye diagram is a field test that can easily be performed to verify that the matched filter output at the sampling times is as expected. It is also a valuable tool for designing the pulse $\psi(t)$.

Because the matched filter output is sampled at a rate of one sample every T seconds, we say that the discrete-time AWGN channel seen by the top layer is used at a rate of one symbol every T seconds. This channel is depicted in Figure 5.12.

Questions that pertain to the encoder/decoder pair (such as the number of bits per symbol, the average energy per symbol, and the error probability) can be answered assuming that the top layer communicates via the discrete-time channel. Questions that pertain to the signal's time/frequency characteristics need to take the waveform former into consideration. Essentially, the top two layers can be designed independently.

5.9 Appendix: $\mathcal{L}_2$, and Lebesgue integral: A primer

A function $g : \mathbb{R} \to \mathbb{C}$ belongs to $\mathcal{L}_2$ if it is Lebesgue measurable and has a finite Lebesgue integral $\int_{-\infty}^{\infty} |g(t)|^2 dt$. Lebesgue measure, Lebesgue integral, and $\mathcal{L}_2$ are technical terms, but the ideas related to these terms are quite natural. In this appendix, our goal is to introduce them informally. The reader can skip this appendix and read "$\mathcal{L}_2$ functions" as "finite-energy functions" and "Lebesgue integral" as "integral".

We can associate the integral of a non-negative function $g : \mathbb{R} \to \mathbb{R}$ to the area between the abscissa and the function's graph. We attempt to determine this area via a sequence of approximations that – we hope – converges to a quantity that can reasonably be identified with the mentioned area. Both the Riemann integral (the one we learn in high school) and the Lebesgue integral (a more general construction) obtain approximations by adding up the area of rectangles.

For the Riemann integral, we think of partitioning the area between the abscissa and the function's graph by means of *vertical* slices of some width (typically the same width for each slice), and we approximate the area of each slice by the area of a rectangle, as shown in Figure 5.13a. The height of the rectangle can be any value taken by the function in the interval defined by the slice. Hence, by adding up the area of the rectangles, we obtain a number that can underestimate or

5.9. Appendix: $\mathcal{L}_2$, and Lebesgue integral: A primer

(a) Riemann integration.

(b) Lebesgue integration.

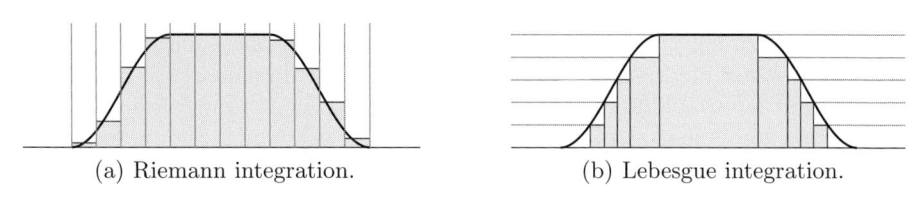

Figure 5.13. Integration.

overestimate the area of interest. We obtain a sequence of estimates by choosing slices of decreasing width. If the sequence converges for any such construction applied to the function being integrated, then the Riemann integral of that function is defined to be the limit of the sequence. Otherwise the function is not Riemann integrable.

The definition of the Lebesgue integral starts with equally spaced *horizontal* lines as shown in Figure 5.13b. From the intersection of these lines with the function's graph, we obtain vertical slices that, unlike for the Riemann's integral, have variable widths. The area of each slice between the abscissa and the function's graph is under-approximated by the area of the highest rectangle that fits *under* the function's graph. By adding the areas of these rectangles, we obtain an under-estimate of the area of interest. If we refine the horizontal partition (a new line halfway between each pair of existing lines) and repeat the construction, we obtain an approximation that is at least as good as the previous one. By repeating the operation, we obtain an increasing – thus convergent – sequence of approximations, the limit of which is the Lebesgue integral of the positive function.

Every non-negative Riemann-integrable function is Lebesgue integrable; and the values of the two integrals agree whenever they are both defined.

Next, we give an example of a function that is Lebesgue integrable but not Riemann integrable, and then we define the Lebesgue integral of general real-valued and complex-valued functions.

EXAMPLE 5.8 *The Dirichlet function $f : [0,1] \to \{0,1\}$ is 0 where its argument is irrational and 1 otherwise. Its Lebesgue integral is well defined and equal to 0. But to see why, we need to introduce the notion of measure. We will do so shortly. The Riemann integral of the Dirichlet function is undefined. The problem is that in each interval of the abscissa, no matter how small, we can find both rational and irrational numbers. So to each approximating rectangle, we can assign height 0 or 1. If we assign height 0 to all approximating rectangles, the integral is approximated by 0; if we assign height 1 to all rectangles, the integral is approximated by 1. And we can choose, no matter how narrow we make the vertical slices. Clearly, we can create approximation sequences that do not converge. This could not happen with a continuous function or a function that has a finite number of discontinuities, but the Dirichlet function is discontinuous everywhere.* □

In our informal construction of the Lebesgue integral, we have implicitly assumed that the area under the function can be approximated using rectangles of a well defined width, which is not necessarily the case with bizarre functions.

The Dirichlet is such a function. When we apply the Lebesgue construction to the Dirichlet function, we effectively partition the domain of the function, namely the unit interval $[0,1]$, into two subsets, say the set $\mathcal{A}$ that contains the rationals and the set $\mathcal{B}$ that contains the irrationals. If we could assign a "total length" $L(\mathcal{A})$ and $L(\mathcal{B})$ to these sets as we do for countable unions of disjoint intervals of the real line, then we could say that the Lebesgue integral of the Dirichlet function is $1 \times L(\mathcal{A}) + 0 \times L(\mathcal{B}) = L(\mathcal{A})$. The *Lebesgue measure* does precisely this. The Lebesgue measure of $\mathcal{A}$ is 0 and that of $\mathcal{B}$ is 1. This is not surprising, given that $[0,1]$ contains a countable number of rationals and an uncountable number of irrationals. Hence, the Lebesgue integral of the Dirichlet function is 0.

Note that it is not possible to assign the equivalent of a "length" to every subset of $\mathbb{R}$. Every attempt to do so leads to contradictions, whereby the measure of the union of certain disjoint sets is not the sum of the individual measures. The subsets of $\mathbb{R}$ to which we can assign the Lebesgue measure are called *Lebesgue measurable sets*. It is hard to come up with non-measurable sets; for an example see the end of Appendix 4.9 of [2]. A function is said to be *Lebesgue measurable* if the Lebesgue's construction partitions the abscissa into Lebesgue measurable sets.

We would not mention the Lebesgue integral if it were just to integrate bizarre functions. The real power of Lebesgue integration theory comes from a number of theorems that give precise conditions under which a limit and an integral can be exchanged. Such operations come up frequently in the study of Fourier transforms and Fourier series. The reader should not be alarmed at this point. We will interchange integrals, swap integrals and sums, swap limits and integrals, all without a fuss: but we do this because we know that the Lebesgue integration theory allows us to do so, in the cases of interest to us.

In introducing the Lebesgue integral, we have assumed that the function being integrated is non-negative. The same idea applies to non-positive functions. (In fact, we can integrate the negative of the function and then take the negative of the result.) If the function takes on positive as well as negative values, we split the function into its positive and negative parts, we integrate each part separately, and we add the two results. This works as long as the two intermediate results are not $+\infty$ and $-\infty$, respectively. If they are, the Lebesgue integral is undefined; otherwise the Lebesgue integral is defined.

A complex-valued function $g : \mathbb{R} \to \mathbb{C}$ is integrated by separately integrating its real and imaginary parts. If the Lebesgue integral over both parts is defined and finite, the function is said to be *Lebesgue integrable*. The set of Lebesgue-integrable functions is denoted by $\mathcal{L}_1$. The notation $\mathcal{L}_1$ comes from the easy-to-verify fact that for a Lebesgue measurable function g, the Lebesgue integral is finite if and only if $\int_{-\infty}^{\infty} |g(x)| dx < \infty$. This integral is the $\mathcal{L}_1$ norm of $g(x)$.

Every bounded Riemann-integrable function of bounded support is Lebesgue integrable and the values of the two integrals agree. This statement does not extend to functions that are defined over the real line. To see why, consider integrating $\text{sinc}(t)$ over the real line. The Lebesgue integral of $\text{sinc}(t)$ is not defined, because the integral of the positive part of $\text{sinc}(t)$ and that over the negative part are $+\infty$ and $-\infty$, respectively. The Riemann integral of $\text{sinc}(t)$ exists because, by definition, Riemann integrates from $-T$ to T and then lets T go to infinity.

5.9. Appendix: $\mathcal{L}_2$, and Lebesgue integral: A primer

All functions that model physical processes are finite-energy and measurable, hence $\mathcal{L}_2$ functions. All finite-energy functions that we encounter in this text are measurable, hence $\mathcal{L}_2$ functions. There are examples of finite-energy functions that are not measurable, but it is hard to imagine an engineering problem where such functions would arise.

The set of $\mathcal{L}_2$ functions forms a complex vector space with the zero vector being the all-zero function. If we modify an $\mathcal{L}_2$ function in a countable number of points, the result is also an $\mathcal{L}_2$ function and the (Lebesgue) integral over the two functions is the same. (More generally, the same is true if we modify the function over a set of measure zero.) The difference between the two functions is an $\mathcal{L}_2$ function $\xi(t)$ such that the Lebesgue integral $\int |\xi(t)|^2 dt = 0$. Two $\mathcal{L}_2$ functions that have this property are said to be $\mathcal{L}_2$ *equivalent*.

Unfortunately, $\mathcal{L}_2$ with the (standard) inner product that maps $a(t), b(t) \in \mathcal{L}_2$ to

$$\langle a, b \rangle = \int a(t) b^*(t) dt \tag{5.19}$$

does not form an inner product space because axiom (c) of Definition 2.36 is not fulfilled. In fact, $\langle a, a \rangle = 0$ implies that $a(t)$ is $\mathcal{L}_2$ equivalent to the zero function, but this is not enough: to satisfy the axiom, $a(t)$ must be *the* all-zero function.

There are two obvious ways around this problem if we want to treat finite-energy functions as vectors of an inner product space. One way is to consider only subspaces $\mathcal{V}$ of $\mathcal{L}_2$ such that there is only one vector $\xi(t)$ in $\mathcal{V}$ that has the property $\int |\xi(t)|^2 dt = 0$. This will always be the case when $\mathcal{V}$ is spanned by a set $\mathcal{W}$ of waveforms that represent electrical signals.

EXAMPLE 5.9 *The set $\mathcal{V}$ that consists of the continuous functions of $\mathcal{L}_2$ is a vector space. Continuity is sufficient to ensure that there is only one function $\xi(t) \in \mathcal{V}$ for which the integral $\int |\xi(t)|^2 dt = 0$, namely the zero function. Hence $\mathcal{V}$ equipped with the standard inner product is an inner product space.*[4] □

Another way is to form equivalence classes. Two signals that are $\mathcal{L}_2$ equivalent cannot be distinguished by means of a physical experiment. Hence the idea of partitioning $\mathcal{L}_2$ into equivalence classes, with the property that two functions are in the same equivalence class if and only if they are $\mathcal{L}_2$ equivalent. With an appropriately defined vector addition and multiplication of a vector by a scalar, the set of equivalence classes forms a complex vector space. We can use (5.19) to define an inner product over this vector space. The inner product between two equivalence classes is the result of applying (5.19) with an element of the first class and an element of the second class. The result does not depend on which element of a class we choose to perform the calculation. This way $\mathcal{L}_2$ can be transformed into an inner product space denoted by L_2. As a "sanity check", suppose that we want to compute the inner product of a vector with itself. Let $a(t)$ be an arbitrary element of the corresponding class. If $\langle a, a \rangle = 0$ then $a(t)$ is in the equivalence class that contains all the functions that have 0 norm. This class *is* the zero vector.

---

[4] However, one can construct a sequence of continuous functions that converges to a discontinuous function. In technical terms, this inner product space is not complete.

For a thorough treatment of measurability, Lebesgue integration, and $\mathcal{L}_2$ functions we refer to a real-analysis book (see e.g. [9]). For an excellent introduction to the Lebesgue integral in a communication engineering context, we recommend the graduate-level text [2, Section 4.3]. Even more details can be found in [3]. A very nice summary of Lebesgue integration can be found on Wikipedia (July 27, 2012).

5.10 Appendix: Fourier transform: A review

In this appendix, we review a few facts about the Fourier transform. They belong to two categories. One category consists of facts that have an operational value. You should know them as they help us in routine manipulations. The other category consists of mathematical subtleties that you should be aware of, even though they do not make a difference in engineering applications. We mention them for awareness, and because in certain cases they affect the language we use in formulating theorems. For an in-depth discussion of this second category we recommend [2, Section 4.5] and [3].

The following two integrals relate a function $g : \mathbb{R} \to \mathbb{C}$ to its Fourier transform $g_\mathcal{F} : \mathbb{R} \to \mathbb{C}$

$$g(u) = \int_{-\infty}^{\infty} g_\mathcal{F}(\alpha) e^{\mathrm{j}2\pi u \alpha} d\alpha \tag{5.20}$$

$$g_\mathcal{F}(v) = \int_{-\infty}^{\infty} g(\alpha) e^{-\mathrm{j}2\pi v \alpha} d\alpha, \tag{5.21}$$

where $\mathrm{j} = \sqrt{-1}$.

Some books use capital letters for the Fourier transform, such as $G(f)$ for the Fourier transform of $g(t)$. Other books use a hat, like in $\hat{g}(f)$. We use the subscript $\mathcal{F}$ because, in this text, capital letters denote random objects (random variables and random processes) and hats are used for estimates in detection-related problems.

Typically we take the Fourier transform of a time-domain function and the inverse Fourier transform of a frequency-domain function. However, from a mathematical point of view it does not matter whether or not there is a physical meaning associated to the variable of the function being transformed. To underline this fact, we use the unbiased dummy variable α in (5.20) and (5.21).

Sometimes the Fourier transform is defined using ω instead of $2\pi f$ in (5.20) and (5.21), but (5.21) also inherits the factor $\frac{1}{2\pi}$, which breaks the nice symmetry between (5.20) and (5.21). Most books on communication define the Fourier transform as we do, because it makes the formulas easier to remember. Lapidoth [3, Section 6.2.1] gives five "good reasons" for defining the Fourier transform as we do.

When we say that the Fourier transform of a function exists, we mean that the integral (5.21) is well defined. It does not imply that the inverse Fourier transform exists. The Fourier integral (5.21) exists and is finite for all $\mathcal{L}_2$ functions defined over a finite-length interval. The technical reason for this is that an $\mathcal{L}_2$ function

5.10. Appendix: Fourier transform: A review

$f(t)$ defined over a finite-length interval is also an $\mathcal{L}_1$ function, i.e. $\int |f(t)|dt$ is finite, which excludes the $\infty - \infty$ problem mentioned in Appendix 5.9. Then also $\int |f(t)e^{j2\pi\alpha t}|dt = \int |f(t)|dt$ is finite. Hence (5.21) is finite.

If the $\mathcal{L}_2$ function is defined over $\mathbb{R}$, then the $\infty - \infty$ problem mentioned in Appendix 5.9 can arise. This is the case with the sinc(t) function. In such cases, we truncate the function to the interval $[-T, T]$, compute its Fourier transform, and let T go to infinity. It is in this sense that the Fourier transform of sinc(t) is defined. An important result of Fourier analysis says that we are allowed to do so (see Plancherel's theorem, [2, Section 4.5.2]). Fortunately we rarely have to do this, because the Fourier transform of most functions of interest to us is tabulated.

Thanks to Plancherel's theorem, we can make the sweeping statement that the Fourier transform is defined for all $\mathcal{L}_2$ functions. The transformed function is in $\mathcal{L}_2$, hence also its inverse is defined. Be aware though, when we compute the transform and then the inverse of the transformed, we do not necessarily obtain the original function. However, what we obtain is $\mathcal{L}_2$ equivalent to the original. As already mentioned, no physical experiment will ever be able to detect the difference between the original and the modified function.

EXAMPLE 5.10 *The function $g(t)$ that has value 1 at $t = 0$ and 0 everywhere else is an $\mathcal{L}_2$ function. Its Fourier transform $g_\mathcal{F}(f)$ is 0 everywhere. The inverse Fourier transform of $g_\mathcal{F}(f)$ is also 0 everywhere.* □

One way to remember whether (5.20) or (5.21) has the minus sign in the exponent is to think of the Fourier transform as a tool that allows us to write a function $g(u)$ as a linear combination of complex exponentials. Hence we are writing $g(u) = \int g_\mathcal{F}(\alpha)\phi(\alpha, u)d\alpha$ with $\phi(\alpha, u) = \exp(j2\pi u\alpha)$ viewed as a function of u with parameter α. Technically this is not an orthonormal expansion, but it looks like one, where $g_\mathcal{F}(\alpha)$ is the coefficient of the function $\phi(\alpha, u)$. Like for an orthonormal expansion, the coefficient is obtained from an expression that takes the form $g_\mathcal{F}(u) = \langle g(\alpha), \phi(\alpha, u)\rangle = \int g(\alpha)\phi^*(\alpha, u)d\alpha$. It is the complex conjugate in the computation of the inner product that brings in the minus sign in the exponent. We emphasize that we are working by analogy here. The complex exponential has infinite energy – hence not a unit-norm function (at least not with respect to the standard inner product).

A useful formula is Parseval's relationship

$$\int a(t)b^*(t)dt = \int a_\mathcal{F}(f)b_\mathcal{F}^*(f)df, \qquad \text{(Parseval)} \qquad (5.22)$$

which states that $\langle a(t), b(t)\rangle = \langle a_\mathcal{F}(f), b_\mathcal{F}(f)\rangle$.

Rectangular pulses and their Fourier transforms often show up in examples and we should know how to go back and forth between them. Two tricks make it easy to relate the rectangle and the sinc as Fourier transform pairs. First, let us recall how the sinc is defined: $\text{sinc}(x) = \frac{\sin(\pi x)}{\pi x}$. (The π makes sinc(x) vanish at all integer values of x except for $x = 0$ where it is 1.)

The first trick is well known: the value of a (time-domain) function at (time) 0 equals the area under its Fourier transform. Similarly, the value of a (frequency-domain) function at (frequency) 0 equals the area under its inverse Fourier transform. These properties follow directly from the definition of Fourier transform, namely

$$g(0) = \int_{-\infty}^{\infty} g_{\mathcal{F}}(\alpha) d\alpha \tag{5.23}$$

$$g_{\mathcal{F}}(0) = \int_{-\infty}^{\infty} g(\alpha) d\alpha. \tag{5.24}$$

Everyone knows how to compute the area of a rectangle. But how do we compute the area under a sinc? Here is where the second and not-so-well-known trick comes in handy. The area under a sinc is the area of the triangle inscribed in its main lobe. Hence the integral under the two curves of Figure 5.14 is identical and equal to ab, and this is true for all positive values a, b.

Let us consider a specific example of how to use the above two tricks. It does not matter if we start from a rectangle or from a sinc and whether we want to find its Fourier transform or its inverse Fourier transform. Let $a, b, c,$ and d be as shown in Figure 5.15.

We want to relate a, b to c, d (or vice versa). Since b must equal the area under the sinc and d the area under the rectangle, we have

$$b = cd$$
$$d = 2ab,$$

which may be solved for a, b or for c, d.

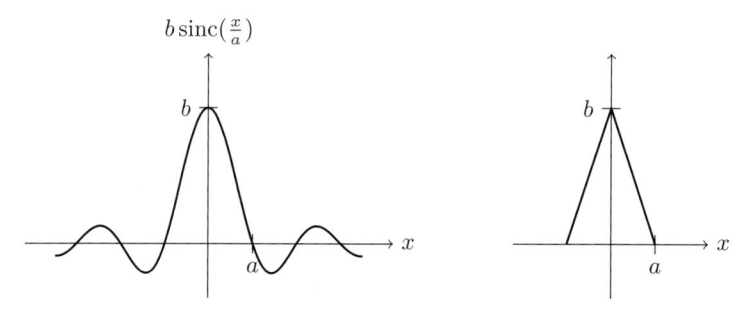

Figure 5.14. $\int_{-\infty}^{\infty} b \operatorname{sinc}(\frac{x}{a}) dx$ equals the area under the triangle on the right.

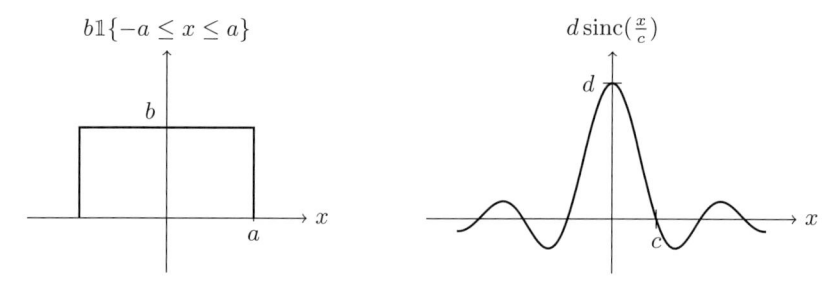

Figure 5.15. Rectangle and sinc to be related as Fourier transform pairs.

5.11. Appendix: Fourier series: A review

Table 5.1 Fourier transform (right-hand column) for a few $\mathcal{L}_2$ functions (left-hand column), where $a > 0$.

$e^{-a\lvert t\rvert}$	$\overset{\mathcal{F}}{\Longleftrightarrow}$	$\frac{2a}{a^2+(2\pi f)^2}$
$e^{-at},\ t \geq 0$	$\overset{\mathcal{F}}{\Longleftrightarrow}$	$\frac{1}{a+j2\pi f}$
$e^{-\pi t^2}$	$\overset{\mathcal{F}}{\Longleftrightarrow}$	$e^{-\pi f^2}$
$\mathbb{1}\{-a \leq t \leq a\}$	$\overset{\mathcal{F}}{\Longleftrightarrow}$	$2a\operatorname{sinc}(2af)$

EXAMPLE 5.11 *The Fourier transform of $b\mathbb{1}\{-a \leq t \leq a\}$ is a sinc that has height $2ab$. Its first zero crossing must be at $\frac{1}{2a}$ so that the area under the sinc becomes b. The result is $2ab\operatorname{sinc}(2af)$.* □

EXAMPLE 5.12 *The Fourier transform of $b\operatorname{sinc}(\frac{t}{a})$ is a rectangle of height ab. Its width must be $\frac{1}{a}$ so that its area is b. The result is $ab\mathbb{1}\{f \in [-\frac{1}{2a}, \frac{1}{2a}]\}$.* □

See Exercise 5.16 for a list of useful Fourier transform relations and see Table 5.1 for the Fourier transform of a few $\mathcal{L}_2$ functions.

5.11 Appendix: Fourier series: A review

We review the Fourier series, focusing on the big picture and on how to remember things. Let $f(x)$ be a periodic function, $x \in \mathbb{R}$. It has period p if $f(x) = f(x+p)$ for all $x \in \mathbb{R}$. Its fundamental period is the smallest positive such p. We are using the "physically unbiased" variable x instead of t, because we want to emphasize that we are dealing with a general periodic function, not necessarily a function of time.

The main idea is that a sufficiently well-behaved function $f(x)$ of fundamental period p can be written as a linear combination of all the complex exponentials of period p. Hence

$$f(x) = \sum_{i \in \mathbb{Z}} A_i\, e^{j2\pi x i/p} \tag{5.25}$$

for some sequence of coefficients $\ldots A_{-1}, A_0, A_1, \ldots$ with values in $\mathbb{C}$.

Two functions of fundamental period p are identical if and only if they coincide over a period. Hence to check if a given series of coefficients $\ldots A_{-1}, A_0, A_1, \ldots$ is the correct series, it suffices to verify that

$$f(x)\mathbb{1}\Big\{-\frac{p}{2} \leq x \leq \frac{p}{2}\Big\} = \sum_{i \in \mathbb{Z}} \sqrt{p}A_i \frac{e^{j2\pi x i/p}}{\sqrt{p}}\mathbb{1}\Big\{-\frac{p}{2} \leq x \leq \frac{p}{2}\Big\},$$

where we have multiplied and divided by $\sqrt{p}$ to make $\phi_i(x) = \frac{e^{j2\pi xi/p}}{\sqrt{p}} \mathbb{1}\{-\frac{p}{2} \leq x \leq \frac{p}{2}\}$ of unit norm, $i \in \mathbb{Z}$. Thus we can write

$$f(x)\mathbb{1}\left\{-\frac{p}{2} \leq x \leq \frac{p}{2}\right\} = \sum_{i \in \mathbb{Z}} \sqrt{p} A_i \phi_i(x). \tag{5.26}$$

The right-hand side of the above expression is an orthonormal expansion. The coefficients of an orthonormal expansion are always computed according to the same expression. The ith coefficient $\sqrt{p}A_i$ equals $\langle f, \phi_i \rangle$. Hence,

$$A_i = \frac{1}{p} \int_{-\frac{p}{2}}^{\frac{p}{2}} f(x) e^{-j2\pi xi/p} dx. \tag{5.27}$$

Notice that the right-hand side of (5.25) is periodic, and that of (5.26) has finite support. In fact we can use the Fourier series for both kind of functions, periodic and finite-support. That Fourier series can be used for finite-support functions is an obvious consequence of the fact that a finite-support function can be seen as one period of a periodic function. In communication, we are more interested in functions that have finite support. (Once we have seen one period of a periodic function, we have seen them all: from that point on there is no information being conveyed.)

In terms of mathematical rigor, there are two things that can go wrong in what we have said. (1) For some i, the integral in (5.27) might not be defined or might not be finite. We can show that neither is the case if $f(x)\mathbb{1}\{-\frac{p}{2} \leq x \leq \frac{p}{2}\}$ is an $\mathcal{L}_2$ function. (2) for a specific x, the truncated series $\sum_{i=-l}^{l} \sqrt{p}A_i\phi_i(x)$ might not converge as l goes to infinity or might converge to a value that differs from $f(x)\mathbb{1}\{-\frac{p}{2} \leq x \leq \frac{p}{2}\}$. It is not hard to show that the norm of the function

$$f(x)\mathbb{1}\left\{-\frac{p}{2} \leq x \leq \frac{p}{2}\right\} - \sum_{i=-l}^{l} \sqrt{p} A_i \phi_i(x)$$

goes to zero as l goes to infinity. Hence the two functions are $\mathcal{L}_2$ equivalent. We write this as follows

$$f(x)\mathbb{1}\left\{-\frac{p}{2} \leq x \leq \frac{p}{2}\right\} = \text{l.i.m.} \sum_{i \in \mathbb{Z}} \sqrt{p} A_i \phi_i(x),$$

where l.i.m. means *limit in mean-square*: it is a short-hand notation for

$$\lim_{l \to \infty} \int_{-\frac{p}{2}}^{\frac{p}{2}} \left| f(x) - \sum_{i=-l}^{l} \sqrt{p} A_i \phi_i(x) \right|^2 dx = 0.$$

We summarize with the following rigorous statement. The details that we have omitted in our "proof" can be found in [2, Section 4.4].

5.12. Appendix: Proof of the sampling theorem

THEOREM 5.13 (Fourier series) Let $g(x) : [-\frac{p}{2}, \frac{p}{2}] \to \mathbb{C}$ be an $\mathcal{L}_2$ function. Then for each $k \in \mathbb{Z}$, the (Lebesgue) integral

$$A_k = \frac{1}{p} \int_{-p/2}^{p/2} g(x) e^{-j2\pi kx/p} dx$$

exists and is finite. Furthermore,

$$g(x) = \text{l.i.m.} \sum_k A_k e^{j2\pi kx/p} \mathbb{1}\left\{-\frac{p}{2} \le x \le \frac{p}{2}\right\}.$$

□

5.12 Appendix: Proof of the sampling theorem

In this appendix, we prove Theorem 5.2. By assumption, for any $b \ge B$, $w_\mathcal{F}(f) = 0$ for $f \notin [-b, b]$. Hence we can write $w_\mathcal{F}(f)$ as a Fourier series (Theorem 5.13):

$$w_\mathcal{F}(f) = \text{l.i.m.} \sum_k A_k e^{j\pi f k/b} \mathbb{1}\{-b \le f \le b\}.$$

Note that $w_\mathcal{F}(f)$ is in $\mathcal{L}_2$ (it is the Fourier transform of an $\mathcal{L}_2$ function) and vanishes outside $[-b, b]$, hence it is $\mathcal{L}_1$ (see e.g. [2, Theorem 4.3.2]). The inverse Fourier transform of an $\mathcal{L}_1$ function is continuous (see e.g. [2, Lemma 4.5.1]). Hence it must be identical to $w(t)$, which is also continuous.

Even though $w_\mathcal{F}(f)$ and $\sum_k A_k e^{j\pi fk/b} \mathbb{1}\{-b \le f \le b\}$ might not agree at every point, they are $\mathcal{L}_2$ equivalent. Hence they have the same inverse Fourier transform which, as argued above, is $w(t)$. Thus

$$w(t) = \sum_k A_k 2b \,\text{sinc}\,(2bt + k) = \sum_k \frac{A_k}{T} \,\text{sinc}\left(\frac{t}{T} + k\right),$$

where $T = \frac{1}{2b}$.

We still need to determine $\frac{A_k}{T}$. It is straightforward to determine A_k from the definition of the Fourier series, but it is even easier to plug $t = nT$ in both sides of the above expression to obtain $w(nT) = \frac{A_{-n}}{T}$. This completes the proof. To see that we can easily obtain A_k from the definition (5.27), we write

$$A_k = \frac{1}{2b} \int_{-b}^{b} w_\mathcal{F}(f) e^{-j\pi kf/b} df = T \int_{-\infty}^{\infty} w_\mathcal{F}(f) e^{-j2\pi Tkf} df = Tw(-kT),$$

where the first equality is the definition of the Fourier coefficient A_k, the second uses the fact that $w_\mathcal{F}(f) = 0$ for $f \notin [-b, b]$, and the third is the inverse Fourier transform evaluated at $t = -kT$.

□

5.13 Appendix: A review of stochastic processes

In the discussion of Section 5.3, we assume familiarity with the following facts about stochastic processes.

Let $(\Omega, \mathcal{F}, P)$ be a *probability space*: Ω is the *sample space*, which consists of all possible outcomes of a random experiment; $\mathcal{F}$ is the set of *events*, which is the set of those subsets of Ω on which a probability is assigned; P is the *probability measure* that assigns a probability to every event. There are technical conditions that $\mathcal{F}$ and P must satisfy to ensure consistency. $\mathcal{F}$ must be a σ-*algebra*, i.e. it must satisfy the following conditions: it contains the empty set $\emptyset$; if $\mathcal{A} \in \mathcal{F}$, then $\mathcal{A}^c \in \mathcal{F}$; the union of every countable collection of sets of $\mathcal{F}$ must be in $\mathcal{F}$.

The probability measure is a function $P : \mathcal{F} \to [0, 1]$ with the properties that $P(\Omega) = 1$ and $P(\cup_{i=1}^{\infty} \mathcal{A}_i) = \sum_{i=1}^{\infty} P(\mathcal{A}_i)$ whenever $\mathcal{A}_1, \mathcal{A}_2, \ldots$ is a collection of disjoint events.

A *random variable* X, defined over a probability space $(\Omega, \mathcal{F}, P)$, is a function $X : \Omega \to \mathbb{R}$, such that for every $x \in \mathbb{R}$, the set $\{\omega \in \Omega : X(\omega) \leq x\}$ is contained in $\mathcal{F}$. This ensures that the *cumulative distribution function* $F_X(x) := P(\{\omega : X(\omega) \leq x\})$ is well defined for every $x \in \mathbb{R}$.

A *stochastic process* (also called *random process*) is a collection of random variables defined over the same probability space $(\Omega, \mathcal{F}, P)$. The stochastic process is *discrete-time* if the collection of random variables is indexed by $\mathbb{Z}$ or a subset thereof, such as in $\{X_i : i \in \mathbb{Z}\}$. It is *continuous-time* if the collection of random variables is indexed by $\mathbb{R}$ or a continuous subset thereof, such as in $\{X_t : t \in \mathbb{R}\}$. Engineering texts often use the short-hand notations $X(t)$ and $X[i]$ to denote continuous-time and discrete-time stochastic processes, respectively.

We continue this review, assuming that the process is a continuous-time process, trusting that the reader can make the necessary changes for discrete-time processes.

Let $t_1, t_2, \ldots, t_k$ be an arbitrary finite collection of indices. The technical condition about $\mathcal{F}$ mentioned above ensures that the cumulative distribution function

$$F_{X_{t_1}, \ldots, X_{t_k}}(x_1, \ldots, x_k) := P(\omega : X_{t_1}(\omega) \leq x_1, \ldots, X_{t_k}(\omega) \leq x_k)$$

is defined for all $(x_1, \ldots, x_k) \in \mathbb{R}^k$. In words, the statistic is defined for every finite collection $X_{t_1}, X_{t_2}, \ldots, X_{t_k}$ of samples of $\{X_t : t \in \mathbb{R}\}$.

The *mean* $m_X(t)$, the *autocorrelation* $R_X(s, t)$, and the *autocovariance* $K_X(s, t)$ of a continuous-time stochastic process $\{X_t : t \in \mathbb{R}\}$ are, respectively,

$$m_X(t) := \mathbb{E}[X_t] \tag{5.28}$$

$$R_X(s, t) := \mathbb{E}[X_s X_t^*] \tag{5.29}$$

$$K_X(s, t) := \mathbb{E}[(X_s - \mathbb{E}[X_s])(X_t - \mathbb{E}[X_t])^*] = R_X(s, t) - m_X(s)m_X^*(t), \tag{5.30}$$

where the "$*$" denotes complex conjugation and can be omitted for real-valued stochastic processes.[5] For a zero-mean process, which is usually the case in our applications, $K_X(s, t) = R_X(s, t)$.

---

[5] To remember that the "$*$" in (5.29) goes on the second random variable, it helps to observe the similarity between the definition of $R_X(s, t)$ and that of an inner product such as $\langle a(t), b(t) \rangle = \int a(t) b^*(t) dt$.

5.13. Appendix: A review of stochastic processes

A continuous-time stochastic process $\{X_t : t \in \mathbb{R}\}$ is said to be *stationary* if, for every finite collection of indices $t_1, t_2, \ldots, t_k$, the statistic of

$$X_{t_1+\tau}, X_{t_2+\tau}, \ldots, X_{t_k+\tau} \quad \text{and that of} \quad X_{t_1}, X_{t_2}, \ldots, X_{t_k}$$

are the same for all $\tau \in \mathbb{R}$, i.e. for every $(x_1, x_2, \ldots, x_k) \in \mathbb{R}^k$,

$$F_{X_{t_1+\tau}, X_{t_2+\tau}, \ldots, X_{t_k+\tau}}(x_1, x_2, \ldots, x_k)$$

is the same for all $\tau \in \mathbb{R}$.

EXAMPLE 5.14 *For a stationary process* $\{X_t : t \in \mathbb{R}\}$,

$$m_X(t) = \mathbb{E}[X_t] = \mathbb{E}[X_0] = m_X(0)$$
$$R_X(s,t) = \mathbb{E}[X_s X_t^*] = \mathbb{E}[X_{s-t} X_0^*] = R_X(s-t, 0). \quad \square$$

We see that a stationary stochastic process has a constant mean $m_X(t)$ and has an autocorrelation $R_X(s,t)$ that depends only on the difference $\tau = s - t$. This is a property that simplifies many results. A stochastic process that has this property is called *wide-sense stationary* (WSS). (An equivalent condition is that $m_X(t)$ does not depend on t and $K_X(s,t)$ depends only on $s - t$.) In this case we can use the short-hand notation $R_X(\tau)$ instead of $R_X(t+\tau, t)$, and $K_X(\tau)$ instead of $K_X(t+\tau, t)$. A stationary process is always WSS, but a WSS process is not necessarily stationary. For instance, the process $X(t)$ of Section 5.3 is WSS but not stationary.

The Fourier transform of the autocovariance $K_X(\tau)$ of a WSS process is the *power spectral density* (or simply *spectral density*) $S_X(f)$. To understand why this name makes sense, suppose for the moment that the process is zero-mean, and recall that the value of a function at the origin equals the integral of its Fourier transform (see (5.23)). Hence,

$$\int_{-\infty}^{\infty} S_X(f) df = K_X(0) = R_X(0) = \mathbb{E}\left[|X_t|^2\right].$$

If the process $\{X_t : t \in \mathbb{R}\}$ represents an electrical signal (voltage or current), then $\mathbb{E}\left[|X_t|^2\right]$ is associated with an average power (see the discussion in Example 3.2), and so is the integral of $S_X(f)$. This partially justifies calling $S_X(f)$ a power spectral density. For the full justification, we need to determine how the spectral density at the output of a linear time-invariant system depends on the spectral density of the input.

Towards this goal, let $X(t)$ be a zero-mean WSS process at the input of a linear time-invariant system of impulse response $h(t)$ and let $Y(t) = \int h(\alpha) X(t-\alpha) d\alpha$ be the output. It is straightforward to determine the mean of $Y(t)$:

$$m_Y(t) = \mathbb{E}\left[\int h(\alpha) X(t-\alpha) d\alpha\right] = \int h(\alpha) \mathbb{E}\left[X(t-\alpha)\right] d\alpha = 0.$$

With slightly more effort we determine

$$K_Y(t+\tau, t) = \int_{-\infty}^{\infty} \int_{-\infty}^{\infty} h(\alpha) h^*(\beta) K_X(\tau + \beta - \alpha) d\alpha d\beta.$$

We see that $Y(t)$ is a zero-mean WSS process. If we write $h(\beta)$ in terms of its Fourier transform $h(\beta) = \int_{-\infty}^{\infty} h_{\mathcal{F}}(f) e^{j2\pi f\beta} df$ and substitute into $K_Y(t+\tau, t)$, henceforth denoted by $K_Y(\tau)$, after a few standard manipulations we obtain

$$K_Y(\tau) = \int_{-\infty}^{\infty} |h_{\mathcal{F}}(f)|^2 S_X(f) e^{j2\pi f\tau} df.$$

This proves that $K_Y(\tau)$ is the inverse Fourier transform of $|h_{\mathcal{F}}(f)|^2 S_X(f)$. Hence,

$$S_Y(f) = |h_{\mathcal{F}}(f)|^2 S_X(f).$$

To understand the physical meaning of $S_X(f)$, we let $X(t)$ be the input to a filter that cuts off all the frequencies of $X(t)$, except for those contained in a small interval $\mathcal{I} = [f_c - \frac{\Delta f}{2}, f_c + \frac{\Delta f}{2}]$ of width Δf around f_c. Suppose that Δf is sufficiently small that $S_X(f)$ is constant over $\mathcal{I}$. Then

$$S_Y(f) = |h_{\mathcal{F}}(f)|^2 S_X(f) = S_X(f)\mathbb{1}\{f \in \mathcal{I}\}$$

and the power of $Y(t)$ is $\int S_Y(f)df = S_X(f_c)\Delta f$. We conclude that $S_X(f_c)\Delta f$ is associated with the power of $X(t)$ contained in a frequency interval of width Δf centered around f_c, which explains why $S_X(f)$ is called the power spectral density of $X(t)$.

$S_X(f)$ is well-defined and is called the power spectral density of $X(t)$ even when $X(t)$ is a WSS process with a *non-vanishing* mean $m_X(t)$, but in this case the integral of $S_X(f)$ is the power of the zero-mean process $\check{X}(t) = X(t) - m$, where $m = m_X(t)$. For this reason, one could argue that *spectral density* is a better name than *power spectral density*. This is not a big issue, because most processes of interest to us are indeed zero-mean. Alternatively, we could be tempted to define $S_X(f)$ as the Fourier transform of $R_X(\tau)$, thinking that in so doing, the integral of $S_X(f)$ is the average power of $X(t)$, even when $X(t)$ has a non-vanishing mean. But this would create a technical problem, because when $X(t)$ has a non-vanishing mean, $R_X(\tau) = K_X(\tau) + |m|^2$ is not an $\mathcal{L}_2$ function. (We have defined the Fourier transform only for $\mathcal{L}_2$ functions.)

5.14 Appendix: Root-raised-cosine impulse response

In this appendix, we derive the inverse Fourier transform of the root-raised-cosine pulse

$$\psi_{\mathcal{F}}(f) = \begin{cases} \sqrt{T}, & |f| \leq \frac{1-\beta}{2T} \\ \sqrt{T}\cos\left[\frac{\pi T}{2\beta}\left(|f| - \frac{1-\beta}{2T}\right)\right], & \frac{1-\beta}{2T} < |f| \leq \frac{1+\beta}{2T} \\ 0, & \text{otherwise.} \end{cases}$$

We write $\psi_{\mathcal{F}}(f) = a_{\mathcal{F}}(f) + b_{\mathcal{F}}(f)$ where $a_{\mathcal{F}}(f) = \sqrt{T}\mathbb{1}\{f \in [-\frac{1-\beta}{2T}, \frac{1-\beta}{2T}]\}$ is the central piece of the root-raised-cosine impulse response and $b_{\mathcal{F}}(f)$ accounts for the two root-raised-cosine edges.

5.15. Appendix: The picket fence "miracle"

The inverse Fourier transform of $a_{\mathcal{F}}(f)$ is

$$a(t) = \frac{\sqrt{T}}{\pi t} \sin\left(\frac{\pi(1-\beta)t}{T}\right).$$

Write $b_{\mathcal{F}}(f) = b_{\mathcal{F}}^-(f) + b_{\mathcal{F}}^+(f)$, where $b_{\mathcal{F}}^{\pm}(f) = b_{\mathcal{F}}(f)\mathbb{1}\{f \geq 0\}$. Let $c_{\mathcal{F}}(f) = b_{\mathcal{F}}^+(f + \frac{1}{2T})$. Specifically,

$$c_{\mathcal{F}}(f) = \sqrt{T}\cos\left[\frac{\pi T}{2\beta}\left(f + \frac{\beta}{2T}\right)\right]\mathbb{1}\left\{f \in \left[-\frac{\beta}{2T}, \frac{\beta}{2T}\right]\right\}.$$

The inverse Fourier transform of $c_{\mathcal{F}}(f)$ is

$$c(t) = \frac{\beta}{2\sqrt{T}}\left[e^{-j\frac{\pi}{4}}\operatorname{sinc}\left(\frac{t\beta}{T} - \frac{1}{4}\right) + e^{j\frac{\pi}{4}}\operatorname{sinc}\left(\frac{t\beta}{T} + \frac{1}{4}\right)\right].$$

Now we use the relationship $b(t) = 2\Re\{c(t)e^{j2\pi\frac{1}{2T}t}\}$ to obtain

$$b(t) = \frac{\beta}{\sqrt{T}}\left[\operatorname{sinc}\left(\frac{t\beta}{T} - \frac{1}{4}\right)\cos\left(\frac{\pi t}{T} - \frac{\pi}{4}\right) + \operatorname{sinc}\left(\frac{t\beta}{T} + \frac{1}{4}\right)\cos\left(\frac{\pi t}{T} + \frac{\pi}{4}\right)\right].$$

After some manipulations of $\psi(t) = a(t) + b(t)$ we obtain the desired expression

$$\psi(t) = \frac{4\beta}{\pi\sqrt{T}}\frac{\cos\left[(1+\beta)\pi\frac{t}{T}\right] + \frac{(1-\beta)\pi}{4\beta}\operatorname{sinc}\left[(1-\beta)\frac{t}{T}\right]}{1 - \left(4\beta\frac{t}{T}\right)^2}.$$

5.15 Appendix: The picket fence "miracle"

In this appendix, we "derive" the *picket fence miracle*, which is a useful tool for informal derivations related to sampling (see Example 5.15 and Exercises 5.1 and 5.8). The "derivation" is not rigorous in that we "handwave" over convergence issues.

Recall that $\delta(t)$ is the (generalized) function defined through its integral against a function $f(t)$ (assumed to be continuous at $t = 0$):

$$\int_{-\infty}^{\infty} f(t)\delta(t)dt = f(0).$$

It follows that $\int_{-\infty}^{\infty} \delta(t)dt = 1$ and that the Fourier transform of $\delta(t)$ is 1 for all $f \in \mathbb{R}$.

The *T-spaced picket fence* is the train of Dirac delta functions

$$\sum_{n=-\infty}^{\infty} \delta(t - nT).$$

The picket fence miracle refers to the fact that the Fourier transform of a picket fence is again a (scaled) picket fence. Specifically,

$$\mathcal{F}\left[\sum_{n=-\infty}^{\infty} \delta(t-nT)\right] = \frac{1}{T} \sum_{n=-\infty}^{\infty} \delta\left(f - \frac{n}{T}\right),$$

where $\mathcal{F}[\cdot]$ stands for the Fourier transform of the enclosed expression. The above relationship can be derived by expanding the periodic function $\sum_n \delta(t-nT)$ as a Fourier series, namely:

$$\sum_{n=-\infty}^{\infty} \delta(t-nT) = \frac{1}{T} \sum_{n=-\infty}^{\infty} e^{j2\pi tn/T}.$$

(The careful reader should wonder in which sense the above equality holds. We are indeed being informal here.)

Taking the Fourier transform on both sides yields

$$\mathcal{F}\left[\sum_{n=-\infty}^{\infty} \delta(t-nT)\right] = \frac{1}{T} \sum_{n=-\infty}^{\infty} \delta\left(f - \frac{n}{T}\right),$$

which is what we wanted to prove.

It is convenient to have a notation for the picket fence. Thus we define[6]

$$\mathrm{E}_T(x) = \sum_{n=-\infty}^{\infty} \delta(x-nT).$$

Using this notation, the relationship that we just proved can be written as

$$\mathcal{F}\left[\mathrm{E}_T(t)\right] = \frac{1}{T} \mathrm{E}_{\frac{1}{T}}(f).$$

The picket fence miracle is a practical tool in engineering and physics, but in the stated form it is not appropriate to obtain results that are mathematically rigorous. An example follows.

EXAMPLE 5.15 *We give an informal proof of the sampling theorem by using the picket fence miracle. Let $s(t)$ be such that $s_\mathcal{F}(f) = 0$ for $f \notin [-B, B]$ and let $T \leq \frac{1}{2B}$. We want to show that $s(t)$ can be reconstructed from the T-spaced samples $\{s(nT)\}_{n\in\mathbb{Z}}$. Define*

$$s|(t) = \sum_{n=-\infty}^{\infty} s(nT)\delta(t-nT).$$

(Note that $s|$ is just a name for the expression on the right-hand side of the equality.) Using the fact that $s(t)\delta(t-nT) = s(nT)\delta(t-nT)$, we can also write

$$s|(t) = s(t)\mathrm{E}_T(t).$$

---

[6] The choice of the letter E is suggested by the fact that it looks like a picket fence when rotated 90 degrees.

5.15. Appendix: The picket fence "miracle"

Taking the Fourier transform on both sides yields

$$\mathcal{F}[s|(t)] = s_\mathcal{F}(f) \star \left(\frac{1}{T}\mathrm{E}_{\frac{1}{T}}(f)\right) = \frac{1}{T}\sum_{n=-\infty}^{\infty} s_\mathcal{F}\left(f - \frac{n}{T}\right).$$

The relationship between $s_\mathcal{F}(f)$ and $\mathcal{F}[s|(t)]$ is depicted in Figure 5.16.

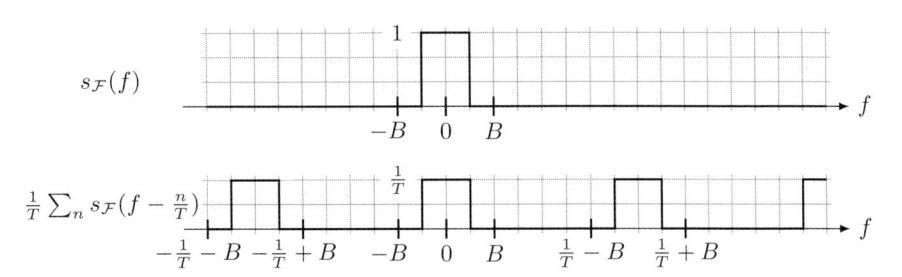

Figure 5.16. Fourier transform of a function $s(t)$ (top) and of $s|(t) = \sum_n s(nT)\delta(t - nT)$ (bottom).

From the figure, it is obvious that we can reconstruct the original signal $s(t)$ by filtering $s|(t)$ with a filter that scales $(1/T)s_\mathcal{F}(f)$ by T and blocks $(1/T)s_\mathcal{F}\left(f - \frac{n}{T}\right)$ for $n \neq 0$. Such a filter exists if, like in the figure, the support of $s_\mathcal{F}(f)$ does not intersect with the support of $s_\mathcal{F}\left(f - \frac{n}{T}\right)$ for $n \neq 0$. This is the case if $T \leq \frac{1}{2B}$. (We allow equality because the output of a filter is unchanged if the filter's input is modified at a countable number of points.) If $h(t)$ is the impulse response of such a filter, the filter output $y(t)$ when the input is $s|(t)$ satisfies

$$y_\mathcal{F}(f) = \left(\frac{1}{T}\sum_{n=-\infty}^{\infty} s_\mathcal{F}\left(f - \frac{n}{T}\right)\right) h_\mathcal{F}(f) = s_\mathcal{F}(f).$$

After taking the inverse Fourier transform, we obtain the reconstruction *(also called* interpolation*) formula*

$$y(t) = \left(\sum_{n=-\infty}^{\infty} s(nT)\delta(t - nT)\right) \star h(t) = \sum_{n=-\infty}^{\infty} s(nT)h(t - nT) = s(t).$$

A specific filter that has the desired properties is the lowpass filter of frequency response

$$h_\mathcal{F}(f) = \begin{cases} T, & f \in [-\frac{1}{2T}, \frac{1}{2T}] \\ 0, & \text{otherwise.} \end{cases}$$

Its impulse response is $\operatorname{sinc}(\frac{t}{T})$. *Inserting into the reconstruction formula yields*

$$s(t) = \sum_{n=-\infty}^{\infty} s(nT) \operatorname{sinc}\left(\frac{t}{T} - n\right), \qquad (5.31)$$

which matches (5.2). □

The picket fence miracle is useful for computing the spectrum of certain signals related to sampling. Examples are found in Exercises 5.1 and 5.8.

5.16 Exercises

Exercises for Section 5.2

EXERCISE 5.1 (Sampling and reconstruction) *Here we use the picket fence miracle to investigate practical ways to approximate sampling and/or reconstruction. We assume that for some positive B, $s(t)$ satisfies $s_\mathcal{F}(f) = 0$ for $f \notin [-B, B]$. Let T be such that $0 < T \leq 1/2B$.*

(a) *As a reference, review Example 5.15 of Appendix 5.15.*
(b) *To generate the intermediate signal $s(t)E_T(t)$ of Example 5.15, we need an electrical circuit that produces δ Diracs. Such a circuit does not exist. As a substitute for $\delta(t)$, we use a rectangular pulse of the form $\frac{1}{T_w}\mathbb{1}\{-\frac{T_w}{2} \leq t \leq \frac{T_w}{2}\}$, where $0 < T_w \leq T$ and the scaling by $\frac{1}{T_w}$ is to ensure that the integral over the substitute pulse and that over $\delta(t)$ give the same result, namely 1. The intermediate signal at the input of the reconstruction filter is then $[s(t)E_T(t)] \star [\frac{1}{T_w}\mathbb{1}\{-\frac{T_w}{2} \leq t \leq \frac{T_w}{2}\}]$. (We can generate this signal without passing through $E_T(t)$.) Express the Fourier transform $y_\mathcal{F}(f)$ of the reconstruction filter output.*
(c) *In the so-called* zero-order interpolator, *the reconstructed approximation is the step-wise signal $[s(t)E_T(t)] \star \mathbb{1}\{-\frac{T}{2} \leq t \leq \frac{T}{2}\}$. This is the intermediate signal of part (b) with $T_w = T$. Express its Fourier transform. Note: There is no interpolation filter in this case.*
(d) *In the* first-order interpolator, *the reconstructed approximation consists of straight lines connecting the values of the original signal at the sampling points. This can be written as $[s(t)E_T(t)] \star p(t)$ where $p(t)$ is the triangular shape waveform*

$$p(t) = \begin{cases} \frac{T-|t|}{T}, & t \in [-T, T] \\ 0, & \text{otherwise.} \end{cases}$$

Express the Fourier transform of the reconstructed approximation.

Compare $s_\mathcal{F}(f)$ to the Fourier transform of the various reconstructions you have obtained.

EXERCISE 5.2 (Sampling and projections) *We have seen that the reconstruction formula of the sampling theorem can be rewritten in such a way that it becomes an orthonormal expansion (expression (5.3)). If $\psi_j(t)$ is the jth element of an orthonormal set of functions used to expand $w(t)$, then the jth coefficient c_j equals the inner product $\langle w, \psi_j \rangle$. Explain why we do not need to explicitly perform an inner product to obtain the coefficients used in the reconstruction formula (5.3).*

5.16. Exercises

Exercises for Section 5.3

EXERCISE 5.3 (Properties of the self-similarity function) *Prove the following properties of the self-similarity function (5.5). Recall that the self-similarity function of an $\mathcal{L}_2$ pulse $\xi(t)$ is $R_\xi(\tau) = \int_{-\infty}^{\infty} \xi(t+\tau)\xi^*(t)dt$.*

(a) Value at zero:
$$R_\xi(\tau) \leq R_\xi(0) = \|\xi\|^2, \qquad \tau \in \mathbb{R}. \tag{5.32}$$

(b) Conjugate symmetry:
$$R_\xi(-\tau) = R_\xi^*(\tau), \qquad \tau \in \mathbb{R}. \tag{5.33}$$

(c) Convolution representation:
$$R_\xi(\tau) = \xi(\tau) \star \xi^*(-\tau), \qquad \tau \in \mathbb{R}. \tag{5.34}$$

Note: The convolution between $a(t)$ and $b(t)$ can be written as $(a \star b)(t)$ or as $a(t) \star b(t)$. Both versions are used in the literature. We prefer the first version, but in the above case the second version does not require the introduction of a name for $\xi^*(-\tau)$.

(d) Fourier relationship:
$$R_\xi(\tau) \text{ is the inverse Fourier transform of } |\xi_\mathcal{F}(f)|^2. \tag{5.35}$$

Note: The fact that $\xi_\mathcal{F}(f)$ is in $\mathcal{L}_2$ implies that $|\xi_\mathcal{F}(f)|^2$ is in $\mathcal{L}_1$. The Fourier inverse of an $\mathcal{L}_1$ function is continuous. Hence $R_\xi(\tau)$ is continuous.

EXERCISE 5.4 (Power spectrum: Manchester pulse) *Derive the power spectral density of the random process*

$$X(t) = \sum_{i=-\infty}^{\infty} X_i \psi(t - iT - \Theta),$$

where $\{X_i\}_{i=-\infty}^{\infty}$ is an iid sequence of uniformly distributed random variables taking values in $\{\pm\sqrt{\mathcal{E}}\}$, Θ is uniformly distributed in the interval $[0, T]$, and $\psi(t)$ is the so-called Manchester pulse shown in Figure 5.17. The Manchester pulse guarantees that $X(t)$ has at least one transition per symbol, which facilitates the clock recovery at the receiver.

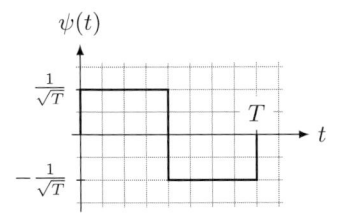

Figure 5.17.

Exercises for Section 5.4

EXERCISE 5.5 (Nyquist's criterion) *For each function $|\psi_\mathcal{F}(f)|^2$ in Figure 5.18, indicate whether the corresponding pulse $\psi(t)$ has unit norm and/or is orthogonal to its time-translates by multiples of T. The function in Figure 5.18d is $\mathrm{sinc}^2(fT)$.*

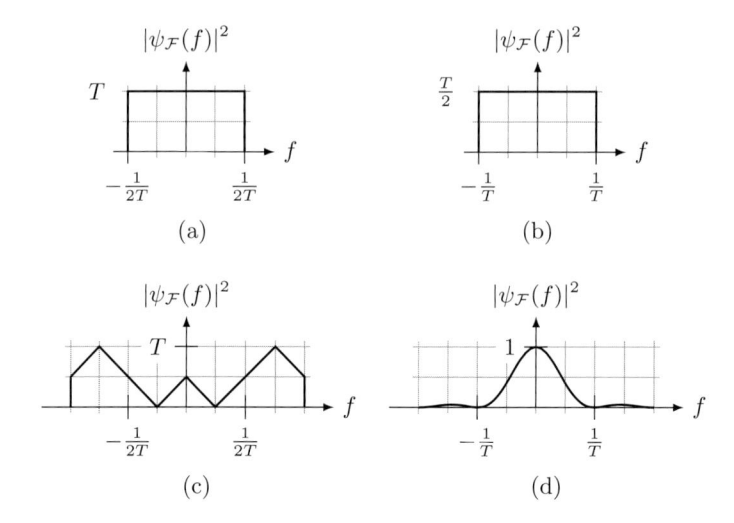

Figure 5.18.

EXERCISE 5.6 (Nyquist pulse) *A communication system uses signals of the form*

$$\sum_{l \in \mathbb{Z}} s_l p(t - lT),$$

where s_l takes values in some symbol alphabet and $p(t)$ is a finite-energy pulse. The transmitted signal is first filtered by a channel of impulse response $h(t)$ and then corrupted by additive white Gaussian noise of power spectral density $\frac{N_0}{2}$. The receiver front end is a filter of impulse response $q(t)$.

(a) Neglecting the noise, show that the front-end filter output has the form

$$y(t) = \sum_{l \in \mathbb{Z}} s_l g(t - lT),$$

where $g(t) = (p \star h \star q)(t)$ and $\star$ denotes convolution.

(b) The necessary and sufficient (time-domain) condition that $g(t)$ has to fulfill so that the samples of $y(t)$ satisfy $y(lT) = s_l$, $l \in \mathbb{Z}$, is

$$g(lT) = \mathbb{1}\{l = 0\}.$$

A function $g(t)$ that fulfills this condition is called a Nyquist pulse of parameter T. Prove the following theorem:

5.16. Exercises

THEOREM 5.16 (Nyquist criterion for Nyquist pulses) *The $\mathcal{L}_2$ pulse $g(t)$ is a Nyquist pulse (of parameter T) if and only if its Fourier transform $g_{\mathcal{F}}(f)$ fulfills Nyquist's criterion (with parameter T), i.e.*

$$\mathop{\mathrm{l.i.m.}}\sum_{l\in\mathbb{Z}} g_{\mathcal{F}}\left(f - \frac{l}{T}\right) = T, \qquad t \in \mathbb{R}.$$

Note: Because of the periodicity of the left-hand side, equality is fulfilled if and only if it is fulfilled over an interval of length $1/T$. Hint: Set $g(t) = \int g_{\mathcal{F}}(f)e^{\mathrm{j}2\pi ft}df$, insert on both sides $t = -lT$ and proceed as in the proof of Theorem 5.6.

(c) Prove Theorem 5.6 as a corollary to the above theorem. Hint: $\mathbb{1}\{l = 0\} = \int \psi(t-lT)\psi^*(t)dt$ if and only if the self-similarity function $R_\psi(\tau)$ is a Nyquist pulse of parameter T.

(d) Let $p(t)$ and $q(t)$ be real-valued with Fourier transform as shown in Figure 5.19, where only positive frequencies are plotted (both functions being even). The channel frequency response is $h_{\mathcal{F}}(f) = 1$. Determine $y(kT)$, $k \in \mathbb{Z}$.

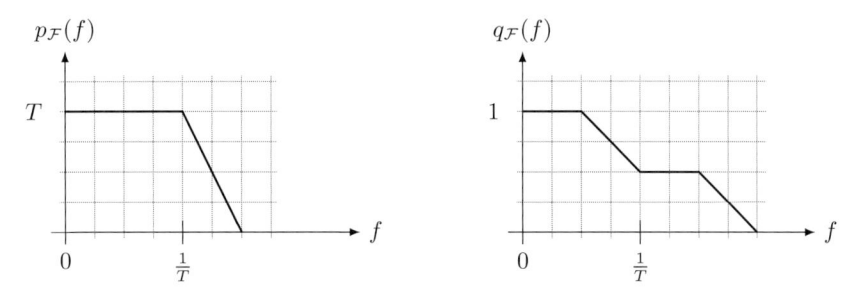

Figure 5.19.

EXERCISE 5.7 (Pulse orthogonal to its T-spaced time translates) *Figure 5.20 shows part of the plot of a function $|\psi_{\mathcal{F}}(f)|^2$, where $\psi_{\mathcal{F}}(f)$ is the Fourier transform of some pulse $\psi(t)$.*

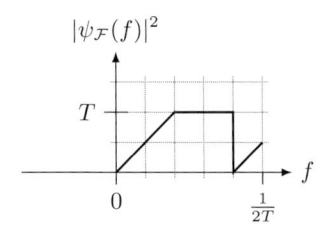

Figure 5.20.

Complete the plot (for positive and negative frequencies) and label the ordinate, knowing that the following conditions are satisfied:

- For every pair of integers k, l, $\int \psi(t - kT)\psi(t - lT)dt = \mathbb{1}\{k = l\}$;
- $\psi(t)$ is real-valued;
- $\psi_{\mathcal{F}}(f) = 0$ for $|f| > \frac{1}{T}$.

EXERCISE 5.8 (Nyquist criterion via picket fence miracle) *Give an informal proof of Theorem 5.6 (Nyquist criterion for orthonormal pulses) using the picket fence miracle (Appendix 5.15). Hint: A function p(t) is a Nyquist pulse of parameter T if and only if $p(t)E_T(t) = \delta(t)$.*

EXERCISE 5.9 (Peculiarity of Nyquist's criterion) *Let*

$$g_{\mathcal{F}}^{(0)}(f) = T\mathbb{1}\left\{-\frac{1}{3T} \leq f \leq \frac{1}{3T}\right\}$$

be the central rectangle in Figure 5.21, and for every positive integer n, let $g_{\mathcal{F}}^{(n)}(f)$ consist of $g_{\mathcal{F}}^{(0)}(f)$ plus 2n smaller rectangles of height $\frac{T}{2n}$ and width $\frac{1}{3T}$, each placed in the middle of an interval of the form $[\frac{l}{T}, \frac{l+1}{T}]$, $l = -n, -n+1, \ldots, n-1$. Figure 5.21 shows $g_{\mathcal{F}}^{(3)}(f)$.

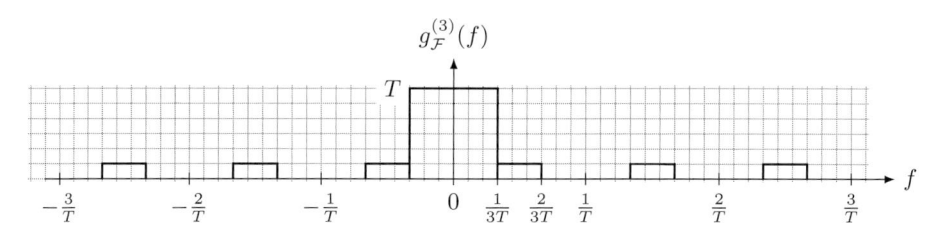

Figure 5.21.

(a) *Show that for every $n \geq 1$, $g_{\mathcal{F}}^{(n)}(f)$ fulfills Nyquist's criterion with parameter T. Hint: It is sufficient that you verify that Nyquist's criterion is fulfilled for $f \in [0, \frac{1}{T}]$. Towards that end, first check what happens to the central rectangle when you perform the operation $\sum_{l \in \mathbb{Z}} g_{\mathcal{F}}^{(n)}(f - \frac{l}{T})$. Then see how the small rectangles fill in the gaps.*
(b) *As n goes to infinity, $g_{\mathcal{F}}^{(n)}(f)$ converges to $g_{\mathcal{F}}^{(0)}(f)$. (It converges for every f and it converges also in $\mathcal{L}_2$, i.e. the squared norm of the difference $g_{\mathcal{F}}^{(n)}(f) - g_{\mathcal{F}}^{(0)}(f)$ goes to zero.) Peculiar is that the limiting function $g_{\mathcal{F}}^{(0)}(f)$ fulfills Nyquist's criterion with parameter $T^{(0)} \neq T$. What is $T^{(0)}$?*
(c) *Suppose that we use symbol-by-symbol on a pulse train to communicate across the AWGN channel. To do so, we choose a pulse $\psi(t)$ such that $|\psi_{\mathcal{F}}(f)|^2 = g_{\mathcal{F}}^{(n)}(f)$ for some n, and we choose n sufficiently large that $\frac{T}{2n}$ is much smaller than the noise power spectral density $\frac{N_0}{2}$. In this case, we can argue that our bandwidth B is only $\frac{1}{3T}$. This means a 30% bandwidth reduction with respect to the minimum absolute bandwidth $\frac{1}{2T}$. This reduction is non-negligible if we pay for the bandwidth we use. How do you explain that such a pulse is not used in practice? Hint: What do you expect $\psi(t)$ to look like?*
(d) *Construct a function $g_{\mathcal{F}}(f)$ that looks like Figure 5.21 in the interval shown by the figure except for the heights of the rectangles. Your function should have infinitely many smaller rectangles on each side of the central rectangle and*

5.16. Exercises

(like $g_\mathcal{F}^{(n)}(f)$) shall satisfy Nyquist's criterion. Hint: One such construction is suggested by the infinite geometric series $\sum_{i=1}^{\infty}(\frac{1}{2})^i$, which adds to 1.

Exercises for Section 5.5

EXERCISE 5.10 (Raised-cosine expression) Let T be a positive number. Following the steps below, derive the raised-cosine function $|\psi_\mathcal{F}(f)|^2$ of roll-off factor $\beta \in (0, 1]$. (It is recommended to plot the various functions.)

(a) Let $p(f) = \cos(f)$, defined over the domain $f \in [0, \pi]$, be the starting point for what will become the right-hand side roll-off edge.
(b) Find constants c and d so that $q(f) = cp(f) + d$ has range $[0, T]$ over the domain $[0, \pi]$.
(c) Find a constant e so that $r(f) = q(ef)$ has domain $[0, \frac{\beta}{T}]$.
(d) Find a constant g so that $s(f) = r(f - g)$ has domain $[\frac{1}{2T} - \frac{\beta}{2T}, \frac{1}{2T} + \frac{\beta}{2T}]$.
(e) Write an expression for the function $|\psi_\mathcal{F}(f)|^2$ that has the following properties:
   (i) it is T for $f \in [0, \frac{1}{2T} - \frac{\beta}{2T})$;
   (ii) it equals $s(f)$ for $f \in [\frac{1}{2T} - \frac{\beta}{2T}, \frac{1}{2T} + \frac{\beta}{2T}]$;
   (iii) it is 0 for $f \in (\frac{1}{2T} + \frac{\beta}{2T}, \infty)$;
   (iv) it is an even function.

Exercises for Section 5.6

EXERCISE 5.11 (Peculiarity of the sinc pulse) Let $\{U_k\}_{k=0}^n$ be an iid sequence of uniformly distributed bits taking value in $\{\pm 1\}$. Prove that for certain values of t and for n sufficiently large, $s(t) = \sum_{k=0}^n U_k \, \text{sinc}(t-k)$ can become larger than any given constant. Hint: The series $\sum_{k=1}^\infty \frac{1}{k}$ diverges and so does $\sum_{k=1}^\infty \frac{1}{k-a}$ for any constant $a \in (0, 1)$. Note: This implies that the eye diagram of $s(t)$ is closed.

Miscellaneous exercises

EXERCISE 5.12 (Matched filter basics) Let

$$w(t) = \sum_{k=1}^{K} d_k \, \psi(t - kT)$$

be a transmitted signal where $\psi(t)$ is a real-valued pulse that satisfies

$$\int_{-\infty}^{\infty} \psi(t)\psi(t - kT)dt = \begin{cases} 0, & k \text{ integer} \neq 0 \\ 1, & k = 0, \end{cases}$$

and $d_k \in \{-1, 1\}$.

(a) Suppose that $w(t)$ is filtered at the receiver by the matched filter with impulse response $\psi(-t)$. Show that the filter output $y(t)$ sampled at mT, m integer, yields $y(mT) = d_m$, for $1 \leq m \leq K$.

(b) Now suppose that the (noiseless) channel outputs the input plus a delayed and scaled replica of the input. That is, the channel output is $w(t) + \rho w(t - T)$ for some T and some $\rho \in [-1, 1]$. At the receiver, the channel output is filtered by $\psi(-t)$. The resulting waveform $\tilde{y}(t)$ is again sampled at multiples of T. Determine the samples $\tilde{y}(mT)$, for $1 \leq m \leq K$.

(c) Suppose that the kth received sample is $Y_k = d_k + \alpha d_{k-1} + Z_k$, where $Z_k \sim \mathcal{N}(0, \sigma^2)$ and $0 \leq \alpha < 1$ is a constant. Note that d_k and d_{k-1} are realizations of independent random variables that take on the values 1 and -1 with equal probability. Suppose that the receiver decides $\hat{d}_k = 1$ if $Y_k > 0$, and decides $\hat{d}_k = -1$ otherwise. Find the probability of error for this receiver.

EXERCISE 5.13 (Communication link design) *Specify the block diagram for a digital communication system that uses twisted copper wires to connect devices that are 5 km apart from each other. The cable has an attenuation of 16 dB/km. You are allowed to use the spectrum between -5 and 5 MHz. The noise at the receiver input is white and Gaussian, with power spectral density $N_0/2 = 4.2 \times 10^{-21}$ W/Hz. The required bit rate is 40 Mbps (megabits per second) and the bit-error probability should be less than 10^{-5}. Be sure to specify the symbol alphabet and the waveform former of the system you propose. Give precise values or bounds for the bandwidth used, the power of the channel input signal, the bit rate, and the error probability. Indicate which bandwidth definition you use.*

EXERCISE 5.14 (Differential encoding) *For many years, telephone companies built their networks on twisted pairs. This is a twisted pair of copper wires invented by Alexander Graham Bell in 1881 as a means to mitigate the effect of electromagnetic interference. In essence, an alternating magnetic field induces an electric field in a loop. This applies also to the loop created by two parallel wires connected at both ends. If the wire is twisted, the electric field components that build up along the wire alternate polarity and tend to cancel out one another. If we swap the two contacts at one end of the cable, the signal's polarity at one end is the opposite of that on the other end. Differential encoding is a technique for encoding the information in such a way that the decoding process is not affected by polarity. The differential encoder takes the data sequence $\{D_i\}_{i=1}^n$, here assumed to have independent and uniformly distributed components taking value in $\{0, 1\}$, and produces the symbol sequence $\{X_i\}_{i=1}^n$ according to the following encoding rule:*

$$X_i = \begin{cases} X_{i-1}, & D_i = 0, \\ -X_{i-1}, & D_i = 1, \end{cases}$$

where $X_0 = \sqrt{\mathcal{E}}$ by convention. Suppose that the symbol sequence is used to form

$$X(t) = \sum_{i=1}^n X_i \psi(t - iT),$$

where $\psi(t)$ is normalized and orthogonal to its T-spaced time-translates. The signal is sent over the AWGN channel of power spectral density $N_0/2$ and at the receiver

5.16. Exercises

is passed through the matched filter of impulse response $\psi^*(-t)$. Let Y_i be the filter output at time iT.

(a) Determine $R_X[k]$, $k \in \mathbb{Z}$, assuming an infinite sequence $\{X_i\}_{i=-\infty}^{\infty}$.
(b) Describe a method to estimate D_i from Y_i and Y_{i-1}, such that the performance is the same if the polarity of Y_i is inverted for all i. We ask for a simple decoder, not necessarily ML.
(c) Determine (or estimate) the error probability of your decoder.

EXERCISE 5.15 (Mixed questions)

(a) Consider the signal $x(t) = \cos(2\pi t)\left(\frac{\sin(\pi t)}{\pi t}\right)^2$. Assume that we sample $x(t)$ with sampling period T. What is the maximum T that guarantees signal recovery?
(b) You are given a pulse $p(t)$ with spectrum $p_{\mathcal{F}}(f) = \sqrt{T(1-|f|T)}$, $|f| \leq \frac{1}{T}$. What is the value of $\int p(t)p(t-3T)dt$?

EXERCISE 5.16 (Properties of the Fourier transform) Prove the following properties of the Fourier transform. The sign $\overset{\mathcal{F}}{\Longleftrightarrow}$ relates Fourier transform pairs, with the function on the right being the Fourier transform of that on the left. The Fourier transforms of $v(t)$ and $w(t)$ are denoted by $v_{\mathcal{F}}(f)$ and $w_{\mathcal{F}}(f)$, respectively.

(a) Linearity:
$$\alpha v(t) + \beta w(t) \overset{\mathcal{F}}{\Longleftrightarrow} \alpha v_{\mathcal{F}}(f) + \beta w_{\mathcal{F}}(f).$$

(b) Time-shifting:
$$v(t-t_0) \overset{\mathcal{F}}{\Longleftrightarrow} v_{\mathcal{F}}(f)e^{-j2\pi f t_0}.$$

(c) Frequency-shifting:
$$v(t)e^{j2\pi f_0 t} \overset{\mathcal{F}}{\Longleftrightarrow} v_{\mathcal{F}}(f-f_0).$$

(d) Convolution in time:
$$(v \star w)(t) \overset{\mathcal{F}}{\Longleftrightarrow} v_{\mathcal{F}}(f)w_{\mathcal{F}}(f).$$

(e) Time scaling by $\alpha \neq 0$:
$$v(\alpha t) \overset{\mathcal{F}}{\Longleftrightarrow} \frac{1}{|\alpha|}v_{\mathcal{F}}\left(\frac{f}{\alpha}\right).$$

(f) Conjugation:
$$v^*(t) \overset{\mathcal{F}}{\Longleftrightarrow} v_{\mathcal{F}}^*(-f).$$

(g) Time-frequency duality:
$$v_{\mathcal{F}}(t) \overset{\mathcal{F}}{\Longleftrightarrow} v(-f).$$

(h) Parseval's relationship:

$$\int v(t)w^*(t)dt = \int v_{\mathcal{F}}(f)w_{\mathcal{F}}^*(f)df.$$

Note: As a mnemonic, notice that the above can be written as $\langle v, w \rangle = \langle v_{\mathcal{F}}, w_{\mathcal{F}} \rangle$.

(i) Correlation:

$$\int v(\lambda + t)w^*(\lambda)d\lambda \quad \overset{\mathcal{F}}{\Longleftrightarrow} \quad v_{\mathcal{F}}(f)w_{\mathcal{F}}^*(f).$$

Hint: Use Parseval's relationship on the expression on the right and interpret the result.

6 Convolutional coding and Viterbi decoding: First layer revisited

6.1 Introduction

In this chapter we shift focus to the encoder/decoder pair. The general setup is that of Figure 6.1, where $N(t)$ is white Gaussian noise of power spectral density $N_0/2$. The details of the waveform former and the n-tuple former are immaterial for this chapter. The important fact is that the channel model from the encoder output to the decoder input is the discrete-time AWGN channel of noise variance $\sigma^2 = N_0/2$.

The study of encoding/decoding methods has been an active research area since the second half of the twentieth century. It is called coding theory. There are many coding techniques, and a general introduction to coding can easily occupy a one-semester graduate-level course. Here we will just consider an example of a technique called convolutional coding. By considering a specific example, we can considerably simplify the notation. As seen in the exercises, applying the techniques learned in this chapter to other convolutional encoders is fairly straightforward. We choose convolutional coding for two reasons: (i) it is well suited in conjunction with the discrete-time AWGN channel; (ii) it allows us to introduce various instructive and useful tools, notably the Viterbi algorithm to do maximum likelihood decoding and generating functions to upper bound the bit-error probability.

6.2 The encoder

The encoder is the device that takes the message and produces the codeword. In this chapter the message consists of a sequence $(b_1, b_2, \ldots, b_k)$ of binary *source symbols*. For comparison with bit-by-bit on a pulse train, we let the codewords consist of symbols that take value in $\{\pm\sqrt{\mathcal{E}_s}\}$. To simplify the description of the encoder, we let the source symbols take value in $\{\pm 1\}$ (rather than in $\{0, 1\}$). For the same reason, we factor out $\sqrt{\mathcal{E}_s}$ from the encoder output. Hence we declare the encoder output to be the sequence $(x_1, x_2, \ldots, x_n)$ with components in $\{\pm 1\}$ and let the codeword be $\sqrt{\mathcal{E}}(x_1, \ldots, x_n)$.

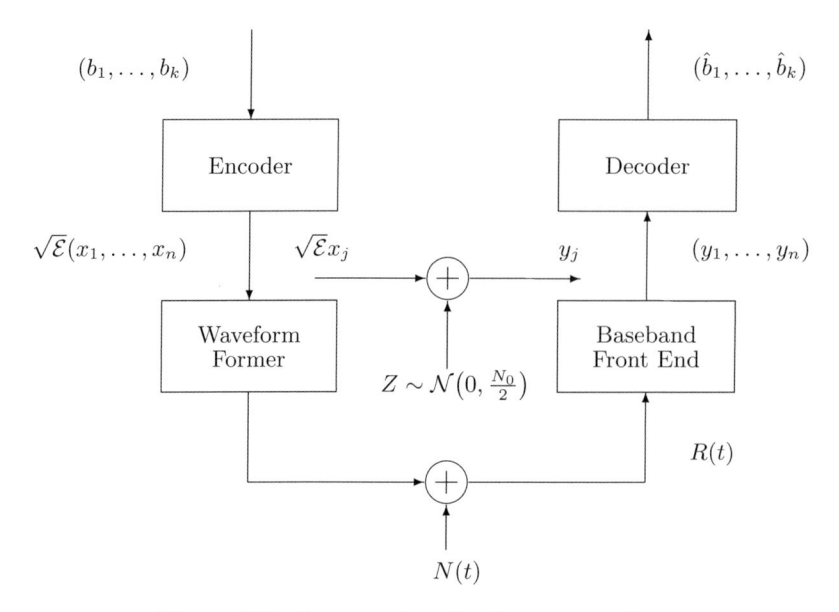

Figure 6.1. System view for the current chapter.

The source symbols enter the encoder sequentially, at regular intervals determined by the encoder clock. During the jth epoch, $j = 1, 2, \ldots$, the encoder takes b_j and produces two output symbols, x_{2j-1} and x_{2j}, according to the *encoding map*[1]

$$x_{2j-1} = b_j b_{j-2}$$
$$x_{2j} = b_j b_{j-1} b_{j-2}.$$

To produce x_1 and x_2 the encoder needs b_{-1} and b_0, which are assumed to be 1 by default.

The circuit that implements the convolutional encoder is depicted in Figure 6.2, where "×" denotes multiplication in $\mathbb{R}$. A shift register stores the past two inputs. As implied by the indices of x_{2j-1}, x_{2j}, the two encoder outputs produced during an epoch are transmitted sequentially.

Notice that the encoder output has length $n = 2k$. The following is an example of a source sequence of length $k = 5$ and the corresponding encoder output sequence of length $n = 10$.

b_j	1	−1	−1	1	1
x_{2j-1}, x_{2j}	1,1	−1,−1	−1,1	−1,1	−1,−1
j	1	2	3	4	5

---

[1] We are choosing this particular encoding map because it is the simplest one that is actually used in practice.

6.2. The encoder

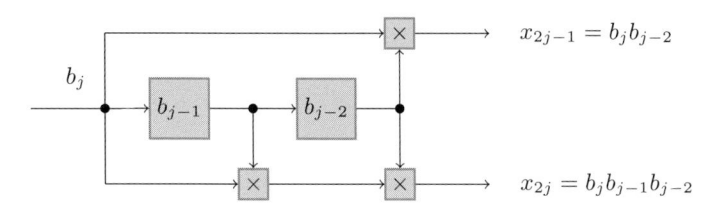

Figure 6.2. Convolutional encoder.

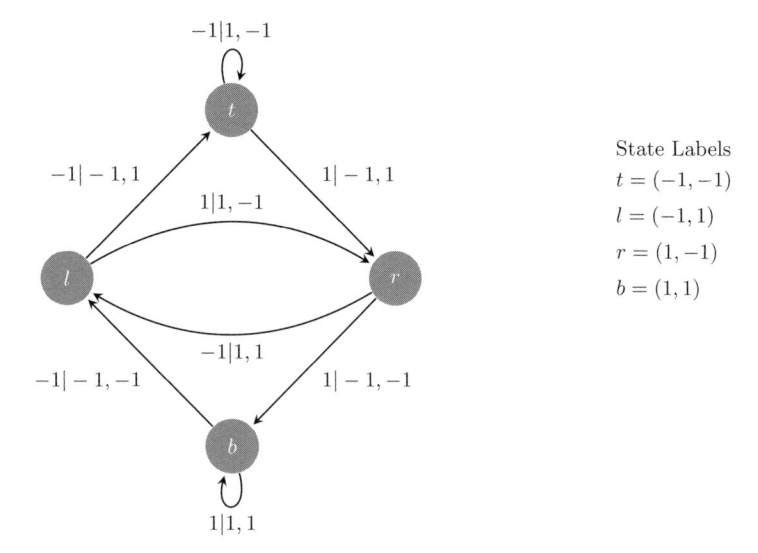

Figure 6.3. State diagram description of the convolutional encoder.

State Labels
$t = (-1, -1)$
$l = (-1, 1)$
$r = (1, -1)$
$b = (1, 1)$

Because the $n = 2k$ encoder output symbols are determined by the k input bits, only 2^k of the 2^n sequences of $\{\pm\sqrt{\mathcal{E}_s}\}^n$ are codewords. Hence we use only a fraction $2^k/2^n = 2^{-k}$ of all possible n-length channel input sequences. Compared with bit-by-bit on a pulse train, we are giving up a factor two in the bit rate to make the signal space much less crowded, hoping that this significantly reduces the probability of error.

We have already seen two ways to describe the encoder (the *encoding map* and the *encoding circuit*). A third way, useful in determining the error probability, is the *state diagram* of Figure 6.3. The diagram describes a *finite state machine*. The state of the convolutional encoder is what the encoder needs to know about past inputs so that the state and the current input determine the current output. For the convolutional encoder of Figure 6.2 the state at time j can be defined to be (b_{j-1}, b_{j-2}). Hence we have four states.

As the diagram shows, there are two possible transitions from each state. The input symbol b_j decides which of the two transitions is taken during epoch j. Transitions are labeled by $b_j | x_{2j-1}, x_{2j}$. To be consistent with the default $b_{-1} = b_0 = 1$, the state is $(1, 1)$ when b_1 enters the encoder.

The choice of letting the encoder input and output symbols be the elements of $\{\pm 1\}$ is not standard. Most authors choose the input/output alphabet to be $\{0, 1\}$ and use addition modulo 2 instead of multiplication over $\mathbb{R}$. In this case, a memoryless mapping at the encoder output transforms the symbol alphabet from $\{0, 1\}$ to $\{\pm\sqrt{\mathcal{E}_s}\}$. The notation is different but the end result is the same. The choice we have made is better suited for the AWGN channel. The drawback of this choice is that it is less evident that the encoder is linear. In Exercise 6.12 we establish the link between the two viewpoints and in Exercise 6.5 we prove from first principles that the encoder is indeed linear.

In each epoch, the convolutional encoder we have chosen has $k_0 = 1$ symbol entering and $n_0 = 2$ symbols exiting the encoder. In general, a convolutional encoder is specified by (i) the number k_0 of source symbols entering the encoder in each epoch; (ii) the number n_0 of symbols produced by the encoder in each epoch, where $n_0 > k_0$; (iii) the constraint length m_0 defined as the number of input k_0-tuples used to determine an output n_0-tuple; and (iv) the encoding function, specified for instance by a $k_0 \times m_0$ matrix of 1s and 0s for each component of the output n_0-tuple. The matrix associated to an output component specifies which inputs are multiplied to obtain that output. In our example, $k_0 = 1$, $n_0 = 2$, $m_0 = 3$, and the encoding function is specified by $[1, 0, 1]$ (for the first component of the output) and $[1, 1, 1]$ (for the second component). (See the connections that determine the top and bottom output in Figure 6.2.) In our case, the elements of the output n_0 tuple are serialized into a single sequence that we consider to be the actual encoder output, but there are other possibilities. For instance, we could take the pair x_{2j-1}, x_{2j} and map it into an element of a 4-PAM constellation.

6.3 The decoder

A maximum likelihood (ML) decoder for the discrete-time AWGN channel decides for (one of) the encoder output sequence x that maximizes

$$\langle c, y \rangle - \frac{\|c\|^2}{2},$$

where y is the channel output sequence and $c = \sqrt{\mathcal{E}_s}x$ is the codeword associated to x. The last term in the above expression is irrelevant as it is $n\mathcal{E}_s/2$, thus the same for all codewords. Furthermore, finding an x that maximizes $\langle c, y \rangle = \langle \sqrt{\mathcal{E}_s}x, y \rangle$ is the same as finding an x that maximizes $\langle x, y \rangle$.

Up to this point the inner product and the norm have been defined for vectors of $\mathbb{C}^n$ written in column form, with n being an arbitrary but fixed positive integer. Considering n-tuples in column form is a standard mathematical practice when matrix operations are involved. (We have used matrix notation to express the density of Gaussian random vectors.) In coding theory, people find it more useful to write n-tuples in row form because it saves space. Hence, we refer to the encoder input and output as *sequences* rather than as k and n-tuples, respectively. This is a minor point. What matters is that the inner products and norms of the previous paragraph are well defined.

6.3. The decoder

To find an x that maximizes $\langle x, y \rangle$, we could in principle compute $\langle x, y \rangle$ for all 2^k sequences that can be produced by the encoder. This *brute-force* approach would be quite unpractical. As already mentioned, if $k = 100$ (which is a relatively modest value for k), $2^k = (2^{10})^{10}$ which is approximately $(10^3)^{10} = 10^{30}$. A VLSI chip that makes 10^9 inner products per second takes 10^{21} seconds to check all possibilities. This is roughly 4×10^{13} years. The universe is "only" roughly 2×10^{10} years old!

We wish for a method that finds a maximizing x with a number of operations that grows linearly (as opposed to exponentially) in k. We will see that the so-called *Viterbi algorithm* achieves this.

To describe the Viterbi algorithm (VA), we introduce a fourth way of describing a convolutional encoder, namely the *trellis*. The trellis is an unfolded transition diagram that keeps track of the passage of time. For our example, if we assume that we start at state $(1, 1)$, that the source sequence is $b_1, b_2, \ldots, b_5$, and that we complete the transmission by feeding the encoder with two "dummy bits" $b_6 = b_7 = 1$ that make the encoder stop in the initial state, we obtain the trellis description shown on the top of Figure 6.4, where an edge (transition) from a state at depth j to a state at depth $j+1$ is labeled with the corresponding encoder output x_{2j-1}, x_{2j}. The encoder input that corresponds to an edge is the first component of the next state.

There is a one-to-one correspondence between an encoder input sequence b, an encoder output sequence x, and a path (or state sequence) that starts at the initial state $(1, 1)$ (left state) and ends at the final state $(1, 1)$ (right state) of the trellis. Hence we can refer to a path by means of an input sequence, an output sequence or a sequence of states.

To decode using the Viterbi algorithm, we replace the label of each edge with the edge metric (also called branch metric) computed as follows. The edge with $x_{2j-1} = a$ and $x_{2j} = b$, where $a, b \in \{\pm 1\}$, is assigned the edge metric $ay_{2j-1} + by_{2j}$. Now if we add up all the edge metrics along a path, we obtain the *path metric* $\langle x, y \rangle$.

EXAMPLE 6.1 *Consider the trellis on the top of Figure 6.4 and let the decoder input sequence be $y = (1, 3), (-2, 1), (4, -1), (5, 5), (-3, -3), (1, -6), (2, -4)$. For convenience, we chose the components of y to be integers, but in reality they are real-valued. Also for convenience, we use parentheses to group the components of y into pairs (y_{2j-1}, y_{2j}) that belong to the same trellis section. The edge metrics are shown on the second trellis (from the top) of Figure 6.4. Once again, by adding the edge metric along a path, we obtain the path metric $\langle x, y \rangle$, where x is the encoder output associated to the path.* □

The problem of finding an x that maximizes $\langle x, y \rangle$ is reduced to the problem of finding a path with the largest path metric. The next example illustrates how the Viterbi algorithm finds such a path.

EXAMPLE 6.2 *Our starting point is the second trellis of Figure 6.4, which has been labeled with the edge metrics. We construct the third trellis in which every state is labeled with the metric of the surviving path to that state obtained as follows. We use $j = 0, 1, \ldots, k+2$ to run over the trellis depth. Depth $j = 0$ refers*

210   6. First layer revisited

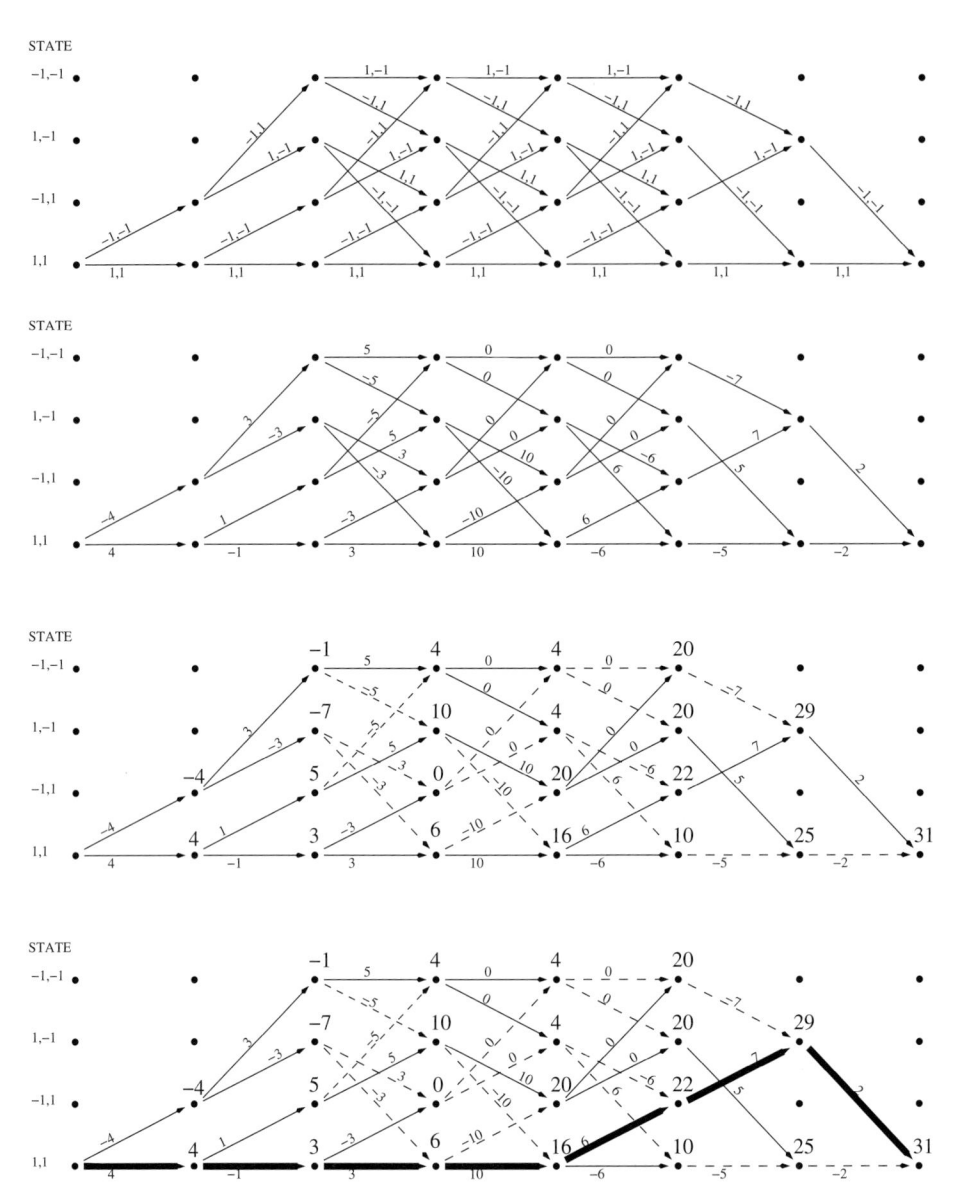

Figure 6.4. The Viterbi algorithm. Top figure: Trellis representing the encoder where edges are labeled with the corresponding output symbols. Second figure: Edges are re-labeled with the edge metric corresponding to the received sequence $(1,3), (-2,1), (4,-1), (5,5), (-3,-3), (1,-6), (2,-4)$ (parentheses have been inserted to facilitate parsing). Third figure: Each state has been labeled with the metric of a survivor to that state and non-surviving edges are pruned (dashed). Fourth figure: Tracing back from the end, we find the decoded path (bold); it corresponds to the source sequence $1, 1, 1, 1, -1, 1, 1$.

6.4. Bit-error probability

to the initial state (leftmost) and depth $j = k + 2$ to the final state (rightmost) after sending the k bits and the 2 "dummy bits". Let $j = 0$ and to the single state at depth j assign the metric 0. Let $j = 1$ and label each of the two states at depth j with the metric of the only subpath to that state. (See the third trellis from the top.) Let $j = 2$ and label the four states at depth j with the metric of the only subpath to that state. For instance, the label to the state $(-1, -1)$ at depth $j = 2$ is obtained by adding the metric of the single state and the single edge that precedes it, namely $-1 = -4 + 3$. From $j = 3$ on the situation is more interesting, because now every state can be reached from two previous states. We label the state under consideration with the largest of the two subpath metrics to that state and make sure to remember to which of the two subpaths it corresponds. In the figure, we make this distinction by dashing the last edge of the other path. (If we were doing this by hand we would not need a third trellis. Rather we would label the states on the second trellis and cross out the edges that are dashed in the third trellis.) The subpath with the highest edge metric (the one that has not been dashed) is called survivor. We continue similarly for $j = 4, 5, \ldots, k + 2$. At depth $j = k + 2$ there is only one state and its label maximizes $\langle x, y \rangle$ over all paths. By tracing back along the non-dashed path, we find the maximum likelihood path. From it, we can read out the corresponding bit sequence. The maximum likelihood path is shown in bold on the fourth and last trellis of Figure 6.4. □

From the above example, it is clear that, starting from the left and working its way to the right, the Viterbi algorithm visits all states and keeps track of the subpath that has the largest metric to that state. In particular, the algorithm finds the path that has the largest metric between the initial state and the final state.

The complexity of the Viterbi algorithm is linear in the number of trellis sections, i.e. in k. Recall that the brute-force approach has complexity exponential in k. The saving of the Viterbi algorithm comes from not having to compute the metric of non-survivors. When we dash an edge at depth j, we are in fact eliminating 2^{k-j} possible extensions of that edge. The brute-force approach computes the metric of all those extensions but not the Viterbi algorithm.

A formal definition of the VA (one that can be programmed on a computer) and a more formal argument that it finds the path that maximizes $\langle x, y \rangle$ is given in the Appendix (Section 6.6).

6.4 Bit-error probability

In this section we derive an upper bound to the bit-error probability P_b. As usual, we fix a signal and we evaluate the error probability conditioned on this signal being transmitted. If the result depends on the chosen signal (which is not the case here), then we remove the conditioning by averaging over all signals.

Each signal that can be produced by the transmitter corresponds to a path in the trellis. The path we condition on is referred to as the *reference* path. We are free to choose the reference path and, for notational convenience, we choose the *all-one* path: it is the one that corresponds to the information sequence being

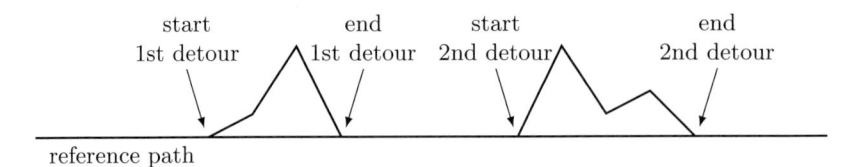

Figure 6.5. Detours.

a sequence of k ones with initial encoder state $(1,1)$. The encoder output is a sequence of 1s of length $n = 2k$.

The task of the decoder is to find (one of) the paths in the trellis that has the largest $\langle x, y \rangle$, where x is the encoder output that corresponds to that path. The encoder input b that corresponds to this x is the maximum likelihood message chosen by the decoder.

The concept of a *detour* plays a key role in upper-bounding the bit-error probability. We start with an analogy. Think of the trellis as a road map, of the path followed by the encoder as the itinerary you have planned for a journey, and of the decoded path as the actual route you follow on your journey. Typically the itinerary and the actual route differ due to constructions that force you to take detours. Similarly, the detours taken by the Viterbi decoder are those segments of the decoded path that share with the reference path only their initial and final state. Figure 6.5 illustrates a reference path and two detours.

Errors are produced when the decoder follows a detour. To the trellis path selected by the decoder, we associate a sequence $\omega_0, \omega_1, \ldots, \omega_{k-1}$ defined as follows. If there is a detour that *starts* at depth j, $j = 0, 1, \ldots, k-1$, we let ω_j be the number of bit errors produced by *that* detour. It is determined by comparing the corresponding segments of the two encoder input sequences and by letting ω_j be the number of positions in which they differ. If depth j does *not* correspond to the start of a detour, then $\omega_j = 0$. Now $\sum_{j=0}^{k-1} \omega_j$ is the number of bits that are incorrectly decoded and $\frac{1}{kk_0} \sum_{j=0}^{k-1} \omega_j$ is the fraction of such bits ($k_0=1$ in our running example; see Section 6.2).

EXAMPLE 6.3 *Consider the example of Figure 6.4 where $k = 5$ bits are transmitted (followed by the two dummy bits $1,1$). The reference path is the all-one path and the decoded path is the one marked by the solid line on the bottom trellis. There is one detour, which starts at depth 4 in the trellis. Hence $\omega_j = 0$ for $j = 0, 1, 2, 3$ whereas $\omega_4 \neq 0$. To determine the value of ω_4, we need to compare the encoder input bits over the span of the detour. The input bits that correspond to the detour are $-1, 1, 1$ and those that correspond to the reference path are $1, 1, 1$. There is one disagreement, hence $\omega_4 = 1$. The fraction of bits that are decoded incorrectly is $\frac{1}{k} \sum_{j=0}^{k-1} \omega_j = 1/5$.* □

Over the ensemble of all possible noise processes, ω_j is modeled by a random variable Ω_j and the bit-error probability is

6.4. Bit-error probability

$$P_b \triangleq \mathbb{E}\left[\frac{1}{kk_0}\sum_{j=0}^{k-1}\Omega_j\right] = \frac{1}{kk_0}\sum_{j=0}^{k-1}\mathbb{E}\left[\Omega_j\right].$$

To upper bound the above expression, we need to learn how many detours of a certain kind there are. We do so in the next section.

6.4.1 Counting detours

In this subsection, we consider the infinite trellis obtained by extending the finite trellis in both directions. Each path of the infinite trellis corresponds to an infinite encoder input sequence $b = \ldots b_{-1}, b_0, b_1, b_2, \ldots$ and an infinite encoder output sequence $x = \ldots x_{-1}, x_0, x_1, x_2, \ldots$. These are sequences that belong to $\{\pm 1\}^\infty$.

Given any two paths in the trellis, we can take one as the reference and consider the other as consisting of a number of detours with respect to the reference. To each of the two paths there corresponds an encoder input and an encoder output sequence. For every detour we can compare the two segments of encoder output sequences and count the number of positions in which they differ. We denote this number by d and call it the output distance (over the span of the detour). Similarly, we can compare the segments of encoder input sequences and call input distance (over the span of the detour) the number i of positions in which they differ.

EXAMPLE 6.4   *Consider again the example of Figure 6.4 and let us choose the all-one path as the reference. Consider the detour that starts at depth $j = 0$ and ends at $j = 3$. From the top trellis, comparing labels, we see that $d = 5$. (There are two disagreements in the first section of the trellis, one in the second, and two in the third.) To determine the input distance i we need to label the transitions with the corresponding encoder input. If we do so and compare we see that $i = 1$. As another example, consider the detour that starts at depth $j = 0$ and ends at $j = 4$. For this detour, $d = 6$ and $i = 2$.*   □

We seek the answer to the following question: For any given reference path and depth $j \in \{0, 1, \ldots\}$, what is the number $a(i, d)$ of detours that start at depth j and have input distance i and output distance d, with respect to the reference path? This number depends neither on j nor on the reference path. It does not depend on j because the encoder is a time-invariant machine, i.e. all the sections of the infinite trellis are identical. (This is the reason why we are considering the infinite trellis in this section.) We will see that it does not depend on the reference path either, because the encoder is linear in a sense that we will discuss.

EXAMPLE 6.5   *Using the top trellis of Figure 6.4 with the all-one path as the reference and $j = 0$, we can verify by inspection that there are two detours that have output distance $d = 6$. One ends at $j = 4$ and the other ends at $j = 5$. The input distance is $i = 2$ in both cases. Because there are two detours with parameters $d = 6$ and $i = 2$, $a(2, 6) = 2$.*   □

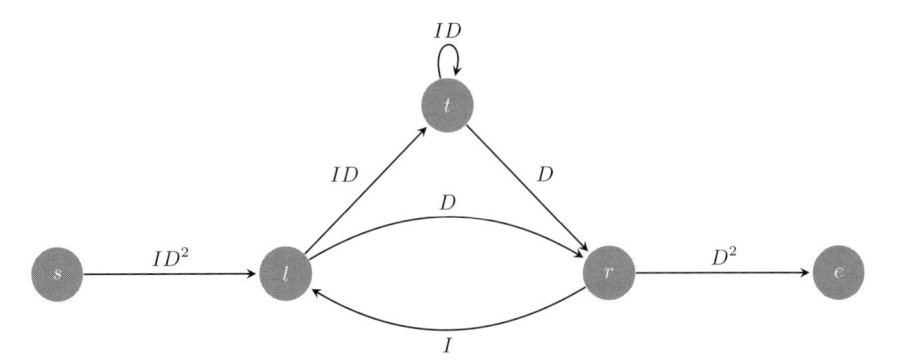

Figure 6.6. Detour flow graph.

To determine $a(i,d)$ in a systematic way, we arbitrarily choose the *all-one* path as the reference and modify the state diagram into a diagram that has a start and an end and for which each path from the start to the end represents a detour with respect to the reference path. This is the *detour flow graph* shown in Figure 6.6. It is obtained from the state diagram by removing the self-loop of state $(1,1)$ and by split opening the state to create two new states denoted by s (for start) and e (for end). For every j, there is a one-to-one correspondence between the set of detours to the all-one path that starts at depth j and the set of paths between state s and state e of the detour flow graph.

The label $I^i D^d$ (where i and d are non-negative integers) on an edge of the detour flow graph indicates that the input and output distances (with respect to the reference path) increase by i and d, respectively, when the detour takes this edge.

In terms of the detour flow graph, $a(i,d)$ is the number of paths between s and e that have path label $I^i D^d$, where the label of a path is the product of all labels along that path.

EXAMPLE 6.6 *In Figure 6.6, the shortest path that connects s to e has length 3. It consists of the edges labeled ID^2, D, and D^2, respectively. The product of these labels is the path label ID^5. This path tells us that there is a detour with $i = 1$ (the exponent of I) and $d = 5$ (the exponent of D). There is no other path with path label ID^5. Hence, as we knew already, $a(1,5) = 1$.* □

Our next goal is to determine the *generating function* $T(I,D)$ of $a(i,d)$ defined as

$$T(I,D) = \sum_{i,d} I^i D^d a(i,d).$$

The letters I and D in the above expression should be seen as "place holders" without any physical meaning. It is like describing a set of coefficients $a_0, a_1, \ldots, a_{n-1}$ by means of the polynomial $p(x) = a_0 + a_1 x + \cdots + a_{n-1} x^{n-1}$. To determine $T(I,D)$, we introduce auxiliary generating functions, one for each intermediate state of the detour flow graph, namely

6.4. Bit-error probability

$$T_l(I,D) = \sum_{i,d} I^i D^d a_l(i,d),$$

$$T_t(I,D) = \sum_{i,d} I^i D^d a_t(i,d),$$

$$T_r(I,D) = \sum_{i,d} I^i D^d a_r(i,d),$$

$$T_e(I,D) = \sum_{i,d} I^i D^d a_e(i,d),$$

where in the first line we define $a_l(i,d)$ as the number of paths in the detour flow graph that start at state s, end at state l, and have path label $I^i D^d$. Similarly, for $x = t, r, e$, $a_x(i,d)$ is the number of paths in the detour flow graph that start at state s, end at state x, and have path label $I^i D^d$. Notice that $T_e(I,D)$ is indeed the $T(I,D)$ of interest to us.

From the detour flow graph, we see that the various generating functions are related as follows, where to simplify notation we drop the two arguments (I and D) of the generating functions:

$$T_l = ID^2 + T_r I$$
$$T_t = T_l ID + T_t ID$$
$$T_r = T_l D + T_t D$$
$$T_e = T_r D^2.$$

To write down the above equations, the reader might find it useful to apply the following rule. *The T_x of a state x is the sum of a product: the sum is over all states y that have an edge into x and the product is T_y times the label on the edge from y to x.* The reader can verify that this rule applies to all of the above equations except the first. When used in an attempt to find the first equation, it yields $T_l = T_s ID^2 + T_r I$, but T_s is not defined because there is no detour starting at s and ending at s. If we define $T_s = 1$ by convention, the rule applies without exception.

The above system can be solved for T_e (hence for T) by pure formal manipulations, like solving a system of equations. The result is

$$T(I,D) = \frac{ID^5}{1 - 2ID}.$$

As we will see shortly, the generating function $T(I,D)$ of $a(i,d)$ is more useful than $a(i,d)$ itself. However, to show that we can indeed obtain $a(i,d)$ from $T(I,D)$ we use the expansion[2] $\frac{1}{1-x} = 1 + x + x^2 + x^3 + \cdots$ to write

---

[2] We do not need to worry about convergence issues at this stage, because for now, x^i is just a "place holder". In other words, we are not adding up the powers of x for some number x.

$$T(I,D) = \frac{ID^5}{1-2ID} = ID^5(1+2ID+(2ID)^2+(2ID)^3+\cdots$$
$$= ID^5 + 2I^2D^6 + 2^2I^3D^7 + 2^3I^4D^8 + \cdots$$

This means that there is one path with parameters $i=1$, $d=5$, that there are two paths with $i=2$, $d=6$, etc. The general expression for $i=1,2,\ldots$ is

$$a(i,d) = \begin{cases} 2^{i-1}, & d=i+4 \\ 0, & \text{otherwise.} \end{cases}$$

By means of the detour flow graph, it is straightforward to verify this expression for small values of i and d.

It remains to be shown that $a(i,d)$ (the number of detours that start at any given depth j and have parameter i and d) does not depend on which reference path we choose. We do this in Exercise 6.6.

6.4.2 Upper bound to P_b

We are now ready to upper bound the bit-error probability. We recapitulate.

We fix an arbitrary encoder input sequence, we let $x=(x_1,x_2,\ldots,x_n)$ be the corresponding encoder output and let $c=\sqrt{\mathcal{E}_s}x$ be the codeword. The waveform signal is

$$w(t) = \sqrt{\mathcal{E}}\sum_{j=1}^{n}x_j\psi_j(t),$$

where $\psi_1(t),\ldots,\psi_n(t)$ forms an orthonormal collection. We transmit this signal over the AWGN channel with power spectral density $N_0/2$. Let $r(t)$ be the received signal, and let

$$y=(y_1,\ldots,y_n), \text{ where } y_i = \int r(t)\psi_i^*(t)dt,$$

be the decoder input.

The Viterbi algorithm labels each edge in the trellis with the corresponding edge metric and finds the path through the trellis with the largest path metric. An edge from depth $j-1$ to j with output symbols x_{2j-1}, x_{2j} is labeled with the edge metric $y_{2j-1}x_{2j-1}+y_{2j}x_{2j}$.

The maximum likelihood path selected by the Viterbi decoder could contain detours. For $j=0,1,\ldots,k-1$, if there is a detour that starts at depth j, we set ω_j to be the number of information-bit errors made on that detour. In all other cases, we set $\omega_j=0$. Let Ω_j be the corresponding random variable (over all possible noise realizations).

For the path selected by the Viterbi algorithm, the total number of incorrect bits is $\sum_{j=0}^{k-1}\omega_j$ and $\frac{1}{kk_0}\sum_{j=0}^{k-1}\omega_j$ is the fraction of errors with respect to the kk_0

6.4. Bit-error probability

source bits. Hence the bit-error probability is

$$P_b = \frac{1}{kk_0} \sum_{j=0}^{k-1} \mathbb{E}[\Omega_j]. \tag{6.1}$$

The expected value $\mathbb{E}[\Omega_j]$ can be written as follows

$$\mathbb{E}[\Omega_j] = \sum_h i(h)\pi(h), \tag{6.2}$$

where the sum is over all detours h that start at depth j with respect to the reference path, $\pi(h)$ stands for the probability that detour h is taken, and $i(h)$ for the input distance between detour h and the reference path.

Next we upper bound $\pi(h)$. If a detour starts at depth j and ends at depth $l = j + m$, then the corresponding encoder-output symbols form a $2m$ tuple $\bar{u} \in \{\pm 1\}^{2m}$. Let $u = x_{2j+1}, \ldots, x_{2l} \in \{\pm 1\}^{2m}$ and $\rho = y_{2j+1}, \ldots, y_{2l}$ be the corresponding sub-sequence of the reference path and of the channel output, respectively, see Figure 6.7.

A necessary (but not sufficient) condition for the Viterbi algorithm to take a detour is that the subpath metric along the detour is at least as large as the corresponding subpath metric along the reference path. An equivalent condition is that ρ is at least as close to $\sqrt{\mathcal{E}_s}\bar{u}$ as it is to $\sqrt{\mathcal{E}_s}u$. Observe that ρ has the statistic of $\sqrt{\mathcal{E}_s}u + Z$ where $Z \sim \mathcal{N}(0, \frac{N_0}{2}I_{2m})$ and $2m$ is the common length of $u, \bar{u}$, and ρ. The probability that ρ is at least as close to $\sqrt{\mathcal{E}_s}\bar{u}$ as it is to $\sqrt{\mathcal{E}_s}u$ is $Q\left(\frac{d_E}{2\sigma}\right)$, where $d_E = 2\sqrt{\mathcal{E}_s d}$ is the Euclidean distance between $\sqrt{\mathcal{E}_s}u$ and $\sqrt{\mathcal{E}_s}\bar{u}$. Using $d_E(h)$ to denote the Euclidean distance of detour h to the reference path, we obtain

$$\pi(h) \leq Q\left(\frac{d_E(h)}{2\sigma}\right) = Q\left(\sqrt{\frac{\mathcal{E}_s d(h)}{\sigma^2}}\right),$$

where the inequality sign is needed because, as mentioned, the event that ρ is at least as close to $\sqrt{\mathcal{E}_s}\bar{u}$ as it is to $\sqrt{\mathcal{E}_s}u$ is only a necessary condition for the Viterbi decoder to take detour $\bar{u}$. Inserting the above bound into (6.2) we obtain the first inequality in the following chain.

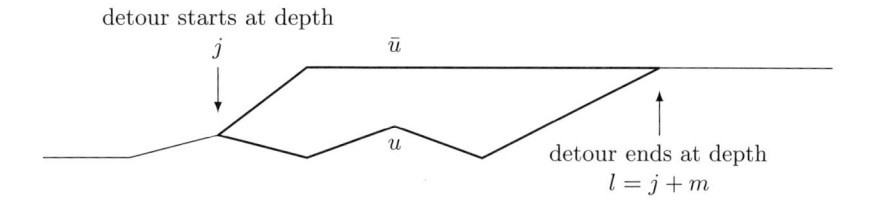

Figure 6.7. Detour and reference path, labeled with the corresponding output subsequences.

$$\mathbb{E}[\Omega_j] = \sum_h i(h)\pi(h)$$

$$\leq \sum_h i(h) Q\left(\sqrt{\frac{\mathcal{E}_s d(h)}{\sigma^2}}\right)$$

$$\stackrel{(a)}{=} \sum_{i=1}^{\infty}\sum_{d=1}^{\infty} iQ\left(\sqrt{\frac{\mathcal{E}_s d}{\sigma^2}}\right) \tilde{a}(i,d)$$

$$\stackrel{(b)}{\leq} \sum_{i=1}^{\infty}\sum_{d=1}^{\infty} iQ\left(\sqrt{\frac{\mathcal{E}_s d}{\sigma^2}}\right) a(i,d)$$

$$\stackrel{(c)}{\leq} \sum_{i=1}^{\infty}\sum_{d=1}^{\infty} iz^d a(i,d).$$

To obtain equality (a) we group the terms of the sum that have the same i and d and introduce $\tilde{a}(i,d)$ to denote the number of such terms in the finite trellis. Note that $\tilde{a}(i,d)$ is the finite-trellis equivalent to $a(i,d)$ introduced in Section 6.4.1. As the infinite trellis contains all the detours of the finite trellis and more, $\tilde{a}(i,d) \leq a(i,d)$. This justifies (b). In (c) we use

$$Q\left(\sqrt{\frac{\mathcal{E}_s d}{\sigma^2}}\right) \leq e^{-\frac{\mathcal{E}_s d}{2\sigma^2}} = z^d, \text{ for } z = e^{-\frac{\mathcal{E}_s}{2\sigma^2}}.$$

For the final step towards the upper bound to P_b, we use the relationship

$$\sum_{i=1}^{\infty} if(i) = \frac{\partial}{\partial I}\sum_{i=1}^{\infty} I^i f(i)\bigg|_{I=1},$$

which holds for any function f and can be verified by taking the derivative of $\sum_{i=1}^{\infty} I^i f(i)$ with respect to I and then setting $I=1$. Hence

$$\mathbb{E}[\Omega_j] \leq \sum_{i=1}^{\infty}\sum_{d=1}^{\infty} iz^d a(i,d) \tag{6.3}$$

$$= \frac{\partial}{\partial I}\sum_{i=1}^{\infty}\sum_{d=1}^{\infty} I^i D^d a(i,d)\bigg|_{I=1, D=z}$$

$$= \frac{\partial}{\partial I} T(I,D)\bigg|_{I=1, D=z}.$$

Plugging into (6.1) and using the fact that the above bound does not depend on j yields

$$P_b = \frac{1}{kk_0}\sum_{j=0}^{k-1} \mathbb{E}[\Omega_j] \leq \frac{1}{k_0}\frac{\partial}{\partial I}T(I,D)\bigg|_{I=1,D=z}. \tag{6.4}$$

6.5. Summary

In our specific example we have $k_0 = 1$ and $T(I,D) = \frac{ID^5}{1-2ID}$, hence $\frac{\partial T}{\partial I} = \frac{D^5}{(1-2ID)^2}$. Thus

$$P_b \leq \frac{z^5}{(1-2z)^2}. \tag{6.5}$$

The bit-error probability depends on the encoder and on the channel. Bound (6.4) nicely separates the two contributions. The encoder is accounted for by $T(I,D)/k_0$ and the channel by z. More precisely, z^d is an upper bound to the probability that a maximum likelihood receiver makes a decoding error when the choice is between two encoder output sequences that have Hamming distance d. As shown in Exercise 2.32(b) of Chapter 2, we can use the Bhattacharyya bound to determine z for any binary-input discrete memoryless channel. For such a channel,

$$z = \sum_y \sqrt{P(y|a)P(y|b)}, \tag{6.6}$$

where a and b are the two letters of the input alphabet and y runs over all the elements of the output alphabet. Hence, the technique used in this chapter is applicable to any binary input discrete memoryless channel.

It should be mentioned that the upper bound (6.5) is valid under the condition that there is no convergence issue associated to the various sums following (6.3). This is the case when $0 \leq z \leq \frac{1}{2}$, which is the case when the numerator and the denominator of (6.5) are non-negative. The z from (6.6) fulfills $0 \leq z \leq 1$. However, if we use the tighter Bhattacharyya bound discussed in Exercise 2.29 of Chapter 2 (which is tighter by a factor $\frac{1}{2}$) then it is guaranteed that $0 \leq z \leq \frac{1}{2}$.

6.5 Summary

To assess the impact of the convolutional encoder, let us compare two situations. In both cases, the transmitted signal looks identical to an observer, namely it has the form

$$w(t) = \sum_{i=1}^{2l} s_i \psi(t - iT)$$

for some positive integer l and some unit-norm pulse $\psi(t)$ that is orthogonal to its T-space translates. In both cases, the symbols take values in $\{\pm\sqrt{\mathcal{E}_s}\}$ for some fixed energy-per-symbol $\mathcal{E}_s$, but the way the symbols are obtained differs in the two cases. In one case, the symbols are obtained from the output of the convolutional encoder studied in this chapter. We call this the coded case. In the other case, the symbols are simply the source bits, which take value in $\{\pm 1\}$, scaled by $\sqrt{\mathcal{E}_s}$. We call this the uncoded case.

For the coded case, the number of symbols is twice the number of bits. Hence, letting R_b, R_s, and $\mathcal{E}_b$ be the bit rate, the symbol rate, and the energy per bit, respectively, we obtain

$$R_b = \frac{R_s}{2} = \frac{1}{2T} \text{ [bits/symbol]},$$
$$\mathcal{E}_b = 2\mathcal{E}_s,$$
$$P_b \le \frac{z^5}{(1-2z)^2},$$

where $z = e^{-\frac{\mathcal{E}_s}{2\sigma^2}}$. As $\frac{\mathcal{E}_s}{2\sigma^2}$ becomes large, the denominator of the above bound for P_b becomes essentially 1 and the bound decreases as z^5.

For the uncoded case, the symbol rate equals the bit rate and the energy per bit equals the energy per symbol. For this case we have an exact expression for the bit-error probability. However, for comparison with the coded case, it is useful to upper bound also the bit-error probability of the uncoded case. Hence,

$$R_b = R_s = \frac{1}{T} \text{ [bits/symbol]},$$
$$\mathcal{E}_b = \mathcal{E}_s,$$
$$P_b = Q\left(\sqrt{\frac{\mathcal{E}_s}{\sigma^2}}\right) \le e^{-\frac{\mathcal{E}_s}{2\sigma^2}} = z,$$

where we have used $Q(x) \le \exp\{-\frac{x^2}{2}\}$. Recall that σ^2 is the noise variance of the discrete-time channel, which equals the power spectral density $\frac{N_0}{2}$ of the continuous-time channel.

Figure 6.8 plots various bit-error probability curves. The dots represent simulated results. From right to left, we see the simulation results for the uncoded system, for the system based on the convolutional encoder, and for a system based on a *low-density parity-check* (LDPC) code, which is a state-of-the-art code used in the DVB-S2 (Digital Video Broadcasting – Satellite – Second Generation) standard. Like the convolutional encoder, the LDPC encoder produces a symbol rate that is twice the bit rate. For the uncoded system, we have plotted also the exact expression for the bit-error probability (dashed curve labeled by the Q function). We see that this expression is in perfect agreement with the simulation results. The upper bound that we have derived for the system that incorporates the convolutional encoder (solid curve) is off by about 1 dB at $P_b = 10^{-4}$ with respect to the simulation results. The dashed curve, which is in excellent agreement with the simulated results for the same code, is the result of a more refined bound (see Exercise 6.11).

Suppose we are to design a system that achieves a target error probability P_b, say $P_b = 10^{-2}$. From the plots in Figure 6.8, we see that the required $\mathcal{E}_s/\sigma^2$ is roughly 7.3 dB for the uncoded system, whereas the convolutional code and the LDPC code require about 2.3 and 0.75 dB, respectively. The gaps become more significant as the target P_b decreases. For $P_b = 10^{-7}$, the required $\mathcal{E}_s/\sigma^2$ is about 14.3, 7.3, and 0.95 dB, respectively. (Recall that a difference of 13 dB means a factor 20 in power.)

Instead of comparing $\mathcal{E}_s/\sigma^2$, we might be interested in comparing $\mathcal{E}_b/\sigma^2$. The conversion is straightforward: for the uncoded system $\mathcal{E}_b/\sigma^2 = \mathcal{E}_s/\sigma^2$, whereas for the two coded systems $\mathcal{E}_b/\sigma^2 = 2\mathcal{E}_s/\sigma^2$.

6.5. Summary

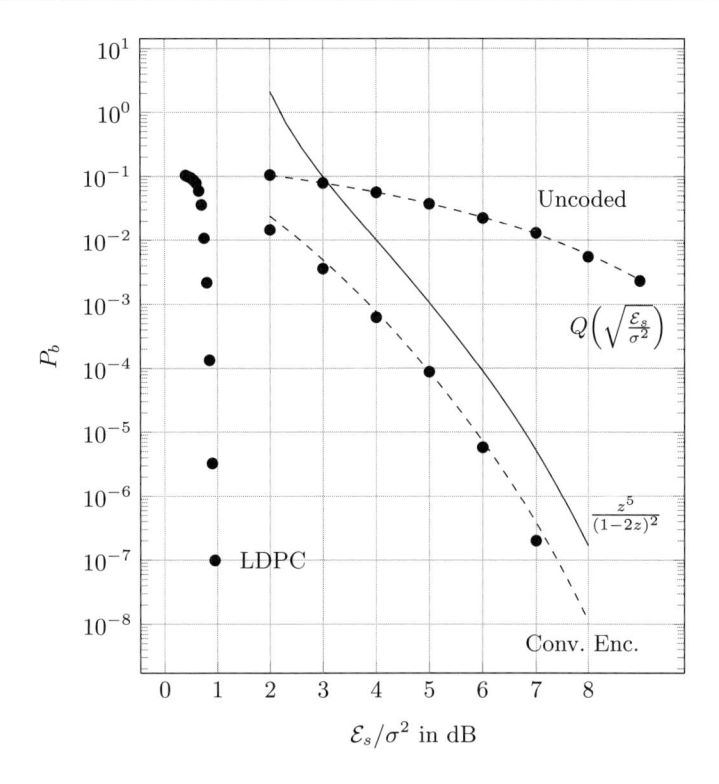

Figure 6.8. Bit-error probabilities.

From a high-level point of view, non-trivial coding is about using only selected sequences to form the codebook. In this chapter, we have fixed the channel-input alphabet to $\{\pm\sqrt{\mathcal{E}_s}\}$. Then our only option to introduce non-trivial coding is to increase the codeword length from $n = k$ to $n > k$. For a fixed bit rate, increasing n implies increasing the symbol rate. To increase the symbol rate we time-compress the pulse $\psi(t)$ by the appropriate factor and the bandwidth expands by the same factor. If we fix the bandwidth, the symbol rate stays the same and the bit rate has to decrease.

It would be wrong to conclude that non-trivial coding always requires reducing the bit rate or increasing the bandwidth. Instead of keeping the channel-input alphabet constant, for the coded system we could have used, say, 4-PAM. Then each pair of binary symbols produced by the encoder can be mapped into a single 4-PAM symbol. In so doing, the bit rate, the symbol rate, and the bandwidth remain unchanged.

The ultimate answer comes from information theory (see e.g. [19]). Information theory tells us that, by means of coding, we can achieve an error probability as small as desired, provided that we send fewer bits per symbol than the channel capacity C, which for the discrete-time AWGN channel is $C = \frac{1}{2}\log_2(1 + \frac{\mathcal{E}_s}{\sigma^2})$ bits/symbol. According to this expression, to send $1/2$ bits per symbol as we do in our example, we need $\mathcal{E}_s/\sigma^2 = 1$, which means 0 dB. We see that the performance of the LDPC code is quite good. Even with the channel-input alphabet restricted to $\{\pm\sqrt{\mathcal{E}_s}\}$ (no such restriction is imposed in the derivation of C), the LDPC code achieves the kind of error probability that we typically want in applications at an $\mathcal{E}_s/\sigma^2$ which is within 1 dB from the ultimate limit of 0 dB required for reliable communication.

Convolutional codes were invented by Elias in 1955 and have been used in many communication systems, including satellite communication, and mobile communication. In 1993, Berrou, Glavieux, and Thitimajshima captured the attention of the communication engineering community by introducing a new class of codes, called turbo codes, that achieved a performance breakthrough by concatenating two convolutional codes separated by an interleaver. Their performance is not far from that of the low-density parity-check codes (LDPC) – today's state-of-the-art in coding.

Thanks to its tremendous success, coding is in every modern communication system. In this chapter we have only scratched the surface. Recommended books on coding are [22] for a classical textbook that covers a broad spectrum of coding techniques and [23] for the reference book on LDPC coding.

6.6 Appendix: Formal definition of the Viterbi algorithm

Let $\Gamma = \{(1,1), (1,-1), (-1,1), (-1,-1)\}$ be the state space and define the *edge metric* $\mu_{j-1,j}(\alpha, \beta)$ as follows. If there is an edge that connects state $\alpha \in \Gamma$ at depth $j - 1$ to state $\beta \in \Gamma$ at depth j let

$$\mu_{j-1,j}(\alpha, \beta) = x_{2j-1} y_{2j-1} + x_{2j} y_{2j},$$

where x_{2j-1}, x_{2j} is the encoder output of the corresponding edge. If there is no such edge, we let $\mu_{j-1,j}(\alpha, \beta) = -\infty$.

Since $\mu_{j-1,j}(\alpha, \beta)$ is the jth term in $\langle x, y \rangle$ for any path that goes through state α at depth $j - 1$ and state β at depth j, $\langle x, y \rangle$ is obtained by adding the edge metrics along the path specified by x.

The *path metric* is the sum of the edge metrics taken along the edges of a path. A *longest path* from state $(1, 1)$ at depth $j = 0$, denoted $(1, 1)_0$, to a state α at depth j, denoted α_j, is one of the paths that has the largest path metric. The Viterbi algorithm works by constructing, for each j, a list of the longest paths to the states at depth j. The following observation is key to understanding the Viterbi algorithm. If $path * \alpha_{j-1} * \beta_j$ is a longest path to state β of depth j, where $path \in \Gamma^{j-2}$ and $*$ denotes concatenation, then $path * \alpha_{j-1}$ must be a longest path to state α of depth $j - 1$, for if another path, say $alternatepath * \alpha_{j-1}$ were shorter for some $alternatepath \in \Gamma^{j-2}$, then $alternatepath * \alpha_{j-1} * \beta_j$ would be shorter than $path * \alpha_{j-1} * \beta_j$. So the longest depth j path to a state can be obtained by checking the extension of the longest depth $(j-1)$ paths by one edge.

The following notation is useful for the formal description of the Viterbi algorithm. Let $\mu_j(\alpha)$ be the metric of a longest path to state α_j and let $B_j(\alpha) \in \{\pm 1\}^j$ be the encoder input sequence that corresponds to this path. We call

$B_j(\alpha) \in \{\pm 1\}^j$ the *survivor* because it is the only path through state α_j that will be extended. (Paths through α_j that have a smaller metric have no chance of extending into a maximum likelihood path.) For each state, the Viterbi algorithm computes two things: a survivor and its metric. The formal algorithm follows, where $B(\beta, \alpha)$ is the encoder input that corresponds to the transition from state β to state α if there is such a transition and is undefined otherwise.

(1) Initially set $\mu_0(1,1) = 0$, $\mu_0(\alpha) = -\infty$ for all $\alpha \neq (1,1)$, $B_0(1,1) = \emptyset$, and $j = 1$.

(2) For each $\alpha \in \Gamma$, find one of the β for which $\mu_{j-1}(\beta) + \mu_{j-1,j}(\beta, \alpha)$ is a maximum. Then set
$$\mu_j(\alpha) \leftarrow \mu_{j-1}(\beta) + \mu_{j-1,j}(\beta, \alpha),$$
$$B_j(\alpha) \leftarrow B_{j-1}(\beta) * B(\beta, \alpha).$$

(3) If $j = k+2$, output the first k bits of $B_j(1,1)$ and stop. Otherwise increment j by one and go to Step 2.

The reader should have no difficulty verifying (by induction on j) that $\mu_j(\alpha)$ as computed by Viterbi's algorithm is indeed the metric of a longest path from $(1,1)_0$ to state α at depth j and that $B_j(\alpha)$ is the encoder input sequence associated to it.

6.7 Exercises

Exercises for Section 6.2

EXERCISE 6.1 (Power spectral density) *Consider the random process*

$$X(t) = \sum_{i=-\infty}^{\infty} X_i \sqrt{\mathcal{E}_s} \psi(t - iT_s - T_0),$$

where T_s and $\mathcal{E}_s$ are fixed positive numbers, $\psi(t)$ is some unit-energy function, T_0 is a uniformly distributed random variable taking values in $[0, T_s)$, and $\{X_i\}_{i=-\infty}^{\infty}$ is the output of the convolutional encoder described by

$$X_{2n} = B_n B_{n-2}$$
$$X_{2n+1} = B_n B_{n-1} B_{n-2}$$

with iid input sequence $\{B_i\}_{i=-\infty}^{\infty}$ taking values in $\{\pm 1\}$.

(a) *Express the power spectral density of $X(t)$ for a general $\psi(t)$.*
(b) *Plot the power spectral density of $X(t)$ assuming that $\psi(t)$ is a unit-norm rectangular pulse of width T_s.*

EXERCISE 6.2 (Power spectral density: Correlative encoding) *Repeat Exercise 6.1 using the encoder:*

$$X_i = B_i - B_{i-1}.$$

Compare this exercise to Exercise 5.4 of Chapter 5.

Exercises for Section 6.3

EXERCISE 6.3 (Viterbi algorithm) *An output sequence $x_1, \ldots, x_{10}$ from the convolutional encoder of Figure 6.9 is transmitted over the discrete-time AWGN channel. The initial and final state of the encoder is $(1, 1)$. Using the Viterbi algorithm, find the maximum likelihood information sequence $\hat{b}_1, \ldots, \hat{b}_4, 1, 1$, knowing that $b_1, \ldots, b_4$ are drawn independently and uniformly from $\{\pm 1\}$ and that the channel output $y_1, \ldots, y_{10} = 1, 2, -1, 4, -2, 1, 1, -3, -1, -2$. (It is for convenience that we are choosing integers rather than real numbers.)*

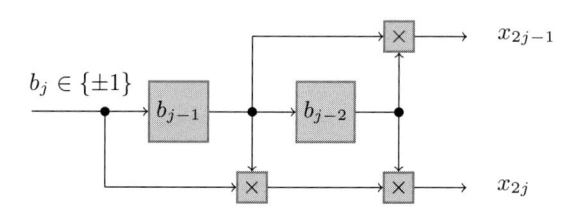

Figure 6.9.

EXERCISE 6.4 (Inter-symbol interference) *From the decoder's point of view, inter-symbol interference (ISI) can be modeled as follows*

$$Y_i = X_i + Z_i$$

$$X_i = \sum_{j=0}^{L} B_{i-j} h_j, \qquad i = 1, 2, \ldots \tag{6.7}$$

where B_i is the ith information bit, $h_0, \ldots, h_L$ are coefficients that describe the inter-symbol interference, and Z_i is zero-mean, Gaussian, of variance σ^2, and statistically independent of everything else. Relationship (6.7) can be described by a trellis, and the ML decision rule can be implemented by the Viterbi algorithm.

(a) *Draw the trellis that describes all sequences of the form $X_1, \ldots, X_6$ resulting from information sequences of the form $B_1, \ldots, B_5, 0$, $B_i \in \{0, 1\}$, assuming*

$$h_i = \begin{cases} 1, & i = 0 \\ -2, & i = 1 \\ 0, & otherwise. \end{cases}$$

To determine the initial state, you may assume that the preceding information sequence terminated with 0. Label the trellis edges with the input/output symbols.

6.7. Exercises

(b) *Specify a metric $f(x_1,\ldots,x_6) = \sum_{i=1}^{6} f(x_i, y_i)$ whose minimization or maximization with respect to the valid $x_1,\ldots,x_6$ leads to a maximum likelihood decision. Specify if your metric needs to be minimized or maximized.*

(c) *Assume $y_1,\ldots,y_6 = 2, 0, -1, 1, 0, -1$. Find the maximum likelihood estimate of the information sequence $B_1,\ldots,B_5$.*

Exercises for Section 6.4

EXERCISE 6.5 (Linearity) *In this exercise, we establish in what sense the encoder of Figure 6.2 is linear.*

(a) *For this part you might want to review the axioms of a field. Consider the set $\mathcal{F}_0 = \{0, 1\}$ with the following addition and multiplication tables.*

+	0	1
0	0	1
1	1	0

×	0	1
0	0	0
1	0	1

(The addition in $\mathcal{F}_0$ is the usual addition over $\mathbb{R}$ with result taken modulo 2. The multiplication is the usual multiplication over $\mathbb{R}$ and there is no need to take the modulo 2 operation because the result is automatically in $\mathcal{F}_0$.) $\mathcal{F}_0$, "+", and "×" form a binary field denoted by $\mathbb{F}_2$. Now consider $\mathcal{F}_- = \{\pm 1\}$ and the following addition and multiplication tables.

+	1	-1
1	1	-1
-1	-1	1

×	1	-1
1	1	1
-1	1	-1

(The addition in $\mathcal{F}_-$ is the usual multiplication over $\mathbb{R}$.) Argue that $\mathcal{F}_-$, "+", and "×" form a binary field as well. Hint: The second set of operations can be obtained from the first set via the transformation $T : \mathcal{F}_0 \to \mathcal{F}_-$ that sends 0 to 1 and 1 to -1. Hence, by construction, for $a, b \in \mathcal{F}_0$, $T(a + b) = T(a) + T(b)$ and $T(a \times b) = T(a) \times T(b)$. Be aware of the double meaning of "+" and "×" in the previous sentence.

(b) *For this part you might want to review the notion of a vector space. Let $\mathcal{F}_0$, "+" and "×" be as defined in (a). Let $\mathcal{V} = \mathcal{F}_0^\infty$. This is the set of infinite sequences taking values in $\mathcal{F}_0$. Does $\mathcal{V}$, $\mathcal{F}_0$, "+" and "×" form a vector space? (Addition of vectors and multiplication of a vector with a scalar is done component-wise.) Repeat using $\mathcal{F}_-$.*

(c) *For this part you might want to review the notion of linear transformation. Let $f : \mathcal{V} \to \mathcal{V}$ be the transformation that sends an infinite sequence $b \in \mathcal{V}$ to an infinite sequence $x \in \mathcal{V}$ according to*

$$x_{2j-1} = b_{j-1} + b_{j-2} + b_{j-3}$$
$$x_{2j} = b_j + b_{j-2},$$

where the "+" is the one defined over the field of scalars implicit in $\mathcal{V}$. Argue that this f is linear. Comment: When $\mathcal{V} = \mathcal{F}_-^\infty$, this encoder is the one used throughout Chapter 6, with the only difference that in the chapter we multiply over $\mathbb{R}$ rather than adding over $\mathcal{F}_-$, but this is just a matter of notation, the result of the two operations on the elements of $\mathcal{F}_-$ being identical. The standard way to describe a convolutional encoder is to choose $\mathcal{F}_0$ and the corresponding addition, namely addition modulo 2. See Exercise 6.12 for the reason we opt for a non-standard description.

EXERCISE 6.6 (Independence of the distance profile from the reference path) We want to show that $a(i, d)$ does not depend on the reference path. Recall that in Section 6.4.1 we define $a(i, d)$ as the number of detours that leave the reference path at some arbitrary but fixed trellis depth j and have input distance i and output distance d with respect to the reference path.

(a) Let b and $\bar{b}$, both in $\{\pm 1\}^\infty$, be two infinite-length input sequences to the encoder of Figure 6.2 and let f be the encoding map. The encoder is linear in the sense that the componentwise product over the reals $b\bar{b}$ is also a valid input sequence and the corresponding output sequence is $f(b\bar{b}) = f(b)f(\bar{b})$ (see Exercise 6.5). Argue that the distance between b and $\bar{b}$ equals the distance between $b\bar{b}$ and the all-one input sequence. Similarly, argue that the distance between $f(b)$ and $f(\bar{b})$ equals the distance between $f(b\bar{b})$ and the all-one output sequence (which is the output to the all-one input sequence).

(b) Fix an arbitrary reference path and an arbitrary detour that splits from the reference path at time 0. Let b and $\bar{b}$ be the corresponding input sequences. Because the detour starts at time 0, $b_i = \bar{b}_i$ for $i < 0$ and $b_0 \neq \bar{b}_0$. Argue that $\bar{b}$ uniquely defines a detour $\tilde{b}$ that splits from the all-one path at time 0 and such that:

   (i) the distance between b and $\bar{b}$ is the same as that between $\tilde{b}$ and the all-one input sequence;
   (ii) the distance between $f(b)$ and $f(\bar{b})$ is the same as that between $f(\tilde{b})$ and the all-one output sequence.

(c) Conclude that $a(i, d)$ does not depend on the reference path.

EXERCISE 6.7 (Rate 1/3 convolutional code) For the convolutional encoder of Figure 6.10 do the following.

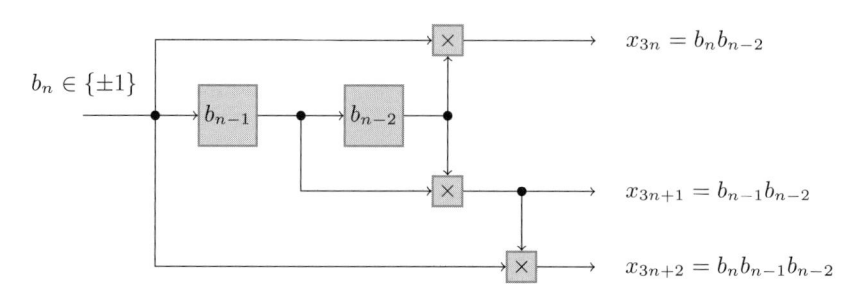

Figure 6.10.

6.7. Exercises

(a) Draw the state diagram and the detour flow graph.

(b) Suppose that the serialized encoder output symbols are scaled so that the resulting energy per bit is $\mathcal{E}_b$ and are sent over the discrete-time AWGN channel of noise variance $\sigma^2 = N_0/2$. Derive an upper bound to the bit-error probability assuming that the decoder implements the Viterbi algorithm.

EXERCISE 6.8 (Rate 2/3 convolutional code) *The following equations describe the output sequence of a convolutional encoder that in each epoch takes $k_0 = 2$ input symbols from $\{\pm 1\}$ and outputs $n_0 = 3$ symbols from the same alphabet.*

$$x_{3n} = b_{2n}b_{2n-1}b_{2n-2}$$
$$x_{3n+1} = b_{2n+1}b_{2n-2}$$
$$x_{3n+2} = b_{2n+1}b_{2n}b_{2n-2}$$

(a) Draw an implementation of the encoder based on delay elements and multipliers.

(b) Draw the state diagram.

(c) Suppose that the serialized encoder output symbols are scaled so that the resulting energy per bit is $\mathcal{E}_b$ and are sent over the discrete-time AWGN channel of noise variance $\sigma^2 = N_0/2$. Derive an upper bound to the bit-error probability assuming that the decoder implements the Viterbi algorithm.

EXERCISE 6.9 (Convolutional encoder, decoder, and error probability) *For the convolutional code described by the state diagram of Figure 6.11:*

(a) draw the encoder;

(b) as a function of the energy per bit $\mathcal{E}_b$, upper bound the bit-error probability of the Viterbi algorithm when the scaled encoder output sequence is transmitted over the discrete-time AWGN channel of noise variance $\sigma^2 = N_0/2$.

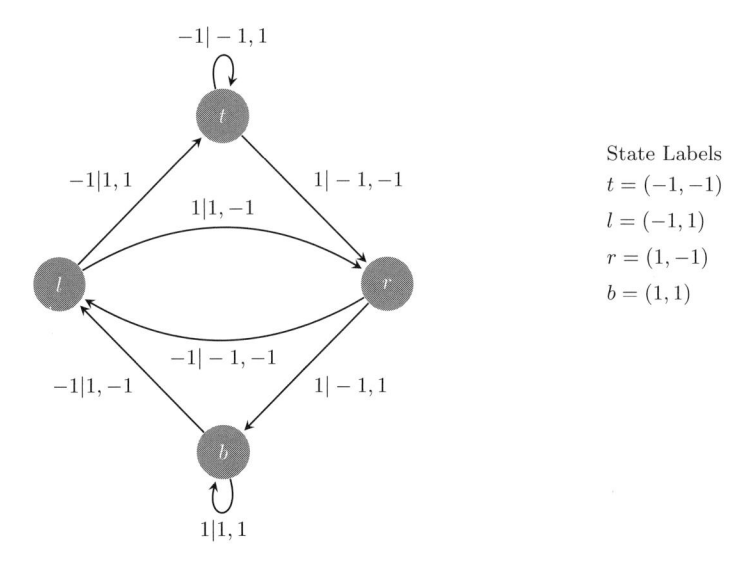

State Labels
$t = (-1, -1)$
$l = (-1, 1)$
$r = (1, -1)$
$b = (1, 1)$

Figure 6.11.

EXERCISE 6.10 (Viterbi for the binary erasure channel) *Consider the convolutional encoder of Figure 6.12 with inputs and outputs over $\{0,1\}$ and addition modulo 2. Its output is sent over the binary erasure channel described by*

$$P_{Y|X}(0|0) = P_{Y|X}(1|1) = 1 - \epsilon,$$
$$P_{Y|X}(?|0) = P_{Y|X}(?|1) = \epsilon,$$
$$P_{Y|X}(1|0) = P_{Y|X}(0|1) = 0,$$

where $0 < \epsilon < 0.5$.

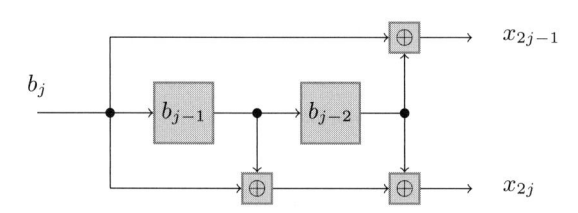

Figure 6.12.

(a) *Draw a trellis section that describes the encoder map.*
(b) *Derive the branch metric and specify whether a maximum likelihood decoder chooses the path with largest or smallest path metric.*
(c) *Suppose that the initial encoder state is $(0,0)$ and that the channel output is $\{0,?,?,1,0,1\}$. What is the most likely information sequence?*
(d) *Derive an upper bound to the bit-error probability.*

EXERCISE 6.11 (Bit-error probability) *In the process of upper bounding the bit-error probability, in Section 6.4.2 we make the following step*

$$\mathbb{E}[\Omega_j] \leq \sum_{i=1}^{\infty} \sum_{d=1}^{\infty} i Q\left(\sqrt{\frac{\mathcal{E}_s d}{\sigma^2}}\right) a(i,d)$$
$$\leq \sum_{i=1}^{\infty} \sum_{d=1}^{\infty} i z^d a(i,d).$$

(a) *Instead of upper bounding the Q function as done above, use the results of Section 6.4.1 to substitute $a(i,d)$ and d with explicit functions of i and get rid of the second sum. You should obtain*

$$P_b \leq \sum_{i=1}^{\infty} i Q\left(\sqrt{\frac{\mathcal{E}_s(i+4)}{\sigma^2}}\right) 2^{i-1}.$$

(b) *Truncate the above sum to the first five terms and evaluate it numerically for $\mathcal{E}_s/\sigma^2$ between 2 and 6 dB. Plot the results and compare to Figure 6.8.*

6.7. Exercises

Miscellaneous exercises

EXERCISE 6.12 (Standard description of a convolutional encoder) *Consider the two encoders of Figure 6.13, where the map $T : \mathcal{F}_0 \to \mathcal{F}_-$ sends 0 to 1 and 1 to -1. Show that the two encoders produce the same output when their inputs are related by $b_j = T(\bar{b}_j)$. Hint: For a and b in $\mathcal{F}_0$, $T(a+b) = T(a) \times T(b)$, where addition is modulo 2 and multiplication is over $\mathbb{R}$.*

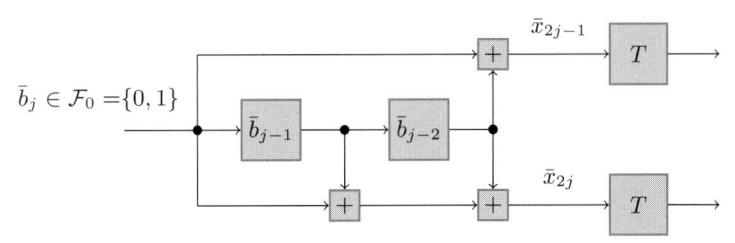

(a) Conventional description. Addition is modulo 2.

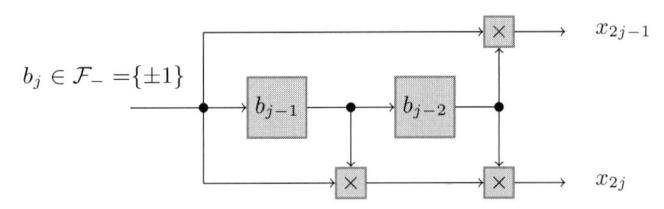

(b) Description used in this text. Multiplication is over $\mathbb{R}$.

Figure 6.13.

Comment: The encoder of Figure 6.13b is linear over the field $\mathcal{F}_-$ (see Exercise 6.5), whereas the encoder of Figure 6.13a is linear over $\mathcal{F}_0$ only if we omit the output map T. The comparison of the two figures should explain why in this chapter we have opted for the description of part (b) even though the standard description of a convolutional encoder is as in part (a).

EXERCISE 6.13 (Trellis with antipodal signals) *Figure 6.14a shows a trellis section labeled with the output symbols x_{2j-1}, x_{2j} of a convolutional encoder. Notice how branches that are the mirror-image of each other have antipodal output symbols (symbols that are the negative of each other). The purpose of this exercise is to see that when the trellis has this particular structure and codewords are sent through the discrete-time AWGN channel, the maximum likelihood sequence detector further simplifies (with respect to the Viterbi algorithm).*

Figure 6.14b shows two consecutive trellis sections labeled with the branch metric. Notice that the mirror symmetry of part (a) implies the same kind of symmetry for part (b). The maximum likelihood path is the one that has the largest path metric. To avoid irrelevant complications we assume that there is only one path that maximizes the path metric.

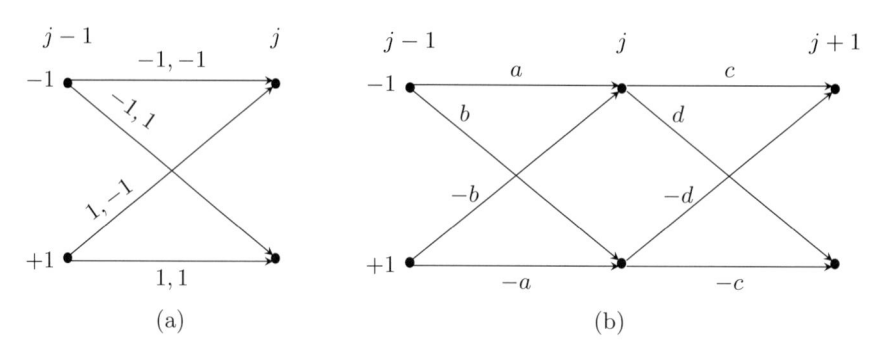

Figure 6.14.

(a) Let $\sigma_j \in \{\pm 1\}$ be the state visited by the maximum likelihood path at depth j. Suppose that a genie informs the decoder that $\sigma_{j-1} = \sigma_{j+1} = 1$. Write down the necessary and sufficient condition for the maximum likelihood path to go through $\sigma_j = 1$.

(b) Repeat for the remaining three possibilities of σ_{j-1} and σ_{j+1}. Does the necessary and sufficient condition for $\sigma_j = 1$ depend on the value of σ_{j-1} and σ_{j+1}?

(c) The branch metric for the branch with output symbols x_{2j-1}, x_{2j} is

$$x_{2j-1} y_{2j-1} + x_{2j} y_{2j},$$

where y_j is x_j plus noise. Using the result of the previous part, specify a maximum likelihood sequence decision for $\sigma_j = 1$ based on the observation $y_{2j-1}, y_{2j}, y_{2j+1}, y_{2j+2}$.

EXERCISE 6.14 (Timing error) *A transmitter sends*

$$X(t) = \sum_i B_i \psi(t - iT),$$

where $\{B_i\}_{i=-\infty}^{\infty}$, $B_i \in \{1, -1\}$, is a sequence of independent and uniformly distributed bits and $\psi(t)$ is a centered and unit-energy rectangular pulse of width T. The communication channel between the transmitter and the receiver is the AWGN channel of power spectral density $\frac{N_0}{2}$. At the receiver, the channel output $Z(t)$ is passed through a filter matched to $\psi(t)$, and the output is sampled, ideally at times $t_k = kT$, k integer.

(a) *Consider that there is a timing error, i.e. the sampling time is $t_k = kT - \tau$ where $\frac{\tau}{T} = 0.25$. Ignoring the noise, express the matched filter output observation w_k at time $t_k = kT - \tau$ as a function of the bit values b_k and b_{k-1}.*

(b) *Extending to the noisy case, let $r_k = w_k + z_k$ be the kth matched filter output observation. The receiver is not aware of the timing error. Compute the resulting error probability.*

6.7. Exercises

(c) Now assume that the receiver knows the timing error τ (same τ as above) but it cannot correct for it. (This could be the case if the timing error becomes known once the samples are taken.) Draw and label four sections of a trellis that describes the noise-free sampled matched filter output for each input sequence b_1, b_2, b_3, b_4. In your trellis, take into consideration the fact that the matched filter is "at rest" before $x(t) = \sum_{i=1}^{4} b_i \psi(t-iT)$ enters the filter.

(d) Suppose that the sampled matched filter output consists of $2, 0.5, 0, -1$. Use the Viterbi algorithm to decide on the transmitted bit sequence.

EXERCISE 6.15 (Simulation) The purpose of this exercise is to determine, by simulation, the bit-error probability of the communication system studied in this chapter. For the simulation, we recommend using MATLAB, as it has high-level functions for the various tasks, notably for generating a random information sequence, for doing convolutional encoding, for simulating the discrete-time AWGN channel, and for decoding by means of the Viterbi algorithm. Although the actual simulation is on the discrete-time AWGN, we specify a continuous-time setup. It is part of your task to translate the continuous-time specifications into what you need for the simulation. We begin with the uncoded version of the system of interest.

(a) By simulation, determine the minimum obtainable bit-error probability P_b of bit-by-bit on a pulse train transmitted over the AWGN channel. Specifically, the channel input signal has the form

$$X(t) = \sum_j X_j \psi(t - jT),$$

where the symbols are iid and take value in $\{\pm\sqrt{\mathcal{E}_s}\}$, the pulse $\psi(t)$ has unit norm and is orthogonal to its T-spaced time translates. Plot P_b as a function of $\mathcal{E}_s/\sigma^2$ in the range from 2 to 6 dB, where σ^2 is the noise variance. Verify your results with Figure 6.8.

(b) Repeat with the symbol sequence being the output of the convolutional encoder of Figure 6.2 multiplied by $\sqrt{\mathcal{E}_s}$. The decoder shall implement the Viterbi algorithm. Also in this case you can verify your results by comparing with Figure 6.8.

7 Passband communication via up/down conversion: Third layer

7.1 Introduction

We speak of *baseband communication* when the signals have their energy in some frequency interval $[-B, B]$ around the origin (Figure 7.1a). Much more common is the situation where the signal's energy is concentrated in $[f_c - B, f_c + B]$ and $[-f_c - B, -f_c + B]$ for some *carrier frequency* f_c greater than B. In this case, we speak of *passband communication* (Figure 7.1b). The carrier frequency f_c is chosen to fulfill regulatory constraints, to avoid interference from other signals, or to make the best possible use of the propagation characteristics of the medium used to communicate.

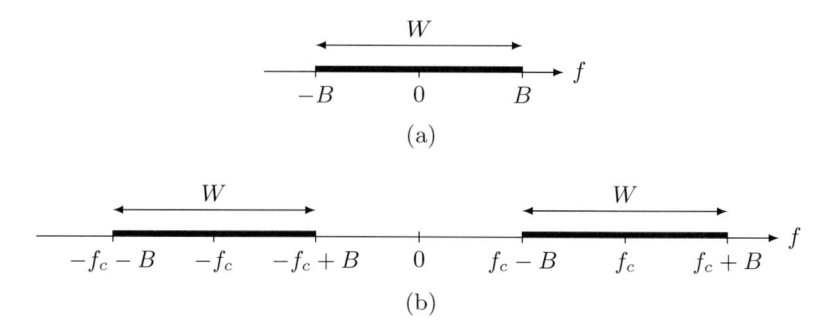

Figure 7.1. Baseband (a) versus passband (b).

The purpose of this chapter is to introduce a third and final layer responsible for passband communication. With this layer in place, the upper layers are designed for baseband communication even when the actual communication happens in passband.

EXAMPLE 7.1 (Regulatory constraints) *Figure 7.2 shows the radio spectrum allocation for the United States (October 2003). To get an idea about its complexity, the chart is presented in its entirety even if it is too small to read. The interested reader can find the original on the website of the (US) National Telecommunications and Information Administration.* □

7.1. Introduction

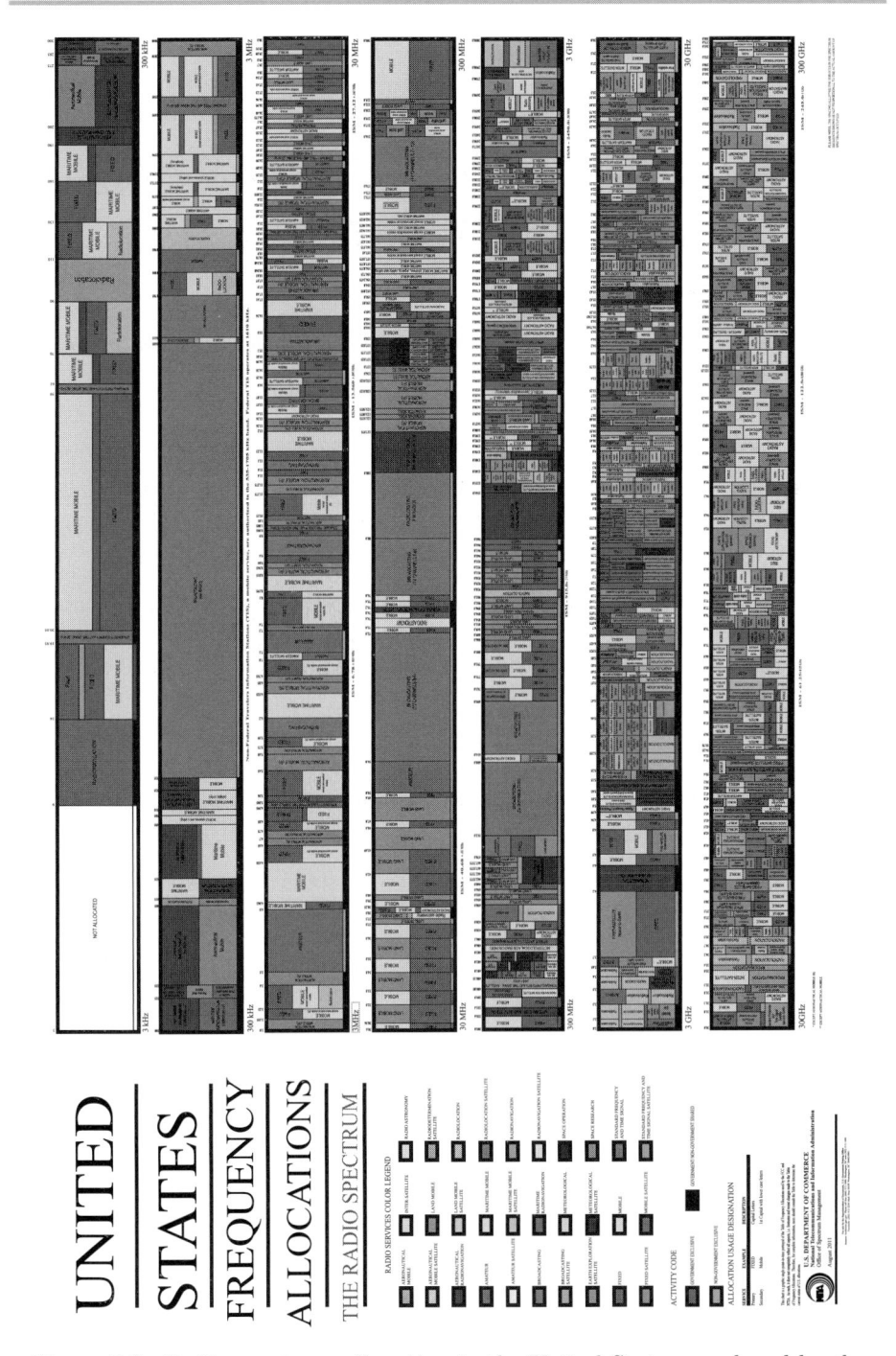

Figure 7.2. Radio spectrum allocation in the United States, produced by the US Department of Commerce, National Telecommunications and Information Administration, Office of Spectrum Management (October 2003).

Electromagnetic waves are subject to reflection, refraction, polarization, diffraction and absorption, and different frequencies experience different amounts of these phenomenons. This makes certain frequency ranges desirable for certain applications and not for others. A few examples follow.

EXAMPLE 7.2 (Reflection/diffraction) *Radio signals reflect and/or diffract off obstacles such as buildings and mountains. For our purpose, we can assume that the signal is reflected if its wavelength is much smaller than the size of the obstacle, whereas it is diffracted if its wavelength is much larger than the obstacle's size. Because of their large wavelengths, very low frequency (VLF) radio waves (see Table 7.1) can be diffracted by large obstacles such as mountains, thus are not blocked by mountain ranges. By contrast, in the UHF range signals propagate mainly by line of sight.* □

Table 7.1 The radio spectrum

Range Name	Frequency Range	Wavelength Range
Extremely Low Frequency (ELF)	3 Hz to 300 Hz	100 000 km to 1,000 km
Ultra Low Frequency (ULF)	300 Hz to 3 kHz	1000 km to 100 km
Very Low Frequency (VLF)	3 kHz to 30 kHz	100 km to 10 km
Low Frequency (LF)	30 kHz to 300 kHz	10 km to 1 km
Medium Frequency (MF)	300 kHz to 3 MHz	1 km to 100 m
High Frequency (HF)	3 MHz to 30 MHz	100 m to 10 m
Very High Frequency (VHF)	30 MHz to 300 MHz	10 m to 1 m
Ultra High Frequency (UHF)	300 MHz to 3 GHz	1 m to 10 cm
Super High Frequency (SHF)	3 GHz to 30 GHz	10 cm to 1 cm
Extremely High Frequency (EHF)	30 GHz to 300 GHz	1 cm to 1 mm
Visible Spectrum	400 THz to 790 THz	750 nm to 390 nm
X-Rays	3×10^{16} to 3×10^{19} Hz	10 nm to 10 pm

EXAMPLE 7.3 (Refraction) *Radio signals are refracted by the ionosphere surrounding the Earth. Different layers of the ionosphere have different ionization densities, hence different refraction indices. As a result, signals can be bent by a layer or can be trapped between layers. This phenomenon concerns mainly the MF and HF range (300 kHz to 30 MHz) but can also affect the MF through the LF and VLF range. As a consequence, radio signals emitted from a ground station can be bent back to Earth, sometimes after traveling a long distance trapped between layers of the ionosphere. This mode of propagation, called* sky wave *(as opposed to* ground wave*) is exploited, for instance, by amateur radio operators to reach locations on Earth that could not be reached if their signals traveled in straight lines. In fact, under particularly favorable circumstances, the communication between any two regions on Earth can be established via sky waves. Although the bending caused by the ionosphere is desirable for certain applications, it is a nuisance for Earth-to-satellite communication. This is why satellites use higher frequencies for which the ionosphere is essentially transparent (typically GHz range).* □

EXAMPLE 7.4 (Absorption) *Because of absorption, electromagnetic waves do not go very far under sea. The lower the frequency the better the penetration. This explains why submarines use the ELF through the VLF range. Because of the very limited bandwidth, communication in the ELF range is limited to a few characters per minute. For this reason, it is mainly used to order a submarine to rise to a shallow depth where it can be reached in the VLF range. Similarly, long waves penetrate the Earth better than short waves. For this reason, communication in mines is done in the ULF band. On the Earth's surface, VLF waves have very little path attenuation (2–3 dB per 1000 km), so they can be used for long-distance communication without repeaters.* □

7.2 The baseband-equivalent of a passband signal

In this section we show how a real-valued *passband signal* $x(t)$ can be represented by a *baseband-equivalent* $x_E(t)$, which is in general complex-valued. The relationship between these two signals is established via the *analytic-equivalent* $\hat{x}(t)$.

Recall two facts from Fourier analysis. If $x(t)$ is a *real-valued* signal, then its Fourier transform $x_{\mathcal{F}}(f)$ is *conjugate symmetric*, that is

$$x_{\mathcal{F}}^*(f) = x_{\mathcal{F}}(-f), \tag{7.1}$$

where $x_{\mathcal{F}}^*(f)$ is the complex conjugate of $x_{\mathcal{F}}(f)$.[1] If $x(t)$ is a *purely imaginary* signal, then its Fourier transform is *conjugate anti-symmetric*, i.e.

$$x_{\mathcal{F}}^*(f) = -x_{\mathcal{F}}(-f).$$

The symmetry and anti-symmetry properties can easily be verified from the definition of the Fourier transform and the fact that the complex conjugate operator commutes with both the integral and the product. For instance, the proof of the symmetry property is

$$x_{\mathcal{F}}^*(f) = \left[\int x(t)e^{-j2\pi ft}dt\right]^*$$

$$= \int \left[x(t)e^{-j2\pi ft}\right]^* dt$$

$$= \int x^*(t)e^{j2\pi ft}dt$$

$$= \int x(t)e^{-j2\pi(-f)t}dt$$

$$= x_{\mathcal{F}}(-f).$$

---

[1] In principle, the notation $x_{\mathcal{F}}^*(f)$ could mean $(x_{\mathcal{F}})^*(f)$ or $(x^*)_{\mathcal{F}}(f)$, but it should be clear that we mean the former because the latter is not useful when $x(t)$ is real-valued, in which case $(x^*)_{\mathcal{F}}(f) = x_{\mathcal{F}}(f)$.

The symmetry property implies that the Fourier transform $x_\mathcal{F}(f)$ of a *real-valued* signal $x(t)$ has redundant information: If we know $x_\mathcal{F}(f)$ for $f \geq 0$, then we know it also for $f < 0$.

If we remove the negative frequencies from $x(t)$ and scale the result by $\sqrt{2}$, we obtain the complex-valued signal $\hat{x}(t)$ called the *analytic-equivalent* of $x(t)$. Intuitively, by removing the negative frequencies of a real-valued signal we reduce its norm by $\sqrt{2}$. The scaling by $\sqrt{2}$ in the definition of $\hat{x}(t)$ is meant to make the norm of $\hat{x}(t)$ identical to that of $x(t)$ (see Corollary 7.6 for a formal proof).

To remove the negative frequencies of $x(t)$ we use the filter of impulse response $h_>(t)$ that has Fourier transform[2]

$$h_{>,\mathcal{F}}(f) = \begin{cases} 1 & \text{for } f \geq 0 \\ 0 & \text{for } f < 0. \end{cases} \tag{7.2}$$

Hence,

$$\hat{x}_\mathcal{F}(f) = \sqrt{2}x_\mathcal{F}(f)h_{>,\mathcal{F}}(f), \tag{7.3}$$

$$\hat{x}(t) = \sqrt{2}(x \star h_>)(t). \tag{7.4}$$

It is straightforward to go from $\hat{x}(t)$ back to $x(t)$. We claim that

$$x(t) = \sqrt{2}\Re\{\hat{x}(t)\}, \tag{7.5}$$

where, once again, we remember the factor $\sqrt{2}$ as being the one that compensates for halving the signal's energy by removing the signal's imaginary part. To prove (7.5), we use the fact that the real part of a complex number is half the sum of the number and its complex conjugate, i.e.

$$\sqrt{2}\Re\{\hat{x}(t)\} = \frac{1}{\sqrt{2}}\left(\hat{x}(t) + \hat{x}^*(t)\right).$$

To complete the proof, it suffices to show that the Fourier transform of the right-hand side, namely

$$x_\mathcal{F}(f)h_{>,\mathcal{F}}(f) + x_\mathcal{F}^*(-f)h_{>,\mathcal{F}}^*(-f)$$

is indeed $x_\mathcal{F}(f)$. For non-negative frequencies, the first term equals $x_\mathcal{F}(f)$ and the second term vanishes. Hence the Fourier transform of $\sqrt{2}\Re\{\hat{x}(t)\}$ and that of $x(t)$ agree for non-negative frequencies. As they are the Fourier transform of real-valued signals, they must agree everywhere. This proves that $\sqrt{2}\Re\{\hat{x}(t)\}$ and $x(t)$ have the same Fourier transform, hence they are $\mathcal{L}_2$ equivalent.

To go from $\hat{x}(t)$ to the baseband-equivalent $x_E(t)$ we use the *frequency-shifting property* of the Fourier transform that we rewrite for reference:

$$x(t)e^{j2\pi f_c t} \longleftrightarrow x_\mathcal{F}(f - f_c).$$

---

[2] Note that $h_{>,\mathcal{F}}(f)$ is not an $\mathcal{L}_2$ function, but it can be made into one by setting it to zero at all frequencies that are outside the support of $x_\mathcal{F}(f)$. Note also that we can arbitrarily choose the value of $h_{>,\mathcal{F}}(f)$ at $f = 0$, because two functions that differ at a single point are $\mathcal{L}_2$ equivalent.

7.2. The baseband-equivalent of a passband signal

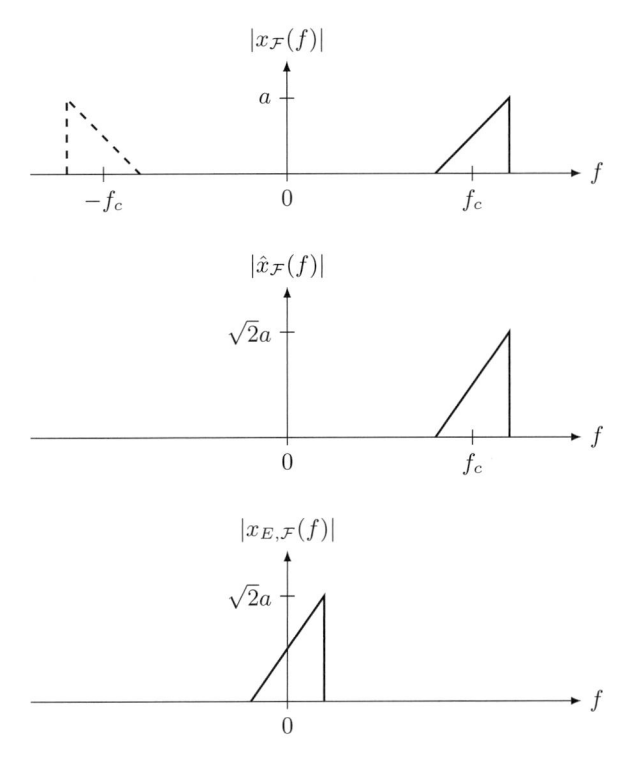

Figure 7.3. Fourier-domain relationship between a real-valued signal, and the corresponding analytic and baseband-equivalent signals.

The *baseband-equivalent* of $x(t)$ with respect to the carrier frequency f_c is defined to be

$$x_E(t) = \hat{x}(t)e^{-j2\pi f_c t}$$

and its Fourier transform is

$$x_{E,\mathcal{F}}(f) = \hat{x}_{\mathcal{F}}(f + f_c).$$

Figure 7.3 depicts the relationship between $|x_{\mathcal{F}}(f)|$, $|\hat{x}_{\mathcal{F}}(f)|$, and $|x_{E,\mathcal{F}}(f)|$. We plot the absolute value to avoid plotting the real and the imaginary components. We use dashed lines to plot $|x_{\mathcal{F}}(f)|$ for $f < 0$ as a reminder that it is completely determined by $|x_{\mathcal{F}}(f)|$, $f > 0$.

The operation that recovers $x(t)$ from its baseband-equivalent $x_E(t)$ is

$$x(t) = \sqrt{2}\Re\{x_E(t)e^{j2\pi f_c t}\}. \tag{7.6}$$

The circuits to go from $x(t)$ to $x_E(t)$ and back to $x(t)$ are depicted in Figure 7.4, where double arrows denote complex-valued signals. Exercises 7.3 and 7.5 derive equivalent circuits that require only operations over the reals.

The following theorem and the two subsequent corollaries are important in that they establish a geometrical link between baseband and passband signals.

(a)

(b)

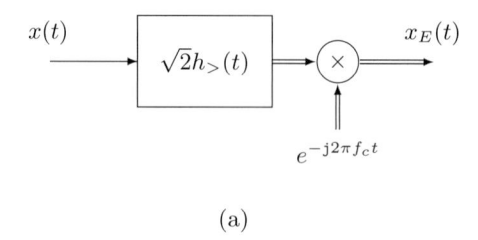

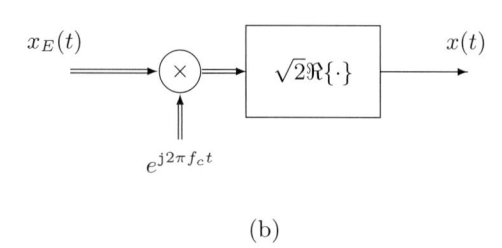

Figure 7.4. From a real-valued signal $x(t)$ to its baseband-equivalent $x_E(t)$ and back.

THEOREM 7.5 (Inner product of passband signals) *Let $x(t)$ and $y(t)$ be (real-valued) passband signals, let $\hat{x}(t)$ and $\hat{y}(t)$ be the corresponding analytic signals, and let $x_E(t)$ and $y_E(t)$ be the baseband-equivalent signals (with respect to a common carrier frequency f_c). Then*

$$\langle x, y \rangle = \Re\{\langle \hat{x}, \hat{y} \rangle\} = \Re\{\langle x_E, y_E \rangle\}.$$

□

Note 1: $\langle x, y \rangle$ is real-valued, whereas $\langle \hat{x}, \hat{y} \rangle$ and $\langle x_E, y_E \rangle$ are complex-valued in general. This helps us see/remember why the theorem cannot hold without taking the real part of the last two inner products. The reader might prefer to remember the more symmetric (and more redundant) form $\Re\{\langle x, y \rangle\} = \Re\{\langle \hat{x}, \hat{y} \rangle\} = \Re\{\langle x_E, y_E \rangle\}$.

Note 2: From the proof that follows, we see that the second equality holds also for the imaginary parts, i.e. $\langle \hat{x}, \hat{y} \rangle = \langle x_E, y_E \rangle$.

Proof Let $\hat{x}(t) = x_E(t)e^{j2\pi f_c t}$. Showing that $\langle \hat{x}, \hat{y} \rangle = \langle x_E, y_E \rangle$ is immediate:

$$\langle \hat{x}, \hat{y} \rangle = \langle x_E(t)e^{j2\pi f_c t}, y_E(t)e^{j2\pi f_c t} \rangle = e^{j2\pi f_c t}e^{-j2\pi f_c t}\langle x_E(t), y_E(t) \rangle = \langle x_E, y_E \rangle.$$

To prove $\langle x, y \rangle = \Re\{\langle \hat{x}, \hat{y} \rangle\}$, we use Parseval's relationship (first and last equality below), we use the fact that the Fourier transform of $x(t) = \frac{1}{\sqrt{2}}[\hat{x}(t) + \hat{x}^*(t)]$ is $x_\mathcal{F}(f) = \frac{1}{\sqrt{2}}[\hat{x}_\mathcal{F}(f) + \hat{x}_\mathcal{F}^*(-f)]$ (second equality), that $\hat{x}_\mathcal{F}(f)\hat{y}_\mathcal{F}(-f) = 0$ because the two functions have disjoint support and similarly $\hat{x}_\mathcal{F}^*(-f)\hat{y}_\mathcal{F}^*(f) = 0$ (third

7.2. The baseband-equivalent of a passband signal

equality), and finally that the integral over a function is the same as the integral over the time-reversed function (fourth equality):

$$\langle x, y \rangle = \int x_{\mathcal{F}}(f) y_{\mathcal{F}}^*(f) df$$

$$= \frac{1}{2} \int \left[\hat{x}_{\mathcal{F}}(f) + \hat{x}_{\mathcal{F}}^*(-f) \right] \left[\hat{y}_{\mathcal{F}}^*(f) + \hat{y}_{\mathcal{F}}(-f) \right] df$$

$$= \frac{1}{2} \int \left[\hat{x}_{\mathcal{F}}(f) \hat{y}_{\mathcal{F}}^*(f) + \hat{x}_{\mathcal{F}}^*(-f) \hat{y}_{\mathcal{F}}(-f) \right] df$$

$$= \frac{1}{2} \int \left[\hat{x}_{\mathcal{F}}(f) \hat{y}_{\mathcal{F}}^*(f) + \hat{x}_{\mathcal{F}}^*(f) \hat{y}_{\mathcal{F}}(f) \right] df$$

$$= \Re \left\{ \int \hat{x}_{\mathcal{F}}(f) \hat{y}_{\mathcal{F}}^*(f) df \right\}$$

$$= \Re \{ \langle \hat{x}, \hat{y} \rangle \}.$$

□

The following two corollaries are immediate consequences of the above theorem. The first proves that the scaling factor $\sqrt{2}$ in (7.4) and (7.5) is what keeps the norm unchanged. We will use the second to prove Theorem 7.13.

COROLLARY 7.6 (Norm preservation) *A passband signal has the same norm as its analytic and its baseband-equivalent signals, i.e.*

$$\|x\|^2 = \|\hat{x}\|^2 = \|x_E\|^2.$$

□

COROLLARY 7.7 (Orthogonality of passband signals) *Two passband signals are orthogonal if and only if the inner product between their baseband-equivalent signals (with respect to a common carrier frequency f_c) is purely imaginary (i.e. the real part vanishes).* □

Typically, we are interested in the baseband-equivalent $x_E(t)$ of a passband signal $x(t)$, but from a mathematical point of view, $x(t)$ need not be passband for $x_E(t)$ to be defined. Specifically, we can feed the circuit of Figure 7.4a with any real-valued signal $x(t)$ and feed the baseband-equivalent output $x_E(t)$ to the circuit of Figure 7.4b to recover $x(t)$.[3]

However, if we reverse the order in Figure 7.4, namely feed the circuit of part (b) with an arbitrary signal $g(t)$ and feed the circuit's output to the input of part (a), we do not necessarily recover $g(t)$ (see Exercise 7.4), unless we set some restriction on $g(t)$. The following lemma sets such a restriction on $g(t)$. (It will be used in the proof of Theorem 7.13.)

LEMMA 7.8 *If $g(t)$ is bandlimited to $[-b, \infty)$ for some $b > 0$ and $f_c > b$, then $g(t)$ is the baseband-equivalent of $\sqrt{2}\Re\{g(t)e^{j2\pi f_c t}\}$ with respect to the carrier frequency f_c.* □

---

[3] It would be a misnomer to call $x_E(t)$ a baseband signal if $x(t)$ is not passband.

Proof If $g(t)$ satisfies the stated condition, then $g(t)e^{\mathrm{j}2\pi f_c t}$ has no negative frequencies. Hence $g(t)e^{\mathrm{j}2\pi f_c t}$ is the analytic signal $\hat{x}(t)$ of $x(t) = \sqrt{2}\Re\{g(t)e^{\mathrm{j}2\pi f_c t}\}$, which implies that $g(t)$ is the baseband-equivalent $x_E(t)$ of $x(t)$. □

Hereafter all passband signals are assumed to be real-valued as they represent actual communication signals. Baseband signals can be signals that we use for baseband communication on real-world channels or can be baseband-equivalents of passband signals. In the latter case, they are complex-valued in general.

7.2.1 Analog amplitude modulations: DSB, AM, SSB, QAM

There is a family of analog modulation techniques, called *amplitude modulations*, that can be seen as a direct application of what we have learned in this section. The well-known AM modulation used in broadcasting is the most popular member of this family.

EXAMPLE 7.9 (Double-sideband modulation with suppressed carrier (DSB-SC)) *Let the source signal be a real-valued baseband signal $b(t)$. The arguably easiest way to convert $b(t)$ into a passband signal $x(t)$ that has the same norm as $b(t)$ is to let $x(t) = \sqrt{2}b(t)\cos(2\pi f_c t)$. This is amplitude modulation in the sense that the carrier $\sqrt{2}\cos(2\pi f_c t)$ is being amplitude modulated by the analog information signal $b(t)$. To see how this relates to what we have learned in the previous section, we write*

$$x(t) = \sqrt{2}b(t)\cos(2\pi f_c t)$$
$$= \sqrt{2}\Re\{b(t)e^{\mathrm{j}2\pi f_c t}\}.$$

If $b(t)$ is bandlimited to $[-B, B]$ and the carrier frequency satisfies $f_c > B$, which we assume to be the case, then $b(t)$ is the baseband-equivalent of the passband signal $x(t)$. Figure 7.5 gives an example of the Fourier-domain relationship between the

(a) Information signal.

(b) DSB-SC modulated signal.

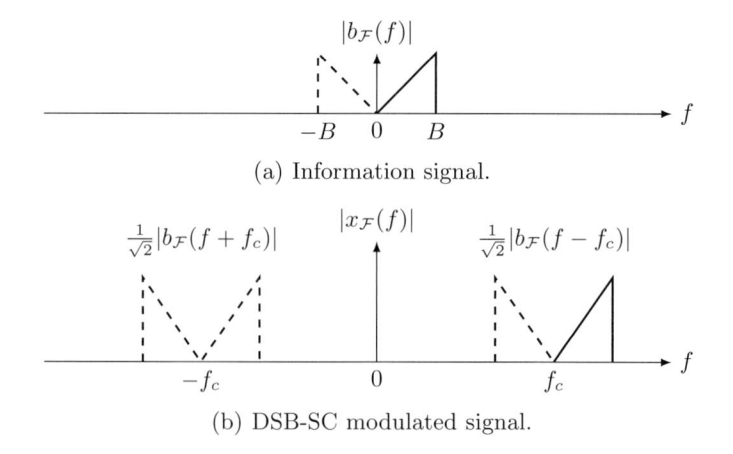

Figure 7.5. Spectrum of a double-sideband modulated signal.

7.2. The baseband-equivalent of a passband signal

baseband information signal $b(t)$ and the modulated signal $x(t)$. The dashed parts of the plots are meant to remind us that they can be determined from the solid parts.

This modulation scheme is called "double-sideband" because of the two bands on the left and right of f_c, only one is needed to recover $b(t)$. Specifically, we could eliminate the sideband below f_c; and to preserve the conjugacy symmetry required by real-valued signals, we would eliminate also the sideband above $-f_c$ and still be able to recover the information signal $b(t)$ from the resulting passband signal. Hence, we could eliminate one of the sidebands and thereby reduce the bandwidth and the energy by a factor 2. (See Example 7.11.) The SC (suppressed carrier) part of the name distinguishes this modulation technique from AM (amplitude modulation, see next example), which is indeed a double-sideband modulation with carrier (at $\pm f_c$). □

EXAMPLE 7.10 (AM modulation) *AM modulation is by far the most popular member of the family of amplitude modulations. Let $b(t)$ be the source signal, and assume that it is zero-mean and $|b(t)| \leq 1$ for all t. AM modulation of $b(t)$ is DSB-SC modulation of $1 + mb(t)$ for some modulation index m such that $0 < m \leq 1$. Notice that $1 + mb(t)$ is always non-negative. By using this fact, the receiver can be significantly simplified (see Exercise 7.7). The possibility of building inexpensive receivers is what made AM modulation the modulation of choice in early radio broadcasting. AM is also a double-sideband modulation but, unlike DSB-SC, it has a carrier at $\pm f_c$. We see the carrier by expanding $x(t) = (1 + mb(t))\sqrt{2}\cos(2\pi f_c t) = mb(t)\sqrt{2}\cos(2\pi f_c t) + \sqrt{2}\cos(2\pi f_c t)$. The carrier consumes energy without carrying any information. It is the "price" that broadcasters are willing to pay to reduce the cost of the receiver. The trade-off seems reasonable given that there is one sender and many receivers.* □

The following two examples are bandwidth-efficient variants of double-sideband modulation.

EXAMPLE 7.11 (Single-sideband modulation (SSB)) *As in the previous example, let $b(t)$ be the real-valued baseband information signal. Let $\hat{b}(t) = (b \star h_>)(t)$ be the analytic-equivalent of $b(t)$. We define $x(t)$ to be the passband signal that has $\hat{b}(t)$ as its baseband-equivalent (with respect to the desired carrier frequency). Figure 7.6 shows the various frequency-domain signals. A comparison with Figure 7.5 should suffice to understand why this process is called single-sideband modulation. Single-sideband modulation is widely used in amateur radio communication. Instead of removing the negative frequencies of the original baseband signal we could remove the positive frequencies. The two alternatives are called SSB-USB (USB stands for upper side-band) and SSB-LSB (lower side-band), respectively. A drawback of SSB is that it requires a sharp filter to remove the negative frequencies. Amateur radio people are willing to pay this price to make efficient use of the limited spectrum allocated to them.* □

EXAMPLE 7.12 (Quadrature amplitude modulation (QAM)) *The idea consists of taking two real-valued baseband information signals, say $b_R(t)$ and $b_I(t)$, and forming the signal $b(t) = b_R(t) + \jmath b_I(t)$. As $b(t)$ is complex-valued, its Fourier*

(a) Information signal.

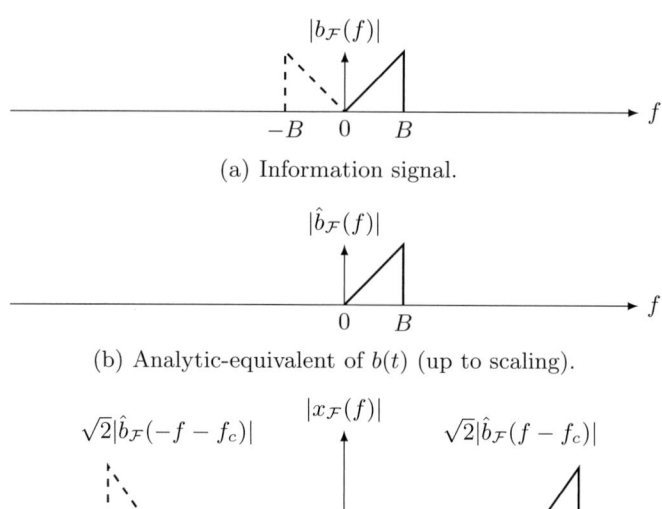

(b) Analytic-equivalent of $b(t)$ (up to scaling).

(c) SSB modulated signal.

Figure 7.6. Spectrum of SSB-USB.

(a) Information signal.

(b) Modulated signal.

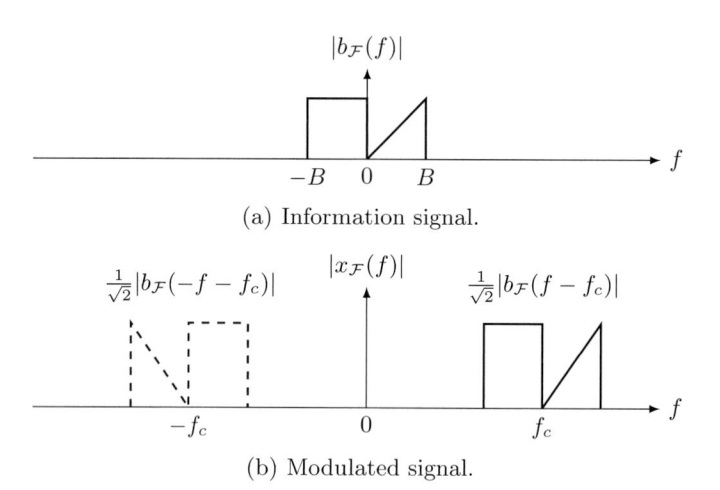

Figure 7.7. Spectrum of QAM.

transform no longer satisfies the conjugacy constraint. Hence the asymmetry of $|b_{\mathcal{F}}(f)|$ on the top of Figure 7.7. We let the passband signal $x(t)$ be the one that has baseband-equivalent $b(t)$ (with respect to the desired carrier frequency). The spectrum of $x(t)$ is shown on the bottom plot of Figure 7.7. QAM is as bandwidth efficient as SSB. Unlike SSB, which achieves the bandwidth efficiency by removing the extra frequencies, QAM doubles the information content. The advantage of

QAM over SSB is that it does not require a sharp filter to remove one of the two sidebands. The drawback is that typically a sender has one, not two, analog signals to send. QAM is not popular as an analog modulation technique. However, it is a very popular technique for digital communication. The idea is to split the bits into two streams, with each stream doing symbol-by-symbol on a pulse train to obtain, say, $b_R(t)$ and $b_I(t)$ respectively, and then proceeding as described above. (See Example 7.15.) □

7.3 The third layer

In this section, we revisit the second layer using the tools that we have learned in Section 7.2. From a structural point of view, the outcome is that the second layer splits into two, giving us the third layer. The third layer enables us to choose the carrier frequency independently from the shape of the power spectral density. It gives us also the freedom to vary the carrier frequency in an easy way without redesigning the system. Various additional advantages are discussed in Section 7.7. We assume passband communication over the AWGN channel.

We start with the ML (or MAP) receiver, following the general approach learned in Chapter 3. Let $\psi_1(t), \ldots, \psi_n(t)$ be an orthonormal basis for the vector space spanned by the passband signals $w_0(t), \ldots, w_{m-1}(t)$. The n-tuple former computes

$$Y = (Y_1, \ldots, Y_n)^\mathsf{T}$$
$$Y_l = \langle R(t), \psi_l(t) \rangle, \quad l = 1, \ldots, n.$$

When $H = i$,

$$Y = c_i + Z,$$

where $c_i \in \mathbb{R}^n$ is the codeword associated to $w_i(t)$, with respect to the orthonormal basis $\psi_1(t), \ldots, \psi_n(t)$ and $Z \sim \mathcal{N}(0, \frac{N_0}{2} I_n)$. When $Y = y$, an ML decoder chooses $\hat{H} = \hat{\imath}$ for one of the $\hat{\imath}$ that minimizes $\|y - c_{\hat{\imath}}\|$ or, equivalently, for one of the $\hat{\imath}$ that maximizes $\langle y, c_{\hat{\imath}} \rangle - \|c_{\hat{\imath}}\|^2 / 2$.

This receiver has the serious drawback that all of its stages depend on the choice of the passband signals $w_0(t), \ldots, w_{m-1}(t)$. For instance, suppose that the sender decides to use a different frequency band. If the signaling method is symbol-by-symbol on a pulse train based on a pulse $\psi(t)$ orthogonal to its T-spaced translates, i.e. such that $|\psi_\mathcal{F}(f)|^2$ fulfills Nyquist's criterion with parameter T, then one way to change the frequency band is to use a different pulse $\tilde{\psi}(t)$ that, as the original pulse, is orthogonal to its T-spaced translates, and such that $|\tilde{\psi}_\mathcal{F}(f)|^2$ occupies the desired frequency band. Such a pulse $\tilde{\psi}(t)$ exists only for certain frequency offsets with respect to the original band (see Exercise 7.14), a fact that makes this approach of limited interest. Still, if we use this approach, then the n-tuple former has to be adapted to the new pulse $\tilde{\psi}(t)$. A much more interesting approach, that applies to any signaling scheme (not only symbol-by-symbol on a pulse train), consists in using the ideas developed in Section 7.2 to frequency-translate $w_0(t), \ldots, w_{m-1}(t)$ into a new set of signals $\tilde{w}_0(t), \ldots, \tilde{w}_{m-1}(t)$ that occupy the desired frequency band. This can be done quite effectively; we will see how. But if we re-design the receiver starting with a new arbitrarily-selected orthonormal basis for the new signal set, then we see that the n-tuple former as well as the decoder could end up being totally different from the original ones. Using the results of Section 7.2, we can find a flexible and elegant solution to this problem, so that we can frequency-translate the signal's band to any desired location without affecting the n-tuple former and the decoder. (The encoder and the waveform former are not affected either.)

Let $w_{E,0}(t), \ldots, w_{E,m-1}(t)$ be the baseband-equivalent signal constellation. We assume that they belong to a complex inner product space and let $\psi_1(t), \ldots, \psi_n(t)$ be an orthonormal basis for this space. Let $c_i = (c_{i,1}, \ldots, c_{i,n})^\mathsf{T} \in \mathbb{C}^n$ be the codeword associated to $w_{E,i}(t)$, i.e.

$$w_{E,i}(t) = \sum_{l=1}^{n} c_{i,l} \psi_l(t),$$

$$w_i(t) = \sqrt{2} \Re\{w_{E,i}(t) e^{j2\pi f_c t}\}.$$

The orthonormal basis for the baseband-equivalent signal set can be lifted up to an orthonormal basis for the passband signal set as follows.

$$w_i(t) = \sqrt{2} \Re\{w_{E,i}(t) e^{j2\pi f_c t}\}$$

$$= \sqrt{2} \Re\{\sum_{l=1}^{n} c_{i,l} \psi_l(t) e^{j2\pi f_c t}\}$$

$$= \sqrt{2} \sum_{l=1}^{n} \Re\{c_{i,l} \psi_l(t) e^{j2\pi f_c t}\}$$

$$= \sqrt{2} \sum_{l=1}^{n} \Re\{c_{i,l}\} \Re\{\psi_l(t) e^{j2\pi f_c t}\}$$

$$- \sqrt{2} \sum_{l=1}^{n} \Im\{c_{i,l}\} \Im\{\psi_l(t) e^{j2\pi f_c t}\}$$

$$= \sum_{l=1}^{n} \Re\{c_{i,l}\} \psi_{1,l}(t) + \sum_{l=1}^{n} \Im\{c_{i,l}\} \psi_{2,l}(t), \qquad (7.7)$$

with $\psi_{1,l}(t)$ and $\psi_{2,l}(t)$ in (7.7) defined as

$$\psi_{1,l}(t) = \sqrt{2} \Re\{\psi_l(t) e^{j2\pi f_c t}\}, \qquad (7.8)$$

$$\psi_{2,l}(t) = -\sqrt{2} \Im\{\psi_l(t) e^{j2\pi f_c t}\}. \qquad (7.9)$$

From (7.7), we see that the set $\{\psi_{1,1}(t), \ldots, \psi_{1,n}(t), \psi_{2,1}(t), \ldots, \psi_{2,n}(t)\}$ spans a vector space that contains the passband signals. As stated by the next theorem, this set forms an orthonormal basis, provided that the carrier frequency is sufficiently high.

7.3. The third layer

THEOREM 7.13 *Let $\{\psi_l(t) : l = 1, 2, \ldots, n\}$ be an orthonormal set of functions that are frequency-limited to $[-B, B]$ for some $B > 0$ and let $f_c > B$. Then the set*

$$\{\psi_{1,l}(t), \psi_{2,l}(t) : l = 1, 2, \ldots, n\} \tag{7.10}$$

defined via (7.8) and (7.9) consists of orthonormal functions. Furthermore, if $w_{E,i}(t) = \sum_{l=1}^{n} c_{i,l} \psi_l(t)$, then

$$w_i(t) = \sqrt{2}\Re\{w_{E,i}(t)e^{\mathrm{j}2\pi f_c t}\} = \sum_{l=1}^{n} \Re\{c_{i,l}\}\psi_{1,l}(t) + \sum_{l=1}^{n} \Im\{c_{i,l}\}\psi_{2,l}(t).$$

□

Proof The last statement is (7.7). Hence (7.10) spans a vector space that contains the passband signals. It remains to be shown that this set is orthonormal. From Lemma 7.8, the baseband-equivalent signal of $\psi_{1,l}(t)$ is $\psi_l(t)$. Similarly, by writing $\psi_{2,l}(t) = \sqrt{2}\Re\{[\mathrm{j}\psi_l(t)]e^{\mathrm{j}2\pi f_c t}\}$, we see that the baseband-equivalent of $\psi_{2,l}(t)$ is $\mathrm{j}\psi_l(t)$. From Corollary 7.7, $\langle\psi_{1,k}(t), \psi_{1,l}(t)\rangle = \Re\{\langle\psi_k(t), \psi_l(t)\rangle\} = \mathbb{1}\{k = l\}$, showing that the set $\{\psi_{1,l}(t) : l = 1, \ldots, n\}$ is made of orthonormal functions. Similarly, $\langle\psi_{2,k}(t), \psi_{2,l}(t)\rangle = \Re\{\langle\mathrm{j}\psi_k(t), \mathrm{j}\psi_l(t)\rangle\} = \Re\{\langle\psi_k(t), \psi_l(t)\rangle\} = \mathbb{1}\{k = l\}$, showing that also $\{\psi_{2,l}(t) : l = 1, \ldots, n\}$ is made of orthonormal functions. To conclude the proof, it remains to be shown that functions from the first set are orthogonal to functions from the second set. Indeed $\langle\psi_{1,k}(t), \psi_{2,l}(t)\rangle = \Re\{\langle\psi_k(t), \mathrm{j}\psi_l(t)\rangle\} = \Re\{-\mathrm{j}\langle\psi_k(t), \psi_l(t)\rangle\} = 0$. The last equality holds for $k \neq l$ because ψ_k and ψ_l are orthogonal and it holds for $k = l$ because $\langle\psi_k(t), \psi_k(t)\rangle = \|\psi_k(t)\|^2$ is real. □

From the above theorem, we see that if the vector space spanned by the baseband-equivalent signals has dimensionality n, the vector space spanned by the corresponding passband signals has dimensionality $2n$. However, the number of real-valued "degrees of freedom" is the same in both spaces. In fact, the coefficients used in the orthonormal expansion of the baseband signals are complex, hence with two degrees of freedom per coefficient, whereas those used in the orthonormal expansion of the passband signals are real.

Next we re-design the receiver using Theorem 7.13 to construct an orthonormal basis for the passband signals. The $2n$-tuple former now computes $Y_1 = (Y_{1,1}, \ldots, Y_{1,n})^\mathsf{T}$ and $Y_2 = (Y_{2,1}, \ldots, Y_{2,n})^\mathsf{T}$, where for $l = 1, \ldots, n$

$$Y_{1,l} = \langle R(t), \psi_{1,l}(t)\rangle \tag{7.11}$$
$$= \langle R(t), \sqrt{2}\Re\{\psi_l(t)e^{\mathrm{j}2\pi f_c t}\}\rangle$$
$$= \Re\{\langle R(t), \sqrt{2}\psi_l(t)e^{\mathrm{j}2\pi f_c t}\rangle\}$$
$$= \Re\{\langle\sqrt{2}e^{-\mathrm{j}2\pi f_c t}R(t), \psi_l(t)\rangle\}, \tag{7.12}$$

and similarly

$$Y_{2,l} = \langle R(t), \psi_{2,l}(t)\rangle \tag{7.13}$$
$$= \Im\{\langle\sqrt{2}e^{-\mathrm{j}2\pi f_c t}R(t), \psi_l(t)\rangle\}. \tag{7.14}$$

To simplify notation, we define the complex random vector $Y \in \mathbb{C}^n$

$$Y = Y_1 + \mathrm{j} Y_2. \tag{7.15}$$

The lth component of Y is

$$Y_l = Y_{1,l} + \mathrm{j} Y_{2,l} \tag{7.16}$$
$$= \langle \sqrt{2} e^{-\mathrm{j} 2\pi f_c t} R(t), \psi_l(t) \rangle. \tag{7.17}$$

This is an important turning point: it is where we introduce complex-valued notation in the receiver design.

Figure 7.8 shows the new architecture, where we use double lines for connections that carry complex-valued quantities (symbols, n-tuples, signals). At the transmitter, the waveform former decomposes into a waveform former for the baseband signal, followed by the *up-converter*. At the receiver, the n-tuple former for passband decomposes into the *down-converter* followed by an n-tuple former for baseband. The significant advantage of this architecture is that f_c affects only the up/down-converter. With this architecture, varying the carrier frequency means turning the knob that varies the frequency of the oscillators that generate $e^{\pm \mathrm{j} 2\pi f_c t}$ in the up/down-converter.

The operations performed by the blocks of Figure 7.8 can be implemented in a digital signal processor (DSP) or with analog electronics. If the operation is done

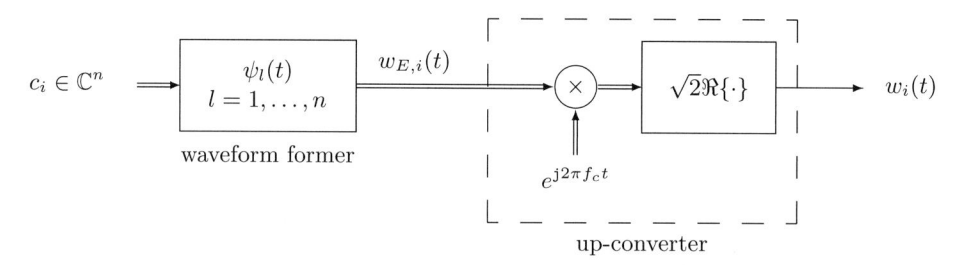

(a) Transmitter back end.

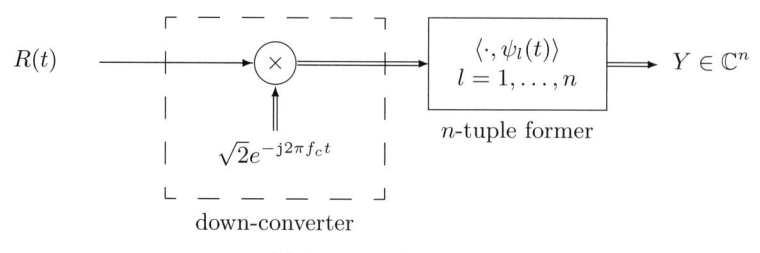

(b) Receiver front end.

Figure 7.8. Baseband waveform former and up-converter at the transmitter back end (a); down-converter and baseband n-tuple former at the receiver front end (b).

7.3. The third layer

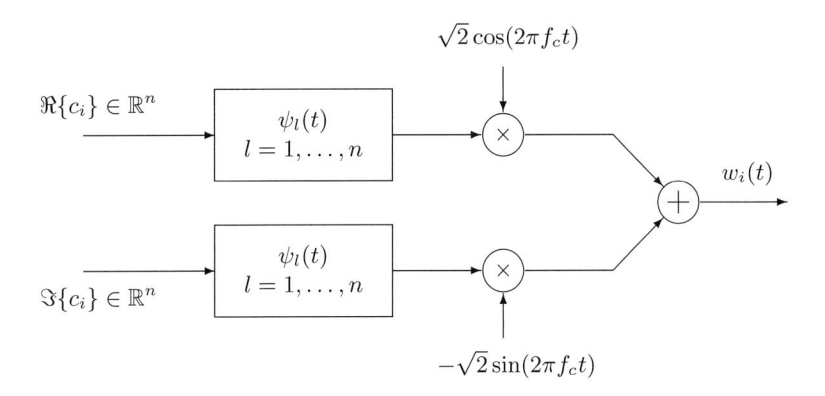

(a) Transmitter back end.

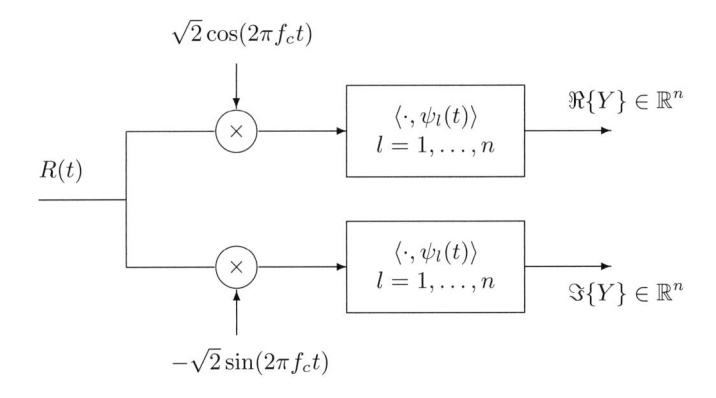

(b) Receiver front end.

Figure 7.9. Real-valued implementation of the diagrams of Figure 7.8 for a real-valued orthonormal basis $\psi_l(t)$, $l = 1, \ldots, n$.

in a DSP, the programmer might be able to rely on functions that can cope with complex numbers. If done with analog electronics, the real and the imaginary parts are kept separate. This is shown in Figure 7.9, for the common situation where the orthonormal basis is real-valued. There is no loss in performance in choosing a real-valued basis and, if we do so, the implementation complexity using analog circuitry is essentially halved (see Exercise 7.9).

We have reached a conceptual milestone, namely the point where working with complex-valued signals becomes natural. It is worth being explicit about how and why we make this important transition. In principle, we are only combining two real-valued vectors of equal length into a single complex-valued vector of the same length (see (7.15)). Because it is a reversible operation, we can always pack a pair of real numbers into a complex number, and we should do so if it provides an advantage. In our case, we can identify a few benefits by doing so. First, the expressions are simplified: compare the pair (7.12) and (7.14) to the single and somewhat simpler (7.17). Similarly, the block diagrams are simplified: compare Figure 7.8 with Figure 7.9, keeping in mind that the former is general, whereas the latter becomes more complicated if the orthonormal basis is complex-valued. (see Exercise 7.9). Finally, as we will see, the expression for the density of the complex random vector Y takes a somewhat simpler form than that of $\Re\{Y\}$ (or of $\Im\{Y\}$), thus simpler than the joint density of $\Re\{Y\}$ and $\Im\{Y\}$, which is what we need if we keep the real and the imaginary parts separate.

EXAMPLE 7.14 (PSK signaling via complex-valued symbols) *Consider the signals*

$$w_E(t) = \sum_l s_l \psi(t - lT)$$

$$w(t) = \sqrt{2}\Re\left\{w_E(t)e^{j2\pi f_c t}\right\},$$

where $\psi(t)$ is real-valued, normalized, and orthogonal to its T-spaced time-translates, and where symbols take value in a 4-ary PSK alphabet, seen as a subset of $\mathbb{C}$. So we can write a symbol s_l as $\sqrt{\mathcal{E}}e^{j\varphi_l}$ with $\varphi_l \in \{0, \frac{\pi}{2}, \pi, \frac{3\pi}{2}\}$ or, alternatively, as $s_l = \Re\{s_l\} + j\Im\{s_l\}$. We work it out both ways.

If we plug $s_l = \sqrt{\mathcal{E}}e^{j\varphi_l}$ into $w_E(t)$ we obtain

$$w(t) = \sqrt{2}\Re\left\{\left(\sum_l \sqrt{\mathcal{E}}e^{j\varphi_l}\psi(t-lT)\right)e^{j2\pi f_c t}\right\}$$

$$= \sqrt{2}\Re\left\{\sum_l \sqrt{\mathcal{E}}e^{j(2\pi f_c t + \varphi_l)}\psi(t-lT)\right\}$$

$$= \sqrt{2}\sum_l \sqrt{\mathcal{E}}\Re\left\{e^{j(2\pi f_c t + \varphi_l)}\right\}\psi(t-lT)$$

$$= \sqrt{2\mathcal{E}}\sum_l \cos(2\pi f_c t + \varphi_l)\psi(t-lT).$$

Figure 7.10 shows a sample $w(t)$ with $\psi(t) = \sqrt{\frac{1}{T}}\mathbb{1}\{0 \leq t < T\}$, $T = 1$, $f_c T = 3$ (there are three periods in a symbol interval T), $\mathcal{E} = \frac{1}{2}$, $\varphi_0 = 0$, $\varphi_1 = \pi$, $\varphi_2 = \frac{\pi}{2}$, and $\varphi_3 = \frac{3\pi}{2}$.

If we plug $s_l = \Re\{s_l\} + j\Im\{s_l\}$ into $w_E(t)$ we obtain

$$w(t) = \sqrt{2}\Re\left\{\left(\sum_l (\Re\{s_l\} + j\Im\{s_l\})\psi(t-lT)\right)e^{j2\pi f_c t}\right\}$$

$$= \sqrt{2}\sum_l \Re\left\{(\Re\{s_l\} + j\Im\{s_l\})e^{j2\pi f_c t}\right\}\psi(t-lT)$$

$$= \sqrt{2}\sum_l \left(\Re\{s_l\}\Re\{e^{j2\pi f_c t}\} - \Im\{s_l\}\Im\{e^{j2\pi f_c t}\}\right)\psi(t-lT)$$

7.3. The third layer

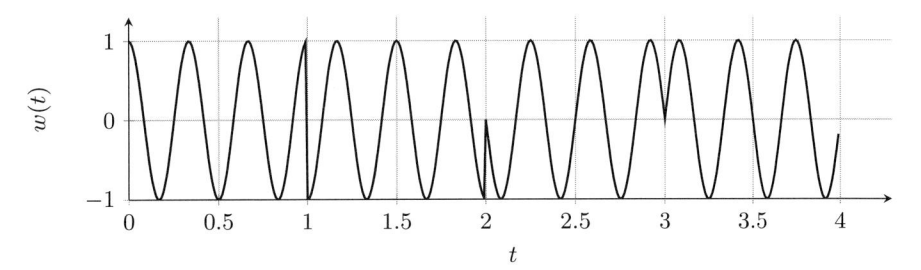

Figure 7.10. Sample PSK modulated signal.

$$= \sqrt{2}\sum_l \Re\{s_l\}\psi(t-lT)\cos(2\pi f_c t)$$

$$- \sqrt{2}\sum_l \Im\{s_l\}\psi(t-lT)\sin(2\pi f_c t). \qquad (7.18)$$

For a rectangular pulse $\psi(t)$, $\sqrt{2}\psi(t-lT)\cos(2\pi f_c t)$ is orthogonal to $\sqrt{2}\psi(t-iT)\sin(2\pi f_c t)$ for all integers l and i, provided that $2f_c T$ is an integer.[4] From (7.18), we see that the PSK signal is the superposition of two PAM signals. This view is not very useful for PSK, because $\Re\{s_l\}$ and $\Im\{s_l\}$ cannot be chosen independently of each other.[5] Hence the two superposed signals cannot be decoded independently. It is more useful for QAM. (See next example.) □

EXAMPLE 7.15 (QAM signaling via complex-valued symbols) *Suppose that the signaling method is as in Example 7.14 but that the symbols take value in a QAM alphabet. As in Example 7.14, it is instructive to write the symbols in two ways. If we write $s_l = a_l e^{j\varphi_l}$, then proceeding as in Example 7.14, we obtain*

$$w(t) = \sqrt{2}\Re\left\{\left(\sum_l a_l e^{j\varphi_l}\psi(t-lT)\right)e^{j2\pi f_c t}\right\}$$

$$= \sqrt{2}\Re\left\{\sum_l a_l e^{j(2\pi f_c t+\varphi_l)}\psi(t-lT)\right\}$$

$$= \sqrt{2}\sum_l a_l \Re\left\{e^{j(2\pi f_c t+\varphi_l)}\right\}\psi(t-lT)$$

$$= \sqrt{2}\sum_l a_l \cos(2\pi f_c t+\varphi_l)\psi(t-lT).$$

---

[4] See the argument in Example 3.10. In practice, the integer condition can be ignored because $2f_c T$ is large, in which case the inner product between the two functions is negligible compared to 1 — the norm of both functions. For a general bandlimited $\psi(t)$, the orthogonality between $\sqrt{2}\psi(t-lT)\cos(2\pi f_c t)$ and $\sqrt{2}\psi(t-iT)\sin(2\pi f_c t)$, for a sufficiently large f_c, follows from Theorem 7.13.

[5] Except for 2-PSK, for which $\Im\{s_l\}$ is always 0.

Figure 7.11 shows a sample $w(t)$ with $\psi(t)$ and f_c as in Example 7.14, with $s_0 = 1 + \mathrm{j} = \sqrt{2}e^{\mathrm{j}\frac{\pi}{4}}$, $s_1 = 3 + \mathrm{j} = \sqrt{10}e^{\mathrm{j}\tan^{-1}(\frac{1}{3})}$, $s_2 = -3 + \mathrm{j} = \sqrt{10}e^{\mathrm{j}(\tan^{-1}(-\frac{1}{3})+\pi)}$, $s_3 = -1 + \mathrm{j} = \sqrt{2}e^{\mathrm{j}\frac{3\pi}{4}}$.

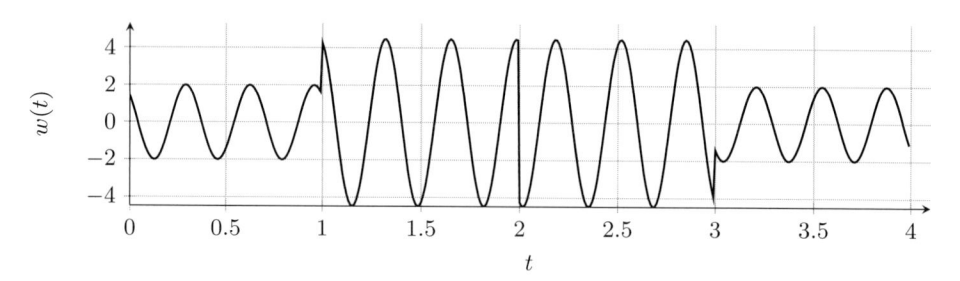

Figure 7.11. Sample QAM signal.

If we write $s_l = \Re\{s_l\} + \mathrm{j}\Im\{s_l\}$, then we obtain the same expression as in Example 7.14

$$w(t) = \sqrt{2}\sum_l \Re\{s_l\}\psi(t - lT)\cos(2\pi f_c t)$$
$$- \sqrt{2}\sum_l \Im\{s_l\}\psi(t - lT)\sin(2\pi f_c t),$$

but unlike for PSK, the $\Re\{s_l\}$ and the $\Im\{s_l\}$ of QAM can be selected independently. Hence, the two superposed PAM signals can be decoded independently, with no interference between the two because $\sqrt{2}\psi(t - lT)\cos(2\pi f_c t)$ is orthogonal to $\sqrt{2}\psi(t - iT)\sin(2\pi f_c t)$. Using (5.10), it is straightforward to verify that the bandwidth of the QAM signal is the same as that of the individual PAM signals. We conclude that the bandwidth efficiency (bits per Hz) of QAM is twice that of PAM. □

Stepping back and looking at the big picture, we now view the physical layer of the OSI model (Figure 1.1) for the AWGN channel as consisting of the three sub-layers shown in Figure 7.12. We are already familiar with all the building blocks of this architecture. The channel models "seen" by the first and second sub-layer, respectively, still need to be discussed. New in these channel models is the fact that the noise is complex-valued. (The signals are complex-valued as well, but we are already familiar with complex-valued signals.)

Under hypothesis $H = i$, the discrete-time channel seen by the first (top) sub-layer has input $c_i \in \mathbb{C}^n$ and output

$$Y = c_i + Z,$$

where, according to (7.16), (7.11) and (7.13), the lth component of Y is $Y_{1,l} + \mathrm{j}Y_{2,l} = c_{i,l} + Z_l$ and $Z_l = Z_{1,l} + \mathrm{j}Z_{2,l}$, where $Z_{1,1}, \ldots, Z_{1,n}, Z_{2,1}, \ldots, Z_{2,n}$ is a collection of iid zero-mean Gaussian random variables of variance $N_0/2$. We have all the ingredients to describe the statistical behavior of Y via the pdf of $Z_{1,1}, \ldots, Z_{1,n}, Z_{2,1}, \ldots, Z_{2,n}$, but it is more elegant to describe the pdf of the *complex-valued* random vector Y. To find the pdf of Y, we introduce the random

7.3. The third layer

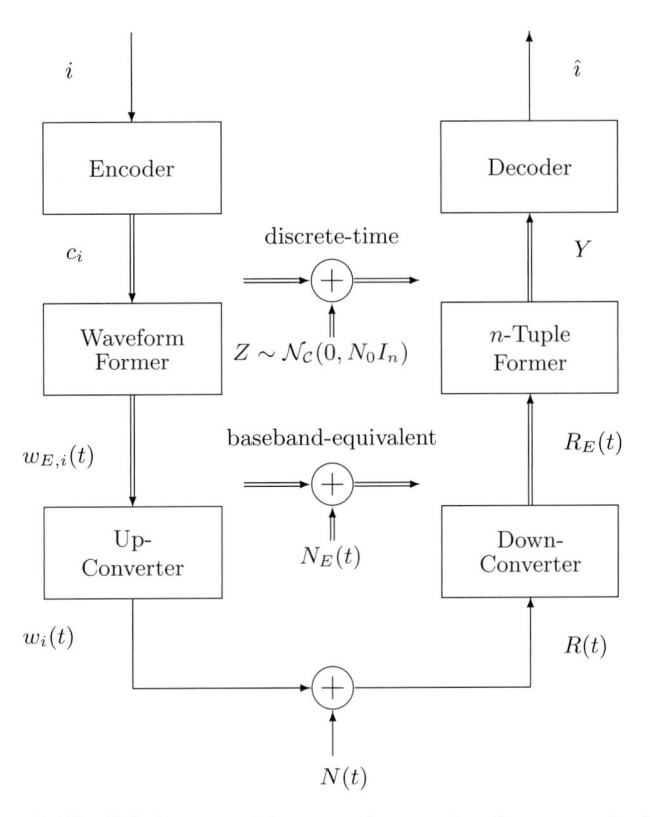

Figure 7.12. Sub-layer architecture for passband communication.

vector $\hat{Y}$ that consists of the (column) n-tuple $Y_1 = \Re\{Y\}$ on top of the (column) n-tuple $Y_2 = \Im\{Y\}$. This notation extends to any complex n-tuple: if $a \in \mathbb{C}^n$ (seen as a column n-tuple), then $\hat{a}$ is the element of $\mathbb{R}^{2n}$ consisting of $\Re\{a\}$ on top of $\Im\{a\}$ (see Appendix 7.8 for an in-depth treatment of the *hat* operator). By definition, the pdf of a complex random vector Y evaluated at y is the pdf of $\hat{Y}$ at $\hat{y}$ (see Appendix 7.9 for a summary on complex-valued random vectors).

Hence,

$$\begin{aligned}
f_{Y|H}(y|i) &= f_{\hat{Y}|H}(\hat{y}|i) \\
&= f_{Y_1,Y_2|H}(\Re\{y\}, \Im\{y\}|i) \\
&= f_{Y_1|H}(\Re\{y\}|i) f_{Y_2|H}(\Im\{y\}|i) \\
&= \frac{1}{(\sqrt{\pi N_0})^n} \exp\left(-\frac{\sum_{l=1}^n (\Re\{y_l\} - \Re\{c_{i,l}\})^2}{N_0}\right) \\
&\quad \times \frac{1}{(\sqrt{\pi N_0})^n} \exp\left(-\frac{\sum_{l=1}^n (\Im\{y_l\} - \Im\{c_{i,l}\})^2}{N_0}\right) \\
&= \frac{1}{(\pi N_0)^n} \exp\left(-\frac{\|y - c_i\|^2}{N_0}\right).
\end{aligned} \qquad (7.19)$$

Naturally, we say that Y is a complex-valued Gaussian random vector with mean c_i and variance N_0 in each (complex-valued) component, and write $Y \sim \mathcal{N}_\mathbb{C}(c_i, N_0 I_n)$, where I_n is the $n \times n$ identity matrix.

In Section 2.4, we assumed that the codebook and the noise are real-valued. If $Y \in \mathbb{C}^n$ as in Figure 7.12, the MAP receiver derived in Section 2.4 applies *as is* to the observable $\hat{Y} \in \mathbb{R}^{2n}$ and codewords $\hat{c}_i \in \mathbb{R}^{2n}$, $i = 0, \ldots, m-1$. But in fact, to describe the decision rule, there is no need to convert $Y \in \mathbb{C}^n$ to $\hat{Y} \in \mathbb{R}^{2n}$ and convert $c_i \in \mathbb{C}^n$ to $\hat{c}_i \in \mathbb{R}^{2n}$. To see why, suppose that we do the conversion. A MAP (or ML) receiver decides based on $\|\hat{y} - \hat{c}_i\|$ or, equivalently, based on $\Re\{\langle \hat{y}, \hat{c}_i \rangle\} - \frac{\|\hat{c}_i\|^2}{2}$. But $\|\hat{y} - \hat{c}_i\|$ is identical to $\|y - c_i\|$. (The squared norm of a complex n-tuple can be obtained by adding the squares of the real components and the squares of the imaginary components.) Similarly, $\langle \hat{y}, \hat{c}_i \rangle = \Re\{\langle y, c_i \rangle\}$. In fact, if $y = y_R + \mathrm{j} y_I$ and $c = c_R + \mathrm{j} c_I$ are (column vectors) in $\mathbb{C}^n$, then $\Re\{\langle y, c \rangle\} = y_R^\mathsf{T} c_R + y_I^\mathsf{T} c_I$, but this is exactly the same as $\langle \hat{y}, \hat{c} \rangle$.[6]

We conclude that an ML decision rule for the complex-valued decoder-input $y \in \mathbb{C}^n$ of Figure 7.12 is

$$\hat{H}_{ML}(y) = \arg\min_i \|y - c_i\|$$
$$= \arg\max_i \Re\{\langle y, c_i \rangle\} - \frac{\|c_i\|^2}{2}.$$

Describing the *baseband-equivalent channel model*, as seen by the second sub-layer of Figure 7.12, requires slightly more work. We do this in the next section for completeness, but it is not needed in order to prove that the receiver structure of Figure 7.12 is completely general (for the AWGN channel) and that it minimizes the error probability. That part is done.

7.4 Baseband-equivalent channel model

We want to derive a *baseband-equivalent channel* that can be used as the channel model seen by the second layer of Figure 7.12. Our goal is to characterize the impulse response $h_0(t)$ and the noise $N_E(t)$ of the baseband-equivalent channel of Figure 7.13b, such that the output U has the same statistic as the output of Figure 7.13a. We assume that every $w_{E,i}(t)$ is bandlimited to $[-B, B]$ where $0 < B < f_c$. Without loss of essential generality, we also assume that every $g_l(t)$ is bandlimited to $[-B, B]$. Except for this restriction, we let $g_l(t)$, $l = 1, \ldots, k$, be any collection of $\mathcal{L}_2$ functions.

We will use the following result,

$$[w(t) \star h(t)] e^{-\mathrm{j} 2\pi f_c t} = [w(t) e^{-\mathrm{j} 2\pi f_c t}] \star [h(t) e^{-\mathrm{j} 2\pi f_c t}], \qquad (7.20)$$

---

[6] For an alternative proof that $\langle \hat{y}, \hat{c}_i \rangle = \Re\{\langle y, c_i \rangle\}$, subtract the two equations $\|y - c_i\|^2 = \|y\|^2 + \|c_i\|^2 - 2\Re\{\langle y, c_i \rangle\}$ and $\|\hat{y} - \hat{c}_i\|^2 = \|\hat{y}\|^2 + \|\hat{c}_i\|^2 - 2\langle \hat{y}, \hat{c}_i \rangle$ and use the fact that the hat on a vector has no effect on the vector's norm.

7.4. Baseband-equivalent channel model

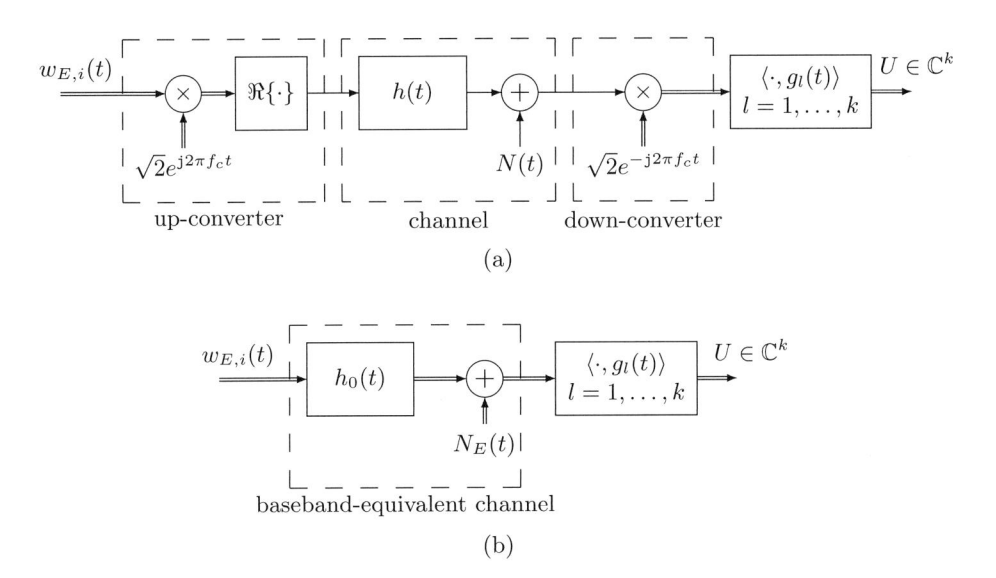

Figure 7.13. Baseband-equivalent channel.

which says that if a signal $w(t)$ is passed through a filter of impulse response $h(t)$ and the filter output is multiplied by $e^{-j2\pi f_c t}$, we obtain the same as passing the signal $w(t)e^{-j2\pi f_c t}$ through the filter with impulse response $h(t)e^{-j2\pi f_c t}$. A direct (time-domain) proof of this result is a simple exercise,[7] but it is more insightful if we take a look at what it means in the frequency domain. In fact, in the frequency domain, the convolution on the left becomes $w_\mathcal{F}(f)h_\mathcal{F}(f)$, and the subsequent multiplication by $e^{-j2\pi f_c t}$ leads to $w_\mathcal{F}(f+f_c)h_\mathcal{F}(f+f_c)$. On the right side we multiply $w_\mathcal{F}(f+f_c)$ with $h_\mathcal{F}(f+f_c)$.

The above relationship should not be confused with the following equalities that hold for any constant $c \in \mathbb{C}$

$$[w(t) \star h(t)]c = [w(t)c] \star h(t) = w(t) \star [h(t)c]. \qquad (7.21)$$

This holds because the left-hand side at an arbitrary time t is c times the integral of the product of two functions. If we bring the constant inside the integral and use it to scale the first function, we obtain the expression in the middle; whereas we obtain the expression on the right if we use c to scale the second function. In the derivation that follows, we use both relationships.

The up-converter, the actual channel, and the down-converter perform *linear* operations, in the sense that their action on the sum of two signals is the sum of the individual actions. Linearity implies that we can consider the signal and the noise separately. We start with the signal part (assuming that there is no noise).

---

[7] Relationship (7.20) is a form of distributivity law, like $[a+b]c = [ac] + [bc]$.

$$\begin{aligned}
U_l &= \langle [w_i(t) \star h(t)]\sqrt{2}e^{-\mathrm{j}2\pi f_c t}, g_l(t)\rangle \\
&= \langle [(w_i(t)\sqrt{2}) \star h(t)]e^{-\mathrm{j}2\pi f_c t}, g_l(t)\rangle \\
&= \langle (w_i(t)\sqrt{2}e^{-\mathrm{j}2\pi f_c t}) \star (h(t)e^{-\mathrm{j}2\pi f_c t}), g_l(t)\rangle \\
&= \langle (w_i(t)\sqrt{2}e^{-\mathrm{j}2\pi f_c t}) \star h_0(t), g_l(t)\rangle \\
&= \langle (w_{E,i}(t) + w_{E,i}^*(t)e^{-\mathrm{j}4\pi f_c t}) \star h_0(t), g_l(t)\rangle \\
&= \langle w_{E,i}(t) \star h_0(t), g_l(t)\rangle, \quad (7.22)
\end{aligned}$$

where in the second line we use (7.21), in the third we use (7.20), in the fourth we introduce the notation

$$h_0(t) = h(t)e^{-\mathrm{j}2\pi f_c t},$$

in the fifth we use

$$w_i(t) = \frac{1}{\sqrt{2}}\left[w_{E,i}(t)e^{\mathrm{j}2\pi f_c t} + w_{E,i}^*(t)e^{-\mathrm{j}2\pi f_c t}\right],$$

and in the sixth we remove the term

$$w_{E,i}^*(t)e^{-\mathrm{j}4\pi f_c t} \star h_0(t),$$

which is bandlimited to $[-2f_c - B, -2f_c + B]$ and therefore has no frequencies in common with $g_l(t)$. By Parseval's relationship, the inner product of functions that have disjoint frequency support is zero.

From (7.22), for all $w_{E,i}(t)$ and all $g_l(t)$ that are bandlimited to $[-B, B]$, the noiseless output of Figure 7.13a is identical to that of Figure 7.13b.

Notice that, not surprisingly, the Fourier transform of $h_0(t)$ is $h_\mathcal{F}(f+f_c)$, namely $h_\mathcal{F}(f)$ frequency-shifted to the left by f_c.

The reader might wonder if $h_0(t)$ is the same as the baseband-equivalent $h_E(t)$ of $h(t)$ (with respect to f_c). In fact it is not, but we can use $\frac{h_E(t)}{\sqrt{2}}$ instead of $h_0(t)$. The two functions are not the same, but it is straightforward to verify that their Fourier transforms agree for $f \in [-B, B]$.

Next we consider the noise alone. To specify $N_E(t)$, we need the following notion of independent noises.[8]

DEFINITION 7.16 *(Independent white Gaussian noises)* $N_R(t)$ *and* $N_I(t)$ *are independent white Gaussian noises if the following two conditions are satisfied.*

(i) *$N_R(t)$ and $N_I(t)$ are white Gaussian noises in the sense of Definition 3.4.*
(ii) *For any two real-valued functions $h_1(t)$ and $h_2(t)$ (possibly the same), the Gaussian random variables $\int N_R(t)h_1(t)dt$ and $\int N_I(t)h_2(t)dt$ are independent.* □

The noise at the output of the down-converter has the form

$$\tilde{N}_E(t) = \tilde{N}_R(t) + \mathrm{j}\tilde{N}_I(t) \quad (7.23)$$

---

[8] The notion of independence is well-defined for stochastic processes, but we do not model the noise as a stochastic process (see Definition 3.4).

7.4. Baseband-equivalent channel model

with

$$\tilde{N}_R(t) = N(t)\sqrt{2}\cos(2\pi f_c t) \tag{7.24}$$
$$\tilde{N}_I(t) = -N(t)\sqrt{2}\sin(2\pi f_c t). \tag{7.25}$$

$\tilde{N}_R(t)$ and $\tilde{N}_I(t)$ are *not* independent white Gaussian noises in the sense of Definition 7.16 (as can be verified by setting $f_c = 0$), but we now show that they do fulfill the conditions of Definition 7.16 when the functions used in the definition are bandlimited to $[-B, B]$ and $B < f_c$.

Let $h_i(t)$, $i = 1, 2$, be real-valued $\mathcal{L}_2$ functions that are bandlimited to $[-B, B]$ and define

$$Z_i = \int \tilde{N}_R(t) h_i(t) dt.$$

Z_i, $i = 1, 2$, is Gaussian, zero-mean, and of variance $\frac{N_0}{2} \|\sqrt{2}\cos(2\pi f_c t) h_i(t)\|^2$. The function $\sqrt{2}\cos(2\pi f_c t) h_i(t)$ is passband with baseband-equivalent $h_i(t)$. By Definition 3.4 and Theorem 7.5,

$$\begin{aligned}
\text{cov}(Z_1, Z_2) &= \frac{N_0}{2} \langle \sqrt{2}\cos(2\pi f_c t) h_1(t), \sqrt{2}\cos(2\pi f_c t) h_2(t) \rangle \\
&= \frac{N_0}{2} \Re\{\langle h_1(t), h_2(t) \rangle\} \\
&= \frac{N_0}{2} \langle h_1(t), h_2(t) \rangle.
\end{aligned}$$

This proves that under the stated bandwidth limitation, $\tilde{N}_R(t)$ behaves as white Gaussian noise of power spectral density $\frac{N_0}{2}$. The proof that the same is true for $\tilde{N}_I(t)$ follows similar patterns, using the fact that $-\sqrt{2}\sin(2\pi f_c t) h_i(t)$ is passband with baseband-equivalent $jh_i(t)$. It remains to be shown that $\tilde{N}_R(t)$ and $\tilde{N}_I(t)$ are independent noises in the sense of Definition 7.16. Let

$$Z_3 = \int \tilde{N}_I(t) h_3(t) dt.$$

Z_3 is zero-mean and jointly Gaussian with Z_2 with

$$\begin{aligned}
\text{cov}(Z_2, Z_3) &= \frac{N_0}{2} \langle \sqrt{2}\cos(2\pi f_c t) h_2(t), -\sqrt{2}\sin(2\pi f_c t) h_3(t) \rangle \\
&= \frac{N_0}{2} \Re\{\langle h_2(t), jh_3(t) \rangle\} \\
&= 0.
\end{aligned}$$

Let us summarize the noise contribution. The noise at the output of the downconverter has the form (7.23)–(7.25). However, as long as we are taking inner products of the noise with functions that are bandlimited to $[-B, B]$ for $B < f_c$, which is the case for $g_l(t)$, $l = 1, \ldots, k$, we can model the equivalent noise $N_E(t)$ of Figure 7.13b as

$$N_E(t) = N_R(t) + jN_I(t) \tag{7.26}$$

where $N_R(t)$ and $N_I(t)$ are independent white Gaussian noises of spectral density $N_0/2$.

This last characterization of $N_E(t)$ suffices to describe the statistic of $U \in \mathbb{C}^k$, even when the $g_l(t)$ are complex-valued, provided they are bandlimited as specified. For the statistical description of a complex random vector, the reader is referred to Appendix 7.9 where, among other things, we introduce and discuss circularly symmetric Gaussian random vectors (which are complex-valued) and prove that the U at the output of Figure 7.13b is always circularly symmetric (even when the $g_l(t)$ are not bandlimited to $[-B, B]$).

7.5 Parameter estimation

In Section 5.7 we have discussed the symbol synchronization problem. The down-converter should also be (phase) synchronized with the up-converter, and the attenuation between the transmitter and the n-tuple former should be estimated.

We assume that the down-converter and the n-tuple former use independent clocks. This is a valid assumption because it is not uncommon that they be implemented in independent modules, possibly made by different manufacturers. The same is true, of course, for the corresponding blocks of the transmitter.

The down-converter uses an oscillator that, when switched on, starts with an arbitrary initial phase. Hence, the received signal $r(t)$ is initially multiplied by $e^{-\mathrm{j}(2\pi f_c t - \varphi)} = e^{\mathrm{j}\varphi} e^{-\mathrm{j}2\pi f_c t}$ for some $\varphi \in [0, 2\pi)$. As a consequence, the n-tuple former input and output have an extra factor of the form $e^{\mathrm{j}\varphi}$. Accounting for this factor goes under the name of *phase synchronization*.

Finally, there is a scaling by some positive number a due to the channel attenuation and the receiver front end amplification. For notational convenience, we sometimes combine the scaling by a and that by $e^{\mathrm{j}\varphi}$ into a single complex scaling factor $\alpha = a e^{\mathrm{j}\varphi}$.

This results in the n-tuple former input

$$R_E(t) = \alpha s_E(t - \theta) + N_E(t),$$

where $N_E(t)$ is complex white Gaussian noise of power spectral density N_0 ($N_0/2$ in both real and imaginary parts) and $\theta \in [0, \theta_{max}]$ accounts for the channel delay and the time offset. For this section, the function $s_E(t)$ represents a training signal known to the receiver, used to estimate θ, a, φ. Once estimated, these channel parameters are used as the true values for the communication that follows. Next we derive the joint ML estimates of θ, a, φ. The good news is that the solution to this joint estimation problem essentially decomposes into three separate ML estimation problems.

The derivation that follows is a straightforward generalization of what we have done in Section 5.7, with the main difference being that signals are now complex-valued. Accordingly, let $Y = (Y_1, \ldots, Y_n)^\mathsf{T}$ be the random vector obtained by

7.5. Parameter estimation

projecting $R_E(t)$ onto the elements of an orthonormal basis[9] for an inner product space that contains $s_E(t - \hat{\theta})$ for all possible values of $\hat{\theta} \in [0, \theta_{max}]$. The *likelihood function* with parameters $\hat{\theta}, \hat{a}, \hat{\varphi}$ is

$$f(y; \hat{\theta}, \hat{a}, \hat{\varphi}) = \frac{1}{(\pi N_0)^n} e^{-\frac{\|y - \hat{a}e^{j\hat{\varphi}}m(\hat{\theta})\|^2}{N_0}},$$

where $m(\hat{\theta})$ is the n-tuple of coefficients of $s_E(t - \hat{\theta})$ with respect to the chosen orthonormal basis.

A joint maximum likelihood estimation of θ, a, φ is a choice of $\hat{\theta}, \hat{a}, \hat{\varphi}$ that maximizes the likelihood function or, equivalently, that maximizes any of the following three expressions

$$-\|y - \hat{a}e^{j\hat{\varphi}}m(\hat{\theta})\|^2,$$

$$-(\|y\|^2 + \|\hat{a}e^{j\hat{\varphi}}m(\hat{\theta})\|^2 - 2\Re\{\langle y, \hat{a}e^{j\hat{\varphi}}m(\hat{\theta})\rangle\}),$$

$$\Re\{\langle y, \hat{a}e^{j\hat{\varphi}}m(\hat{\theta})\rangle\} - \frac{|\hat{a}e^{j\hat{\varphi}}|^2}{2}\|m(\hat{\theta})\|^2. \qquad (7.27)$$

Notice that $\|m(\hat{\theta})\|^2 = \|s_E(t - \hat{\theta})\|^2 = \|s_E(t)\|^2$. Hence, for a fixed $\hat{a}$, the second term in (7.27) is independent of $\hat{\theta}, \hat{\varphi}$. Thus, for any fixed $\hat{a}$, we can maximize over $\hat{\theta}, \hat{\varphi}$ by maximizing any of the following three expressions

$$\Re\{\langle y, \hat{a}e^{j\hat{\varphi}}m(\hat{\theta})\rangle\},$$

$$\Re\{e^{-j\hat{\varphi}}\langle y, m(\hat{\theta})\rangle\},$$

$$\Re\{e^{-j\hat{\varphi}}\langle r_E(t), s_E(t - \hat{\theta})\rangle\}, \qquad (7.28)$$

where the last line is justified by the argument preceding (5.18). The maximum of

$$\Re\{e^{-j\hat{\varphi}}\langle r_E(t), s_E(t - \hat{\theta})\rangle\}$$

is achieved when $\hat{\theta}$ is such that the absolute value of $\langle r_E(t), s_E(t - \hat{\theta})\rangle$ is maximized and $\hat{\varphi}$ is such that $e^{-j\hat{\varphi}}\langle r_E(t), s_E(t - \hat{\theta})\rangle$ is real-valued and positive. The latter happens when $\hat{\varphi}$ equals the phase of $\langle r_E(t), s_E(t - \hat{\theta})\rangle$. Thus

$$\hat{\theta}_{ML} = \arg\max_{\hat{\theta}} |\langle r_E(t), s_E(t - \hat{\theta})\rangle|, \qquad (7.29)$$

$$\hat{\varphi}_{ML} = \angle\langle r_E(t), s_E(t - \hat{\theta}_{ML})\rangle. \qquad (7.30)$$

Finally, for $\hat{\theta} = \hat{\theta}_{ML}$ and $\hat{\varphi} = \hat{\varphi}_{ML}$, the maximum of (7.27) with respect to $\hat{a}$ is achieved by

$$\hat{a}_{ML} = \arg\max_{\hat{a}} \langle y, \hat{a}e^{j\hat{\varphi}_{ML}}m(\hat{\theta}_{ML})\rangle - \frac{|\hat{a}e^{j\hat{\varphi}_{ML}}|^2}{2}\|m(\hat{\theta}_{ML})\|^2$$

---

[9] As in Section 5.7.1, for notational simplicity we assume that the orthonormal basis has finite dimension n. The final result does not depend on the choice of the orthonormal basis.

$$= \arg\max_{\hat{a}} \hat{a} e^{-\mathrm{j}\hat{\varphi}_{ML}} \langle r_E(t), s_E(t - \hat{\theta}_{ML}) \rangle - \hat{a}^2 \frac{\mathcal{E}}{2}$$

$$= \arg\max_{\hat{a}} \hat{a} |\langle r_E(t), s_E(t - \hat{\theta}_{ML}) \rangle| - \hat{a}^2 \frac{\mathcal{E}}{2},$$

where $\mathcal{E}$ is the energy of $s_E(t)$, and in the last line we use the fact that

$$e^{-\mathrm{j}\hat{\varphi}_{ML}} \langle r_E(t), s_E(t - \hat{\theta}_{ML}) \rangle$$

is real-valued and positive (by the choice of $\hat{\varphi}_{ML}$). Taking the derivative of $\hat{a}|\langle r_E(t), s_E(t - \hat{\theta}_{ML}) \rangle| - \hat{a}^2 \frac{\mathcal{E}}{2}$ with respect to $\hat{a}$ and equating to zero yields

$$\hat{a}_{ML} = \frac{|\langle r_E(t), s_E(t - \hat{\theta}_{ML}) \rangle|}{\mathcal{E}}.$$

Until now we have not used any particular signaling method. Now assume that the training signal takes the form

$$s_E(t) = \sum_{l=0}^{K-1} c_l \psi(t - lT), \tag{7.31}$$

where the symbols $c_0, \ldots, c_{K-1}$ and the pulse $\psi(t)$ can be real- or complex-valued and $\psi(t)$ has unit norm and is orthogonal to its T-spaced translates.

The essence of what follows applies whether the n-tuple former incorporates a correlator or a matched filter. For the sake of exposition, we assume that it incorporates the matched filter of impulse response $\psi^*(-t)$.

Once we have determined $\hat{\theta}_{ML}$ according to (7.29), we sample the matched filter output at times $t = \hat{\theta}_{ML} + kT$, k integer. The kth sample is

$$y_k = \langle r_E(t), \psi(t - kT - \hat{\theta}_{ML}) \rangle \tag{7.32}$$

$$= \langle \alpha \sum_{l=0}^{K-1} c_l \psi(t - lT - \theta) + N_E(t), \psi(t - kT - \hat{\theta}_{ML}) \rangle$$

$$= \alpha \sum_{l=0}^{K-1} c_l \langle \psi(t - lT - \theta), \psi(t - kT - \hat{\theta}_{ML}) \rangle + Z_k$$

$$= \alpha \sum_{l=0}^{K-1} c_l R_\psi(\hat{\theta}_{ML} - \theta + (k - l)T) + Z_k$$

$$= \alpha \sum_{l=0}^{K-1} c_l h_{k-l} + Z_k,$$

where $\{Z_k\}$ is iid $\sim \mathcal{N}_\mathbb{C}(0, N_0)$ and we have defined $h_i = R_\psi(\hat{\theta}_{ML} - \theta + iT)$, where R_ψ is ψ's self-similarity function. In particular, if $\hat{\theta}_{ML} = \theta$, then $h_i = \mathbb{1}\{i = 0\}$ and

$$y_k = \alpha c_k + Z_k. \tag{7.33}$$

7.5. Parameter estimation

If N_0 is not too large compared to the signal's power, $\hat{\theta}_{ML}$ should be sufficiently close to θ for (7.33) to be a valid model.

Next we re-derive the ML estimates of φ and a in terms of the matched filter output samples. We do so because it is easier to implement the estimator in a DSP that operates on the matched filter output samples rather than by analog technology operating on the continuous-time signals. Using (7.31) and the linearity of the inner product, we obtain $\langle r_E(t), s_E(t-\hat{\theta}_{ML})\rangle = \sum_{l=0}^{K-1} c_l^* \langle r_E(t), \psi(t-lT-\hat{\theta}_{ML})\rangle$, and using (7.32) we obtain

$$\langle r_E(t), s_E(t-\hat{\theta}_{ML})\rangle = \sum_{l=0}^{K-1} y_l c_l^*.$$

Thus, from (7.30),

$$\hat{\varphi}_{ML} = \angle \sum_{l=0}^{K-1} y_l c_l^*.$$

It is instructive to interpret $\hat{\varphi}_{ML}$ without noise. In the absence of noise, $\hat{\theta}_{ML} = \theta$ and $y_k = a c_k$. Hence $\sum_{l=0}^{K-1} y_l c_l^* = a e^{j\varphi} \sum_{l=0}^{K-1} c_l c_l^* = a e^{j\varphi} \mathcal{E}$, where $\mathcal{E}$ is the energy of the training sequence. From (7.30), we see that $\hat{\varphi}_{ML}$ is the angle of $e^{j\varphi} a \mathcal{E}$, i.e. $\hat{\varphi}_{ML} = \varphi$.

Proceeding similarly, we obtain

$$\hat{a}_{ML} = \frac{|\sum_{l=0}^{K-1} y_l c_l^*|}{\mathcal{E}}.$$

It is immediate to check that if there is no noise and $\hat{\varphi}_{ML} = \varphi$, then $\hat{a}_{ML} = a$.

Notice that both $\hat{\varphi}_{ML}$ and $\hat{a}_{ML}$ depend on the observation $y_0, \ldots, y_{K-1}$ only through the inner product $\sum_{l=0}^{K-1} y_l c_l^*$.

Depending on various factors and, in particular, on the duration of the transmission, the stability of the oscillators and the possibility that the delay and/or the attenuation vary over time, a one-time estimate of θ, a, and φ might not be sufficient.

In Section 5.7.2, we have presented the delay locked loop to track θ, assuming real-valued signals. The technique can be adapted to the situation of this section. In particular, if the symbol sequence $c_0, \ldots, c_{K-1}$ that forms the training signal is as in Section 5.7.2, once φ has been estimated and accounted for, the imaginary part of the matched filter output contains only noise and the real part is as in Section 5.7.2. Thus, once again, θ can be tracked with a delay locked loop.

The most critical parameter is φ because it is very sensitive to channel delay variations and to instabilities of the up/down-converter oscillators.

EXAMPLE 7.17 *A communication system operates at a symbol rate of* 10 *Msps (mega symbols per second) with a carrier frequency* $f_c = 1$ *GHz. The* local oscillator *that produces* $e^{j2\pi f_c t}$ *is based on a crystal oscillator and a phase locked loop (PLL). The frequency of the crystal oscillator can only be guaranteed up to a certain precision and it is affected by the temperature. Typical precisions are in the range* *of 10–100 parts per million (ppm). If we take a 50 ppm precision as a mid-range value, then the frequency used at the transmitter and that used at the receiver are guaranteed to be within 100 ppm from each other, i.e. the carrier frequency difference could be as large as $\Delta f_c = 10^{-4} f_c = 10^5$ Hz. Over the symbol time, this difference accumulates a phase offset $\Delta \varphi = 2\pi \Delta f_c T = 2\pi \times 10^{-2}$, which constitutes a rotation of 3.6 degrees. For 16-QAM, a one-time rotation of 3.6 degrees might have a negligible effect on the error probability, but clearly we cannot let the phase drift accumulate over several symbols.* □

Example 7.17 is quite representative. We are in the situation where there is a small phase drift that creates a rotation of the sampled matched filter output. From one sample to the next, the rotation is by some small angle $\Delta\varphi$. An effective technique to deal with this situation is to let the decoder decide on symbol c_k and then, assuming that the decision $\hat{c}_k$ is correct, use it to estimate φ_k. This general approach is denoted *decision-directed* phase synchronization.

A popular technique for doing decision-directed phase synchronization is based on the following idea. Suppose that after the sample indexed by $k-1$, the rotation has been correctly estimated and corrected. Because of the frequency offset, the kth sample is $y_k = e^{j\Delta\varphi} c_k$ plus noise. Neglecting for the moment the noise, $y_k c_k^* = e^{j(\Delta\varphi)} |c_k|^2$. Hence

$$\Im\{y_k c_k^*\} = \sin(\Delta\varphi)|c_k|^2 \approx \Delta\varphi |c_k|^2,$$

where the approximation holds for small values of $|\Delta\varphi|$. Assuming that $|\Delta\varphi|$ is indeed small, the idea is to decode y_k ignoring the rotation by $\Delta\varphi$. With high probability the decoded symbol $\hat{c}_k$ equals c_k and $\Im\{y_k \hat{c}_k^*\} \approx \Delta\varphi |c_k|^2$. The feedback signal $\Im\{y_k \hat{c}_k^*\}$ can be used by the local oscillator to correct the phase error. Alternatively, the decoder can use the feedback signal to find an estimate $\hat{\Delta\varphi}$ of $\Delta\varphi$ and to rotate y_k by $-\hat{\Delta\varphi}$. This method works well also in the presence of noise, assuming that the noise is zero-mean and independent from sample to sample. Averaging over subsequent samples helps to mitigate the effect of the noise.

Another possibility of tracking φ is to use a phase locked loop – a technique similar to the delay locked loop discussed in Section 5.7.2 to track θ.

Differential encoding is a different technique to deal with a constant or slowly changing phase. It consists in encoding the information in the phase difference between consecutive symbols.

When the phase φ_k is either constant or varies slowly, as assumed in this section, we say that the phase comes through coherently. In the next section, we will see what we can do when this is not the case.

7.6 Non-coherent detection

Sometimes it is not realistic to assume that the phase comes through coherently. We are then in the domain of *non-coherent detection*.

7.6. Non-coherent detection

EXAMPLE 7.18 *Frequency hopping (FH) is a communication technique that consists in varying the carrier frequency according to a predetermined schedule known to the receiver. An example is symbol-by-symbol on a pulse train with non-overlapping pulses and a carrier frequency that changes after each pulse. It would be hard to guarantee a specified initial phase after each change of frequency. So a model for the sampled matched filter output is*

$$y_k = e^{j\varphi_k} c_k + Z_k,$$

with φ_k uniformly distributed in $[0, 2\pi)$ and independent from sample to sample and from anything else. Frequency hopping is sometimes used in wireless military communication, because it makes it harder for someone that does not know the carrier-frequency schedule to detect that a communication is taking place. In particular, it is harder for an enemy to locate the transmitter and/or to jam the signal. Frequency hopping is also used in wireless civil applications when there is no guarantee that a fixed band is free of interference. This is often the case in unlicensed bands. In this case, rather than choosing a fixed frequency that other applications might choose – risking that a long string of symbols are hit by interfering signals – it is better to let the carrier frequency hop and use coding to deal with the occasional symbols that are hit by interfering signals. Frequency hopping is a form of spread spectrum (SS) communication.[10] We speak of spread spectrum when the signal's bandwidth is much larger (typically by several orders of magnitude) than the symbol rate. □

We consider the following fairly general situation. When $H = i \in \mathcal{H} = \{0, 1, \ldots, m-1\}$, the baseband-equivalent channel output is

$$R_E(t) = ae^{j\varphi} w_{E,i}(t) + N_E(t),$$

where $w_{E,i}(t)$ is the baseband-equivalent signal, $N_E(t)$ is the complex-valued white Gaussian noise of power spectral density N_0, $a \in \mathbb{R}_+$ is an arbitrary scaling factor unknown to the receiver, and $\varphi \in [0, 2\pi)$ an unknown phase. We want to do hypothesis testing in the presence of the unknown parameters a and φ. (We are not considering a delay θ in our model, because a delay, unlike a scaling factor $ae^{j\varphi}$, can be assumed to be constant over several symbols, hence it can be estimated via a training sequence as described in Section 7.5.)

One approach is to extend the maximum likelihood (ML) decision rule by maximizing the likelihood function not only over the message i but also over the unknown parameters a and φ. Another approach assumes that the unknown parameters are samples of random variables of known statistic. In this case, we can obtain the distribution of the channel-output observations given the hypothesis, by marginalizing over the distribution of a and φ, at which point we are back to the familiar situation (but with a new channel model). We work out the ML approach, which has the advantage of not requiring the distributions associated to a and φ.

---

[10] There is much literature on spread spectrum. The interested reader can find introductory articles on the Web.

The steps to maximize the likelihood function mimic what we have done in the previous section. Let c_i be the codeword associated with $w_{E,i}(t)$ (with respect to some orthonormal basis). Let y be the n-tuple former output. The likelihood function is

$$f(y;\hat{i},\hat{a},\hat{\varphi}) = \frac{1}{(\pi N_0)^n} e^{-\frac{\|y-\hat{a}e^{j\hat{\varphi}}c_{\hat{i}}\|^2}{N_0}}.$$

We seek the $\hat{i}$ that maximizes

$$g(c_{\hat{i}}) = \max_{\hat{a},\hat{\varphi}} \Re\{\langle y, \hat{a}e^{j\hat{\varphi}}c_{\hat{i}}\rangle\} - \frac{1}{2}\|\hat{a}e^{j\hat{\varphi}}c_{\hat{i}}\|^2$$

$$= \max_{\hat{a},\hat{\varphi}} \hat{a}\Re\{e^{-j\hat{\varphi}}\langle y, c_{\hat{i}}\rangle\} - \frac{\hat{a}^2}{2}\|c_{\hat{i}}\|^2.$$

The $\hat{\varphi}$ that achieves the maximum is the one that makes $e^{-j\hat{\varphi}}\langle y, c_{\hat{i}}\rangle$ real-valued and positive. Let $\hat{\varphi}_{ML}$ be the maximizing $\hat{\varphi}$ and observe that $\Re\{e^{-j\hat{\varphi}_{ML}}\langle y, c_{\hat{i}}\rangle\} = |\langle y, c_{\hat{i}}\rangle|$. Hence,

$$g(c_{\hat{i}}) = \max_{\hat{a}} \hat{a}|\langle y, c_{\hat{i}}\rangle| - \frac{\hat{a}^2}{2}\|c_{\hat{i}}\|^2. \tag{7.34}$$

By taking the derivative of $\hat{a}|\langle y, c_{\hat{i}}\rangle| - \frac{\hat{a}^2}{2}\|c_{\hat{i}}\|^2$ with respect to $\hat{a}$ and setting to zero, we obtain the maximizing $\hat{a}$

$$\hat{a}_{ML} = \frac{|\langle y, c_{\hat{i}}\rangle|}{\|c_{\hat{i}}\|^2}.$$

Inserting into (7.34) yields

$$g(c_{\hat{i}}) = \frac{1}{2}\frac{|\langle y, c_{\hat{i}}\rangle|^2}{\|c_{\hat{i}}\|^2}.$$

Hence

$$\hat{i}_{ML} = \arg\max_{\hat{i}} g(c_{\hat{i}})$$

$$= \arg\max_{\hat{i}} \frac{1}{2}\frac{|\langle y, c_{\hat{i}}\rangle|^2}{\|c_{\hat{i}}\|^2}$$

$$= \arg\max_{\hat{i}} \frac{|\langle y, c_{\hat{i}}\rangle|}{\|c_{\hat{i}}\|}. \tag{7.35}$$

Notice that for every vector $c \in \mathbb{C}^n$ and every scalar $b \in \mathbb{C}$, $\frac{|\langle y,c\rangle|}{\|c\|} = \frac{|\langle y,bc\rangle|}{\|bc\|}$. Hence, it would be a bad idea if the codebook contains two or more codewords that are collinear with respect to the complex inner product space $\mathbb{C}^n$.

The decoding rule (7.35) is not surprising. Recall that, without phase and amplitude uncertainty and with signals of equal energy, the maximum likelihood decision is the one that maximizes $\Re\{\langle y, c_{\hat{i}}\rangle\}$. If the channel scales the signal by an arbitrary real-value a unknown to the decoder, then the decoder should not take the signal's energy into account. This can be accomplished by a decoder

7.6. Non-coherent detection

that maximizes $\Re\{\langle y, c_{\hat{i}}\rangle\}/\|c_{\hat{i}}\|$. Next, assume that the channel can also rotate the signal by an arbitrary phase φ (i.e. the channel multiplies the signal by $e^{j\varphi}$). As we increase the phase by $\pi/2$, the imaginary part of the new inner product becomes the real part of the old (with a possible sign change). One way to make the decoder insensitive to the phase, is to substitute $\Re\{\langle y, c_{\hat{i}}\rangle\}$ with $|\langle y, c_{\hat{i}}\rangle|$. The result is the decoding rule (7.35).

EXAMPLE 7.19 (A bad choice of signals) *Consider m-ary phase-shift keying, i.e.* $w_{E,i}(t) = c_i\psi(t)$, *where* $c_i = \sqrt{\mathcal{E}_s}e^{j2\pi i/m}$, $i = 0, \ldots, m-1$, *and* $\psi(t)$ *is a unit-norm pulse. If we plug into (7.35), we obtain*

$$\hat{i}_{ML} = \arg\max_{\hat{i}} \frac{|\langle y, \sqrt{\mathcal{E}_s}e^{j2\pi\hat{i}/m}\rangle|}{\sqrt{\mathcal{E}_s}}$$

$$= \arg\max_{\hat{i}} \left|e^{-j2\pi\hat{i}/m}\langle y, 1\rangle\right|$$

$$= \arg\max_{\hat{i}} |\langle y, 1\rangle|$$

$$= \arg\max_{\hat{i}} |y|,$$

which means that the decoder has no preference among the $\hat{i} \in \mathcal{H}$, *i.e. the error probability is the same independently of the decoder's choice. In fact, a PSK constellation is a bad choice for a codebook, because it conveys information in the phase and the phase information is destroyed by the channel.* □

More generally, any one-dimensional codebook, like a PAM, a QAM, or a PSK alphabet, is a bad choice, because in that case all codewords are collinear to each other. (However, for $n > 1$, we could use codebooks that consist of n-length sequences with components (symbols) taking value in a PAM, QAM, or PSK alphabet.)

EXAMPLE 7.20 (A good choice) *Two vectors in* $\mathbb{C}^n$ *that are orthogonal to each other cannot be made equal by multiplying one of the two by a scalar* $ae^{j\varphi}$, *which was the underlying issue in Example 7.19. Complex-valued orthogonal signals remain orthogonal after we multiply them by* $ae^{j\varphi}$. *This suggests that they are a good choice for the channel model assumed in this section. Specifically, suppose that the ith codeword* $c_i \in \mathbb{C}^m$, $i = 1, \ldots, m$, *has* $\sqrt{\mathcal{E}_s}$ *at position i and is zero elsewhere. In this case,* $|\langle y, c_i\rangle| = \sqrt{\mathcal{E}_s}|y_i|$ *and*

$$\hat{i}_{ML} = \arg\max_{i} \frac{|\langle y, c_i\rangle|}{\sqrt{\mathcal{E}_s}}$$

$$= \arg\max_{i} |y_i|.$$

(Compare this rule to the decision rule of Example 4.6, where the signaling scheme is the same but there is neither amplitude nor phase uncertainty.) □

7.7 Summary

The fact that each passband signal (real-valued by definition) has an equivalent baseband signal (complex-valued in general) makes it possible to separate the communication system into two parts: a part (top two layers) that processes baseband signals and a part (bottom layer) that implements the conversion to/from passband. With the bottom layer in place, the top two layers are designed to communicate over a complex-valued baseband AWGN channel. This separation has several advantages: (i) it simplifies the design and the analysis of the top two layers, where most of the system complexity lies; (ii) it reduces the implementation costs; and (iii) it provides greater flexibility by making it possible to choose the carrier frequency, simply by changing the frequency of the oscillator in the up/down-converter. For instance, for frequency hopping (Example 7.18), as long as the down-converter is synchronized with the up-converter, the top two layers are unaware that the carrier frequency is hopping. Without the third layer in place, we change the carrier frequency by changing the pulse, and the options that we have in choosing the carrier frequency are limited. (If the Nyquist criterion is fulfilled for $|\psi_{\mathcal{F}}(f + f_1)|^2 + |\psi_{\mathcal{F}}(f - f_1)|^2$, it is not necessarily fulfilled for $|\psi_{\mathcal{F}}(f + f_2)|^2 + |\psi_{\mathcal{F}}(f - f_2)|^2$, $f_1 \neq f_2$.)

Theorem 7.13 tells us how to transform an orthonormal basis of size n for the baseband-equivalent signals into an orthonormal basis of size $2n$ for the corresponding passband signals. The factor 2 is due to the fact that the former is used with complex-valued coefficients, whereas the latter is used with real-valued coefficients.

For mathematical convenience, we assume that neither the up-converter nor the down-converter modifies the signal's norm. This is not what happens in reality, but the system-level designer (as opposed to the hardware designer) can make this assumption because all the scaling factors can be accounted for by a single factor in the channel model. Even this factor can be removed (i.e. it can be made to be 1) without affecting the system-level design, provided that the power spectral density of the noise is adjusted accordingly so as to keep the signal-energy to noise-power-density ratio unchanged.

In practice, the up-converter, as well as the down-converter, amplifies the signals, and the down-converter contains a noise-reduction filter that removes the out-of-band noise (see Section 3.6). The transmitter back end (the physical embodiment of the up-converter) deals with high power, high frequencies, and a variable carrier frequency f_c. The skills needed to design it are quite specific. It is very convenient that the transmitter back end can essentially be designed and built separately from the rest of the system and can be purchased as an off-the-shelf device.

With the back end in place, the earlier stages of the transmitter, which perform the more sophisticated signal processing, can be implemented under the most favorable conditions, namely in baseband and using voltages and currents that are in the range of standard electronics, rather than being tied to the power of the transmitted signal. The advantage of working in baseband is two-fold: the carrier frequency is fixed and working with low frequencies is less tricky.

7.8. Real- vs. complex-valued operations

A similar discussion applies to the receiver. Contrary to the transmitter back end, the receiver front end (the embodiment of the down-converter) processes very small signals, but the conclusions are the same: it can be an off-the-shelf device designed and built separately from the rest of the receiver.[11]

There is one more non-negligible advantage to processing the high-frequency signals of a device (transmitter or receiver) in a confined space that can be shielded from the rest of the device. Specifically, the powerful high-frequency signal at the transmitter output can feed back over the air to the earlier stages of the transmitter. If picked up and amplified by them, the signal can turn the transmitter into an oscillator, similarly to what happens to an audio amplifier when the microphone is in the proximity of the speaker. This cannot happen if the earlier stages are designed for baseband signals. A similar discussion applies for the receiver: the highly sensitive front end could pick up the stronger signals of later stages, if they operate in the same frequency band.

We have seen that a delay in the signal path translates into a rotation of the complex-valued symbols at the output of the n-tuple former. We say that the phase comes through coherently when the rotation is constant or varies slowly from symbol to symbol. Then, the phase can be estimated and its effect can be removed. In this case the decision process is called coherent detection. Sometimes the phase comes through incoherently, meaning that each symbol can be rotated by a different amount. In this case PSK, which carries information in the symbol's phase, becomes useless; whereas PAM, which carries information in the symbol's amplitude, is still a viable possibility. When the phase comes through incoherently, the decision process is referred to as non-coherent detection.

7.8 Appendix: Relationship between real- and complex-valued operations

In this appendix we establish the relationship between complex-valued operations involving n-tuples and matrices and their real-valued counterparts. We use these results in Appendix 7.9 to derive the probability density function of circularly symmetric Gaussian random vectors.

To every complex n-tuple $u \in \mathbb{C}^n$, we associate a real $2n$-tuple $\hat{u}$ defined as follows:

$$\hat{u} = \begin{bmatrix} u_R \\ u_I \end{bmatrix} := \begin{bmatrix} \Re[u] \\ \Im[u] \end{bmatrix}. \tag{7.36}$$

Now let A be an $m \times n$ complex matrix and define $u = Av$. We are interested in the real matrix $\hat{A}$ so that

$$\hat{u} = \hat{A}\hat{v}. \tag{7.37}$$

---

[11] Realistically, a specific back/front end implementation has certain characteristics that limit its usage to certain applications. In particular, for the back end we consider its gain, output power, and bandwidth. For the front end, we consider its bandwidth, sensitivity, gain, and noise temperature.

It is straightforward to verify that

$$\hat{A} = \begin{bmatrix} A_R & -A_I \\ A_I & A_R \end{bmatrix} := \begin{bmatrix} \Re[A] & -\Im[A] \\ \Im[A] & \Re[A] \end{bmatrix}. \qquad (7.38)$$

To remember the form of $\hat{A}$, observe that the top half of $\hat{u}$ is the real part of Av, i.e. $A_R v_R - A_I v_I$. This explains the top half of $\hat{A}$. Similarly, the bottom half of $\hat{u}$ is the imaginary part of Av, i.e. $A_R v_I + A_I v_R$, which explains the bottom half of $\hat{A}$. The following lemma summarizes a number of useful properties.

LEMMA 7.21 *The following properties hold*

$$\widehat{Au} = \hat{A}\hat{u} \qquad (7.39\text{a})$$
$$\widehat{u+v} = \hat{u} + \hat{v} \qquad (7.39\text{b})$$
$$\Re\{u^\dagger v\} = \hat{u}^\mathsf{T} \hat{v} \qquad (7.39\text{c})$$
$$\|u\|^2 = \|\hat{u}\|^2 \qquad (7.39\text{d})$$
$$\widehat{AB} = \hat{A}\hat{B} \qquad (7.39\text{e})$$
$$\widehat{A+B} = \hat{A} + \hat{B} \qquad (7.39\text{f})$$
$$\widehat{A^\dagger} = (\hat{A})^\mathsf{T} \qquad (7.39\text{g})$$
$$\widehat{I_n} = I_{2n} \qquad (7.39\text{h})$$
$$\widehat{A^{-1}} = \hat{A}^{-1} \qquad (7.39\text{i})$$
$$\det(\hat{A}) = |\det(A)|^2 = \det(AA^\dagger) \qquad (7.39\text{j})$$

□

Proof Property (7.39a) is (7.37) restated; (7.39b) is immediate from (7.36); (7.39c) follows from the fact that $\Re\{u^\dagger v\}$ is $(u_R)^\mathsf{T} v_R + (u_I)^\mathsf{T} v_I = \hat{u}^\mathsf{T} \hat{v}$; (7.39d) follows from (7.39c) with $v = u$ and using the fact that $\|u\|^2 = u^\dagger u$ is already real-valued; (7.39e) follows from the observation that if $v = ABu$, we can write $\hat{v} = \widehat{AB}\hat{u}$ but also $\hat{v} = \widehat{AB}u = \hat{A}\hat{B}\hat{u}$. By comparing terms we obtain $\widehat{AB} = \hat{A}\hat{B}$; (7.39f), (7.39g), and (7.39h) follow immediately from (7.38); (7.39i) follows from (7.39e) and (7.39h); finally, to prove (7.39j) we use the fact that the determinant of a product is the product of the determinants and the determinant of a block triangular matrix is the product of the determinants of the diagonal blocks. Hence

$$\det(\hat{A}) = \det\left(\begin{bmatrix} I & jI \\ 0 & I \end{bmatrix} \hat{A} \begin{bmatrix} I & -jI \\ 0 & I \end{bmatrix}\right) = \det\left(\begin{bmatrix} A & 0 \\ \Im(A) & A^* \end{bmatrix}\right) = \det(A)\det(A)^*,$$

where A^* is A with each element complex conjugated. □

COROLLARY 7.22 *If $U \in \mathbb{C}^{n \times n}$ is unitary then $\hat{U} \in \mathbb{R}^{2n \times 2n}$ is orthonormal.* □

Proof From $U^\dagger U = I_n$, applying the hat operator on both sides we obtain

$$(\hat{U})^\mathsf{T} \hat{U} = \hat{I}_n = I_{2n}.$$

□

7.9. Appendix: Complex-valued random vectors

To remember (7.39c), observe that $u^\dagger v$ is complex-valued in general whereas $\hat{u}^\mathsf{T}\hat{v}$ is always real-valued. To remember (7.39h), recall that the hat operator doubles the size of a matrix. In (7.39j) we need to take the absolute value of $\det(A)$ because $\det(A)$ can be a complex number, and we need to square because the hat operator doubles the size of a matrix. In doubt, it helps to do a "sanity check" using a scalar $\mathsf{j}a$, $a \in \mathbb{R}$ as a special case of a matrix A. The determinant of a scalar is the scalar itself. Hence $\det(\mathsf{j}a) = \mathsf{j}a$. From

$$\widehat{\mathsf{j}a} = \begin{pmatrix} 0 & -a \\ a & 0 \end{pmatrix},$$

$\det(\widehat{\mathsf{j}a}) = a^2 = |\det(\mathsf{j}a)|^2$. The remaining relationships of Lemma 7.21 are natural and easy to remember. However, be careful that $\hat{u}^\mathsf{T} \neq \widehat{u^\dagger}$. To see this, consider a scalar u as a special case of an n-tuple. Then $u^\dagger$ is still a scalar, hence $\widehat{u^\dagger}$ has dimension 2×1, whereas $\hat{u}^\mathsf{T}$ has dimension 1×2. Finally, notice that when we apply the hat operator to a scalar, the result depends on whether we consider the scalar as a special case of an n-tuple or of a matrix.

COROLLARY 7.23 *If $Q \in \mathbb{C}^{n \times n}$ is non-negative definite, then so is $\hat{Q} \in \mathbb{R}^{2n \times 2n}$. Moreover, $u^\dagger Q u = \hat{u}^\mathsf{T} \hat{Q} \hat{u}$.* □

Proof If Q is non-negative definite, $u^\dagger Q u$ is a non-negative real-valued number for all $u \in \mathbb{C}^n$. Hence,

$$u^\dagger Q u = \Re\{u^\dagger(Qu)\} = \hat{u}^\mathsf{T} \widehat{(Qu)}$$
$$= \hat{u}^\mathsf{T} \hat{Q} \hat{u},$$

where in the last two equalities we use (7.39c) and (7.39a), respectively. □

7.9 Appendix: Complex-valued random vectors

7.9.1 General statements

A complex-valued random variable Z (hereafter simply called complex random variable) is an object of the form

$$Z = X + \mathsf{j}Y,$$

where X and Y are real random variables.

Recall that a real random variable X is specified by its cumulative distribution function $F_X(x) = Pr\{X \leq x\}$. For a complex random variable Z, as there is no natural ordering in the complex plane, the event $Z \leq z$ does not make sense. Instead, we specify a complex random variable by giving the joint distribution of its real and imaginary parts $F_{\Re\{Z\},\Im\{Z\}}(x,y) = Pr\{\Re\{Z\} \leq x, \Im\{Z\} \leq y\}$. Since the pair of real numbers (x,y) can be identified with a complex number $z = x + \mathsf{j}y$, we will write the joint distribution $F_{\Re\{Z\},\Im\{Z\}}(x,y)$ as $F_Z(z)$. Just as we do for real-valued random variables, if the function $F_{\Re\{Z\},\Im\{Z\}}(x,y)$ is differentiable in x and y, we will call the function

$$f_{\Re\{Z\},\Im\{Z\}}(x,y) = \frac{\partial^2}{\partial x \partial y} F_{\Re\{Z\},\Im\{Z\}}(x,y)$$

the joint density of $(\Re\{Z\}, \Im\{Z\})$, and again associating with (x,y) the complex number $z = x + \mathrm{j}y$, we will call the function

$$f_Z(z) = f_{\Re\{Z\},\Im\{Z\}}(\Re\{z\}, \Im\{z\})$$

the density of the random variable Z.

A complex random vector $Z = (Z_1, \ldots, Z_n)$ is specified by the joint distribution of its real and imaginary parts

$$F_Z(z) = \Pr(\Re\{Z_1\} \leq \Re\{z_1\}, \ldots, \Re\{Z_n\} \leq \Re\{z_n\}, \Im\{Z_1\} \leq \Im\{z_1\}, \ldots, \Im\{Z_n\} \leq \Im\{z_n\}),$$

and if this function is differentiable in $\Re\{z_1\}, \ldots, \Re\{z_n\}, \Im\{z_1\}, \ldots, \Im\{z_n\}$, then we define the density of Z as

$$f_Z(x_1 + \mathrm{j}y_1, \ldots, x_n + \mathrm{j}y_n) = \frac{\partial^{2n}}{\partial x_1 \cdots \partial x_n \partial y_1 \cdots \partial y_n} F_Z(x_1 + \mathrm{j}y_1, \ldots, x_n + \mathrm{j}y_n).$$

If X is a real random vector with density f_X, and A is a nonsingular matrix, then the density of $Y = AX$ is given by (see Appendix 2.9)

$$f_Y(y) = |\det(A)|^{-1} f_X(A^{-1}y).$$

If W is a complex random vector with density f_W and if A is a complex nonsingular matrix, then $Z = AW$ is again a complex random vector with

$$\begin{bmatrix} \Re\{Z\} \\ \Im\{Z\} \end{bmatrix} = \begin{bmatrix} \Re\{A\} & -\Im\{A\} \\ \Im\{A\} & \Re\{A\} \end{bmatrix} \begin{bmatrix} \Re\{W\} \\ \Im\{W\} \end{bmatrix}$$

and thus the density of Z will be given by

$$f_Z(z) = \left| \det\left(\begin{bmatrix} \Re\{A\} & -\Im\{A\} \\ \Im\{A\} & \Re\{A\} \end{bmatrix} \right) \right|^{-1} f_W(A^{-1}z).$$

From (7.39j) we know that

$$\det\left(\begin{bmatrix} \Re\{A\} & -\Im\{A\} \\ \Im\{A\} & \Re\{A\} \end{bmatrix} \right) = |\det(A)|^2,$$

and thus the transformation formula becomes

$$f_Z(z) = |\det(A)|^{-2} f_W(A^{-1}z). \tag{7.40}$$

7.9. Appendix: Complex-valued random vectors

7.9.2 The Gaussian case

DEFINITION 7.24 *The random vector $Z = X + jY \in \mathbb{C}^n$, where $X \in \mathbb{R}^n$ and $Y \in \mathbb{R}^n$, is a complex Gaussian random vector if X and Y are jointly Gaussian random vectors.* □

The probability density function (pdf) of a complex Gaussian random vector is obtained from the pdf of $\hat{Z}$ (see Appendix 2.10),

$$f_Z(z) := f_{\hat{Z}}(\hat{z}) = \frac{1}{\sqrt{\det(2\pi K_{\hat{Z}})}} e^{-\frac{1}{2}(\hat{z}-\hat{m})^T K_{\hat{Z}}^{-1}(\hat{z}-\hat{m})}, \qquad (7.41)$$

where

$$\hat{m} = \mathbb{E}[\hat{Z}]$$

$$K_{\hat{Z}} = \mathbb{E}[(\hat{Z}-\hat{m})(\hat{Z}-\hat{m})^\mathsf{T}] = \begin{pmatrix} K_X & K_{XY} \\ K_{YX} & K_Y \end{pmatrix}. \qquad (7.42)$$

In writing (7.41), we assume that $K_{\hat{Z}}$ is nonsingular. Example 2.32 shows how to proceed when a Gaussian random vector has a singular covariance matrix.

Define the following complex-valued quantities

$$\mathbb{E}[Z] := \mathbb{E}[X] + j\mathbb{E}[Y] \qquad (7.43)$$

$$K_Z := \mathbb{E}[(Z - \mathbb{E}[Z])(Z - \mathbb{E}[Z])^\dagger] \qquad (7.44)$$

$$J_Z := \mathbb{E}[(Z - \mathbb{E}[Z])(Z - \mathbb{E}[Z])^\mathsf{T}] \qquad (7.45)$$

called the *mean*, the *covariance matrix*, and the *pseudo-covariance matrix* (of the complex random vector Z), respectively.

Clearly $\mathbb{E}[\hat{Z}]$ is in one-to-one correspondence with $\mathbb{E}[Z]$. Furthermore

$$K_Z = K_X + K_Y - j(K_{XY} - K_{YX}) \qquad (7.46)$$

$$J_Z = K_X - K_Y + j(K_{XY} + K_{YX}), \qquad (7.47)$$

showing that we can compute K_Z and J_Z from $K_{\hat{Z}}$. We can also go the other way by using

$$K_X = \frac{1}{2}\Re\{K_Z + J_Z\} \qquad (7.48)$$

$$K_Y = \frac{1}{2}\Re\{K_Z - J_Z\} \qquad (7.49)$$

$$K_{YX} = \frac{1}{2}\Im\{K_Z + J_Z\}, \qquad (7.50)$$

and of course $K_{XY} = K_{YX}^\mathsf{T}$. Hence, K_Z and J_Z are in one-to-one correspondence with $K_{\hat{Z}}$. This implies that, in principle, even though $f_Z(z)$ is defined via $f_{\hat{Z}}(\hat{z})$, we can express it using $\mathbb{E}[Z]$, K_Z and J_Z. Doing so is rather cumbersome for general complex Gaussian random vectors, but the expression simplifies for so-called *proper* Gaussian vectors. Fortunately, all complex random vectors that concern us are proper. The next subsection is dedicated to their study.

We conclude this subsection with a lemma that gives an alternative characterization of a complex Gaussian random vector. In Appendix 2.10 we have defined a zero-mean Gaussian random vector $Z \in \mathbb{R}^n$ as a vector that can be obtained via $Z = AW$ for some $n \times m$ matrix A and $W \sim \mathcal{N}(0, I_m)$. If A has linearly independent rows, then K_Z is nonsingular and the pdf of Z is well defined. We can do the same for zero-mean complex Gaussian random vectors.

LEMMA 7.25 *The random vector $Z \in \mathbb{C}^n$ is a zero-mean complex Gaussian random vector if and only if there exists a matrix $A \in \mathbb{C}^{n \times m}$ such that*

$$Z = AW, \tag{7.51}$$

where $W \sim \mathcal{N}(0, I_m)$. □

Proof The "if" part is obvious. If Z can be written as in (7.51), $X = \Re\{Z\} = \Re\{A\}W$ and $Y = \Im\{Z\} = \Im\{A\}W$ are jointly Gaussian random vectors (Definition 2.30) and Z is a complex Gaussian random vector (Definition 7.24). For the "only if" part, let $Z \in \mathbb{C}^n$ be a complex Gaussian random vector. By Definition 7.24, $\Re\{Z\} \in \mathbb{R}^n$ and $\Im\{Z\} \in \mathbb{R}^n$ are jointly Gaussian random vectors. By Definition 2.30, there exist $n \times m$ real matrices A_R and A_I such that

$$\begin{pmatrix} X \\ Y \end{pmatrix} = \begin{pmatrix} A_R \\ A_I \end{pmatrix} W,$$

where $W \sim \mathcal{N}(0, I_m)$. Hence

$$Z = A_R W + j A_I W = AW$$

with $A = A_R + j A_I$. □

7.9.3 The circularly symmetric Gaussian case

Define $V = X + jY \in \mathbb{C}^n$, where $X \in \mathbb{R}^n$ and $Y \in \mathbb{R}^n$ are independent Gaussian random vectors with iid $\sim \mathcal{N}(0, \frac{1}{2})$ components. Then

$$\begin{aligned}
f_V(v) = f_{\hat{V}}(\hat{v}) &= \frac{1}{\pi^n} e^{-\sum_{i=1}^{2n} \hat{v}_i^2} \\
&= \frac{1}{\pi^n} e^{-\|v\|^2} \\
&= \frac{1}{\pi^n} e^{-v^\dagger v}.
\end{aligned} \tag{7.52}$$

Notice that although $f_V(v)$ is derived via $\hat{v}$, it can be expressed in compact form as a function of v. Notice also that $f_V(v)$ only depends on $\|v\|$. Hence $e^{j\theta} V$, which is V with each component rotated by the angle θ, has the same pdf as V. Gaussian random vectors that have this property are of particular interest to us for two reasons: (i) all the noise vectors of interest to us are of this kind, and (ii) the pdf of a Gaussian random vector that has this property takes on a simplified form. For these two reasons, it is worthwhile investing in the study of such random vectors, called circularly symmetric.

7.9. Appendix: Complex-valued random vectors

DEFINITION 7.26 *A complex random vector $Z \in \mathbb{C}^n$ is* circularly symmetric *if for any $\theta \in [0, 2\pi)$, the distribution of $Ze^{j\theta}$ is the same as the distribution of Z. (If it is the case and $n = 1$, we speak of a circularly symmetric random variable.)* □

A circularly symmetric random vector is always zero-mean. In fact

$$\mathbb{E}[Z] = \mathbb{E}[e^{j\theta} Z] = e^{j\theta} \mathbb{E}[Z],$$

which can be true for all θ only if $\mathbb{E}[Z] = 0$.

DEFINITION 7.27 *A complex random vector Z is called* proper *if its pseudo-covariance J_Z vanishes.*

□

The above two definitions are related. To see how, let Z be circularly symmetric. Then it is zero-mean and, for every $\theta \in [0, 2\pi)$,

$$J_Z = \mathbb{E}[ZZ^T] = \mathbb{E}[(e^{j\theta} Z)(e^{j\theta} Z)^T] = e^{j2\theta} \mathbb{E}[ZZ^T],$$

which, evaluated for $\theta = \pi/2$, yields $J_Z = \mathbb{E}[ZZ^T] = -\mathbb{E}[ZZ^T]$. Hence $J_Z = 0$. We conclude that if a random vector Z is circularly symmetric, then it is zero-mean and proper. Note that Z need not be Gaussian.

If Z is indeed *Gaussian*, then the converse is also true, i.e. if Z is zero-mean and proper, then it is circularly symmetric. To see this, let Z be zero-mean, proper, and Gaussian. Then $e^{j\theta} Z$ is also zero-mean and Gaussian. Hence Z and $e^{j\theta} Z$ have the same density if and only if they have the same covariance and pseudo-covariance matrices. We know that the pseudo-covariance matrix of Z vanishes. From

$$\mathbb{E}[(e^{j\theta} Z)(e^{j\theta} Z)^T] = e^{j2\theta} \mathbb{E}[ZZ^T] = 0$$

we see that the pseudo-covariance matrix of $e^{j\theta} Z$ vanishes as well. Finally, the covariance matrix of $e^{j\theta} Z$ is

$$\mathbb{E}\left[(e^{j\theta} Z)(e^{j\theta} Z)^\dagger\right] = \mathbb{E}\left[ZZ^\dagger\right] = K_Z,$$

proving that the covariance matrices are identical. We summarize this result in a lemma.

LEMMA 7.28 *For a complex-valued Gaussian random vector Z, the following statements are equivalent:*

(i) Z is circularly symmetric;
(ii) Z is zero-mean and proper. □

The covariance matrix of two real-valued random vectors X and Y satisfies $K_{XY} = K_{YX}^T$. Hence, (7.47) can be written as

$$J_Z = (K_X - K_Y) + j(K_{XY} + K_{XY}^T),$$

which proves the following.

LEMMA 7.29   A complex random vector $Z = X + jY \in \mathbb{C}^n$, where $X \in \mathbb{R}^n$ and $Y \in \mathbb{R}^n$, is proper if and only if

$$K_X = K_Y \quad \text{and} \quad K_{XY} = -K_{XY}^T,$$

i.e. J_Z vanishes if and only if X and Y have identical auto-covariance matrices and their cross-covariance matrix is skew-symmetric.[12]   □

COROLLARY 7.30   A real random vector Z is a proper complex random vector if and only if it is constant (with probability 1).   □

Proof   If $Z = X + jY \in \mathbb{C}^n$, with $X \in \mathbb{R}^n$ and $Y = 0 \in \mathbb{R}^n$, then $K_Y = 0$. If Z is proper, Lemma 7.29 implies $K_X = 0$. Hence X is constant (with probability 1). For the other direction, if Z is constant, $J_Z = 0$, hence Z is proper.   □

To gain insight into Lemma 7.29, it is helpful to consider a circularly symmetric random vector $V = X + jY$, with $X, Y \in \mathbb{R}^n$. When we multiply V by $e^{j\pi/2} = j$, we obtain $jV = -Y + jX$. For V and jV to have the same distribution, X and $-Y$ have to have the same distribution. This explains why the real part and the imaginary part of V have to have the same covariance. The covariance between the real and the imaginary part of V is K_{XY}, whereas that between the real and the imaginary part of jV is $K_{(-Y)X} = -K_{YX} = -K_{XY}^T$. This explains why K_{XY} has to be skew-symmetric.

The skew-symmetry of K_{XY} implies that K_{XY} has zeros on the main diagonal, which means that the real and imaginary part of each component Z_k of Z are uncorrelated. However, it does not imply that the real part of Z_k and the imaginary part of Z_l are uncorrelated for $k \neq l$. In the following example, we construct a case where the real part of a component is correlated with the imaginary part of another component.

EXAMPLE 7.31   Let $U = X + jY \in \mathbb{C}$, where $X \in \mathbb{R}$ and $Y \in \mathbb{R}$ are independent, zero-mean, Gaussian random variables of unit variance, and let

$$Z = (U, jU)^\mathsf{T} \in \mathbb{C}^2.$$

Now $Z_R = \Re\{Z\} = (X, -Y)^\mathsf{T}$ and $Z_I = \Im\{Z\} = (Y, X)^\mathsf{T}$ have the same covariance, i.e. $K_{Z_R} = K_{Z_I} = I_2$. Furthermore,

$$K_{Z_R Z_I} = \begin{pmatrix} 0 & 1 \\ -1 & 0 \end{pmatrix}.$$

By Lemma 7.29, Z is proper despite the fact that its real and imaginary parts are correlated.   □

LEMMA 7.32   (Closure under affine transformations) Let Z be a proper n-dimensional random vector, i.e. $J_Z = 0$. Then any vector obtained from Z by an affine transformation, i.e. any vector $\tilde{Z}$ of the form $\tilde{Z} = AZ + b$, where $A \in \mathbb{C}^{m \times n}$ and $b \in \mathbb{C}^m$ are constant, is also proper.   □

---

[12] A real-valued matrix A is skew-symmetric if $A^\mathsf{T} = -A$.

7.9. Appendix: Complex-valued random vectors

Proof From
$$\mathbb{E}[\tilde{Z}] = A\mathbb{E}[Z] + b,$$
it follows that
$$\tilde{Z} - \mathbb{E}[\tilde{Z}] = A(Z - \mathbb{E}[Z]).$$
Hence we have
$$\begin{aligned} J_{\tilde{Z}} &= \mathbb{E}[(\tilde{Z} - \mathbb{E}[\tilde{Z}])(\tilde{Z} - \mathbb{E}[\tilde{Z}])^\mathsf{T}] \\ &= \mathbb{E}\{A(Z - \mathbb{E}[Z])(Z - \mathbb{E}[Z])^\mathsf{T} A^\mathsf{T}\} \\ &= A J_Z A^\mathsf{T} = 0. \end{aligned}$$
$\square$

The pdf of a proper Gaussian random vector Z is described by the mean m_Z and covariance matrix K_Z. We denote it by $\mathcal{N}_\mathcal{C}(m_Z, K_Z)$.

The complex Gaussian random vector V introduced at the beginning of this subsection is characterized by
$$\begin{aligned} \mathbb{E}[V] &= 0 \\ K_V &= I_n \\ J_V &= 0. \end{aligned}$$
Hence $V \sim \mathcal{N}_\mathcal{C}(0, I_n)$.

We have seen that every zero-mean complex Gaussian random vector $Z \in \mathbb{C}^n$ can be written as $Z = AW$ for some complex $n \times m$ matrix A and $W \sim \mathcal{N}(0, I_m)$. There is a similar result for circularly symmetric Gaussian random vectors.

LEMMA 7.33 *The random vector $Z \in \mathbb{C}^n$ is a circularly symmetric Gaussian random vector if and only if there exists a matrix $A \in \mathbb{C}^{n \times m}$ such that*
$$Z = AV, \tag{7.53}$$
with $V \sim \mathcal{N}_\mathcal{C}(0, I_m)$. Furthermore, if Z has nonsingular covariance matrix, then we can write (7.53) with $V \sim \mathcal{N}_\mathcal{C}(0, I_n)$ and a nonsingular matrix $A \in \mathbb{C}^{n \times n}$. $\square$

Proof Clearly any Z of the form (7.53) for some $A \in \mathbb{C}^{n \times m}$ and $V \sim \mathcal{N}_\mathcal{C}(0, I_m)$ is Gaussian, zero-mean, and proper (Lemma 7.32), hence circularly symmetric (Lemma 7.28). To prove the other direction, let Z be a circularly symmetric Gaussian random vector specified by its $n \times n$ covariance matrix K_Z. A covariance matrix is Hermitian, hence K_Z can be written in the form $K_Z = U\Lambda U^\dagger$, where $U \in \mathbb{C}^{n \times n}$ is unitary and $\Lambda \in \mathbb{R}^{n \times n}$ is diagonal (Appendix 2.8). If K_Z is nonsingular, then all the diagonal elements of Λ are positive. (In fact, they are non-negative because a covariance matrix is positive semidefinite, and they cannot vanish, because the product of the eigenvalues equals the determinant of the matrix, which is non-zero by assumption.) Define $V = \Lambda^{-\frac{1}{2}} U^\dagger Z$, where for $\alpha \in \mathbb{R}$, Λ^α is the diagonal matrix obtained by raising to the power α the diagonal elements of Λ. Clearly V is a zero-mean Gaussian random vector and $Z = U\Lambda^{\frac{1}{2}} V$ has the form (7.53) for the nonsingular matrix $A = U\Lambda^{\frac{1}{2}} \in \mathbb{C}^{n \times n}$. The covariance matrix of V is

$$\begin{aligned} K_V &= \mathbb{E}\left[VV^\dagger\right] \\ &= \mathbb{E}\left[\Lambda^{-\frac{1}{2}}U^\dagger ZZ^\dagger U\Lambda^{-\frac{1}{2}}\right] \\ &= \Lambda^{-\frac{1}{2}}U^\dagger \mathbb{E}\left[ZZ^\dagger\right] U\Lambda^{-\frac{1}{2}} \\ &= \Lambda^{-\frac{1}{2}}U^\dagger K_Z U\Lambda^{-\frac{1}{2}} \\ &= \Lambda^{-\frac{1}{2}}U^\dagger U\Lambda U^\dagger U\Lambda^{-\frac{1}{2}} \\ &= \Lambda^{-\frac{1}{2}}\Lambda\Lambda^{-\frac{1}{2}} \\ &= I_n. \end{aligned}$$

Finally, V is proper (by Lemma 7.32) and circularly symmetric (by Lemma 7.28). This completes the proof for the case that K_Z is nonsingular. If K_Z is singular, then some of its components are linearly dependent on other components. In this case, we can write $Z = B\tilde{Z}$ for some $B \in \mathbb{C}^{n \times m}$, where $\tilde{Z} \in \mathbb{C}^m$ consists of linearly independent components of Z. The covariance matrix of $\tilde{Z}$ is nonsingular. Hence we can find a nonsingular matrix $\tilde{A} \in \mathbb{C}^{m \times m}$ such that $\tilde{Z} = \tilde{A}V$ with $V \sim \mathcal{N}_\mathcal{C}(0, I_m)$. Finally, $Z = B\tilde{Z} = B\tilde{A}V = AV$ has the desired form with $A = B\tilde{A} \in \mathbb{C}^{n \times m}$. $\square$

We are now in the position to derive a general expression for a circularly symmetric Gaussian random vector Z of nonsingular covariance matrix.

THEOREM 7.34 *The probability density function of a circularly symmetric Gaussian random vector $Z \in \mathbb{C}^n$ of nonsingular covariance matrix K_Z can be written as*

$$f_Z(z) = \frac{1}{\pi^n \det(K_Z)} e^{-z^\dagger K_Z^{-1} z}. \tag{7.54}$$

$\square$

Proof Let $Z \in \mathbb{C}^n$ be circularly symmetric and Gaussian with nonsingular covariance matrix K_Z. By Lemma 7.33, we can write $Z = AV$ where A is a nonsingular $n \times n$ matrix and $V \sim \mathcal{N}_\mathcal{C}(0, I_n)$. From (7.40),

$$f_Z(z) = \frac{1}{\pi^n |\det(A)|^2} e^{-(A^{-1}z)^\dagger (A^{-1}z)}. \tag{7.55}$$

Now

$$\begin{aligned} K_Z &= \mathbb{E}\left[ZZ^\dagger\right] \\ &= \mathbb{E}\left[AVV^\dagger A^\dagger\right] \\ &= A\mathbb{E}\left[VV^\dagger\right] A^\dagger \\ &= AA^\dagger, \end{aligned}$$

hence

$$\det(K_Z) = \det(AA^\dagger) = |\det(A)|^2. \tag{7.56}$$

Furthermore,

$$\begin{aligned}(A^{-1}z)^\dagger(A^{-1}z) &= z^\dagger (A^{-1})^\dagger A^{-1} z \\ &= z^\dagger (AA^\dagger)^{-1} z \\ &= z^\dagger K_Z^{-1} z, \end{aligned} \qquad (7.57)$$

where in the second equality we use the fact that, for nonsingular $n \times n$ matrices, $(AB)^{-1} = B^{-1}A^{-1}$ and $(A^\dagger)^{-1} = (A^{-1})^\dagger$. Inserting (7.56) and (7.57) into (7.55) yields (7.54). □

The above theorem justifies one of the two claims we have made at the beginning of this appendix, specifically that the pdf of a circularly symmetric Gaussian random vector takes on a simplified form. (Compare (7.54) and (7.41) when $\hat{m} = 0$, keeping in mind (7.42) to compute $K_{\hat{Z}}$ from K_X, K_Y, and K_{XY}.) The next theorem justifies the other claim: that the complex-valued noise vectors of interest to us, those at the output of Figure 7.13b, are Gaussian and circularly symmetric.

THEOREM 7.35 *Let $N_E(t) = N_R(t) + \mathrm{j} N_I(t)$, where $N_R(t)$ and $N_I(t)$ are independent white Gaussian noises of spectral density $N_0/2$. For any collection of $\mathcal{L}_2$ functions $g_l(t)$, $l = 1, \ldots, k$ that belong to a finite-dimensional inner product space $\mathcal{V}$, the complex-valued random vector $Z = (Z_1, \ldots, Z_k)^\mathsf{T}$, $Z_l = \langle N_E(t), g_l(t)\rangle$, is circularly symmetric and Gaussian.* □

Proof Let $\psi_1(t), \ldots, \psi_n(t)$ be an orthonormal basis for $\mathcal{V}$. First consider the random vector $V = (V_1, \ldots, V_n)^\mathsf{T}$, where $V_i = \langle N_E(t), \psi_i(t)\rangle$. It is straightforward to check that $V \sim \mathcal{N}_\mathcal{C}(0, I_n)$. Every $g_l(t)$ can be written as $g_l(t) = \sum_{j=1}^n c_{l,j} \psi_j(t)$, where $c_{l,j} \in \mathbb{C}$. By the linearity of the inner product, $Z = AV$, where $A \in \mathbb{C}^{k \times n}$ is the matrix that has $(c_{l,1}^*, \ldots, c_{l,n}^*)$ in its lth row. By Lemma 7.33, Z is circularly symmetric with $K_Z = AA^\dagger$. □

7.10 Exercises

Exercises for Section 7.2

EXERCISE 7.1 (Lifting up) *Let $p(t)$ be real-valued and frequency-limited to $[-B, B]$, where $0 < B < f_c$ for some f_c. Without making any calculations, argue that $p(t)\sqrt{2}\cos(2\pi f_c t)$ and $p(t)\sqrt{2}\sin(2\pi f_c t)$ are orthogonal to each other and have the same norm as $p(t)$.*

EXERCISE 7.2 (Bandpass filtering in baseband) *We want to implement a passband filter of impulse response $h(t) = \sqrt{2}\Re\{h_E(t) e^{\mathrm{j}2\pi f_c t}\}$ using baseband filters, where $h_E(t)$ is frequency-limited to $[-B, B]$ and $0 < B < f_c$.*

(a) *Draw the block diagram of an implementation of the filter of impulse response $h(t)$, based on a filter of impulse response $h_E(t)$ (possibly scaled). Your implementation can use an up-converter, a down-converter, and shall behave like the filter of impulse response $h(t)$ for all (passband) input signals of bandwidth not exceeding $2B$ and center frequency f_c.*

(b) *Draw the box diagram of an implementation that uses only real-valued signals.*

EXERCISE 7.3 (Equivalent representations) *A real-valued passband signal $x(t)$ can be written as $x(t) = \sqrt{2}\Re\{x_E(t)e^{j2\pi f_c t}\}$, where $x_E(t)$ is the baseband-equivalent signal (complex-valued in general) with respect to the carrier frequency f_c. Also, a general complex-valued signal $x_E(t)$ can be written in terms of two real-valued signals, either as $x_E(t) = u(t) + jv(t)$ or as $\alpha(t)\exp(j\beta(t))$.*

(a) *Show that a real-valued passband signal $x(t)$ can always be written as*

$$x_{EI}(t)\cos(2\pi f_c t) - x_{EQ}(t)\sin(2\pi f_c t)$$

and relate $x_{EI}(t)$ and $x_{EQ}(t)$ to $x_E(t)$. Note: This formula can be used at the sender to produce $x(t)$ without doing complex-valued operations. The signals $x_{EI}(t)$ and $x_{EQ}(t)$ are called the in-phase *and the* quadrature *components, respectively.*

(b) *Show that a real-valued passband signal $x(t)$ can always be written as*

$$a(t)\cos[2\pi f_c t + \theta(t)]$$

and relate $x_E(t)$ to $a(t)$ and $\theta(t)$. Note: This explains why sometimes people make the claim that a passband signal is modulated in amplitude and in phase.

(c) *Use part (b) to find the baseband-equivalent of the signal*

$$x(t) = A(t)\cos(2\pi f_c t + \varphi),$$

where $A(t)$ is a real-valued lowpass signal. Verify your answer with Example 7.9 where we assumed $\varphi = 0$.

EXERCISE 7.4 (Passband) *Let f_c be a positive carrier frequency and consider an arbitrary real-valued function $w(t)$. You can visualize its Fourier transform as shown in Figure 7.14.*

(a) *Argue that there are two different functions, $a_1(t)$ and $a_2(t)$, such that, for $i = 1, 2$,*

$$w(t) = \sqrt{2}\Re\{a_i(t)\exp(j2\pi f_c t)\}.$$

This shows that, without some constraint on the input signal, the operation performed by the circuit of Figure 7.4b is not reversible, even in the absence of noise. This was already pointed out in the discussion preceding Lemma 7.8.

(b) *Argue that if we limit the input of Figure 7.4b to signals $a(t)$ such that $a_{\mathcal{F}}(f) = 0$ for $f < -f_c$, then the circuit of Figure 7.4a will retrieve $a(t)$ when fed with the output of Figure 7.4b.*

(c) *Find an example showing that the condition of part (b) is necessary. (Can you find an example with a real-valued $a(t)$?)*

(d) *Argue that if we limit the input of Figure 7.4b to signals $a(t)$ that are real-valued, then the input of Figure 7.4b can be retrieved from the output. Hint 1: we are not claiming that the circuit of Figure 7.4a will retrieve $a(t)$. Hint 2: You may argue in the time domain or in the frequency domain.*

7.10. Exercises

If you argue in the time domain, you can assume that $a(t)$ is continuous. In the frequency-domain argument, you can assume that $a(t)$ has finite bandwidth.

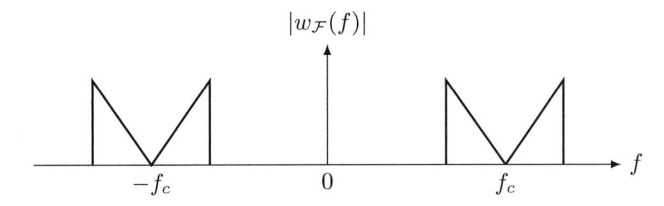

Figure 7.14.

EXERCISE 7.5 (From passband to baseband via real-valued operations) *Let the signal $x_E(t)$ be bandlimited to $[-B, B]$ and let $x(t) = \sqrt{2}\Re\{x_E(t)e^{j2\pi f_c t}\}$, where $0 < B < f_c$. Show that the circuit of Figure 7.15, when fed with $x(t)$, recovers the real and imaginary part of $x_E(t)$. (The two boxes are ideal lowpass filters of cutoff frequency B.) Note: The circuit uses only real-valued operations.*

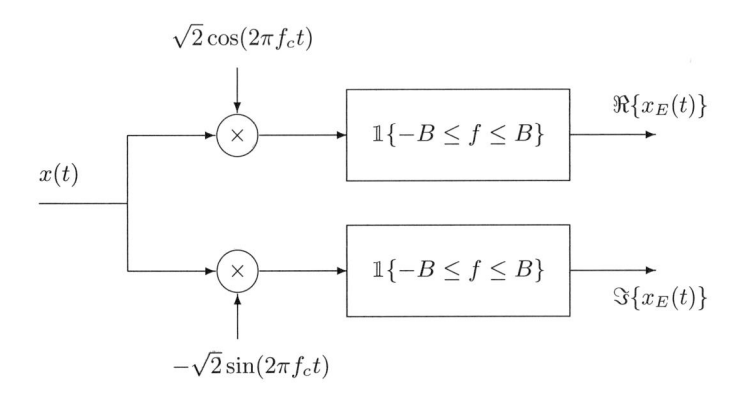

Figure 7.15.

EXERCISE 7.6 (Reverse engineering) *Figure 7.16 shows a toy passband signal. (Its carrier frequency is unusually the horizontal time scale, which is 1 ms per square and the vertical scale is 1 unit per square. Specify the three layers of a transmitter that generates the given signal, namely the following.*

(a) *The carrier frequency f_c used by the up-converter.*
(b) *The orthonormal basis used by the waveform former to produce the baseband-equivalent signal $w_E(t)$.*
(c) *The symbol alphabet, seen as a subset of $\mathbb{C}$.*
(d) *An encoding map, the encoder input sequence that leads to $w(t)$, the bit rate, the encoder output sequence, and the symbol rate.*

w(t)

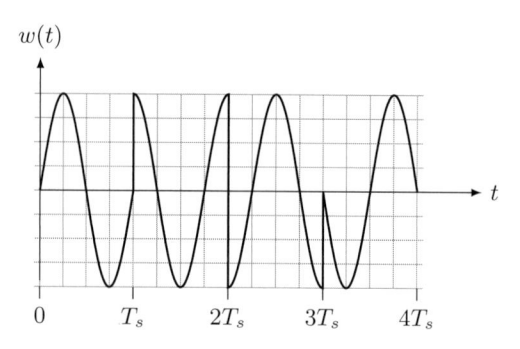

Figure 7.16.

EXERCISE 7.7 (AM receiver) Let $x(t) = (1 + mb(t))\sqrt{2}\cos(2\pi f_c t)$ be an AM modulated signal as described in Example 7.10. We assume that $1 + mb(t) > 0$, that $b(t)$ is bandlimited to $[-B, B]$, and that $f_c > 2B$.

(a) Argue that the envelope of $|x(t)|$ is $(1 + mb(t))\sqrt{2}$ (a drawing will suffice).
(b) Argue that with a suitable choice of components, the output in Figure 7.17 is essentially $b(t)$. Hint: Draw, qualitatively, the voltage on top of R_1 and that on top of R_2.
(c) As an alternative approach, prove that if we pass the signal $|x(t)|$ through an ideal lowpass filter of cutoff frequency f_0, we obtain $1 + mb(t)$ scaled by some factor. Specify a suitable interval for f_0. Hint: Expand $|\cos(2\pi f_c t)|$ as a Fourier series. No need to find explicit values for the Fourier series coefficients.

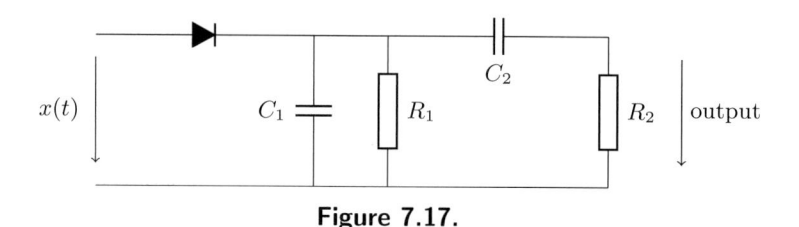

Figure 7.17.

Exercises for Section 7.3

EXERCISE 7.8 (Alternative down-converter) Assuming that all the $\psi_l(t)$ are bandlimited to $[-B, B]$ and that $0 < B < f_c$, show that the n-tuple former output remains unchanged if we substitute the down-converter of Figure 7.8b with the block diagram of Figure 7.4a.

EXERCISE 7.9 (Real-valued implementation) Draw a block diagram for the implementation of the transmitter and receiver of Figure 7.8 by means of real-valued operations. Unlike in Figure 7.9, do not assume that the orthonormal basis is real-valued.

Exercises for Section 7.9

EXERCISE 7.10 (Circular symmetry)

(a) Suppose X and Y are real-valued iid random variables with probability density function $f_X(s) = f_Y(s) = c\exp(-|s|^\alpha)$, where α is a parameter and $c = c(\alpha)$ is the normalizing factor.
  (i) Draw the contour of the joint density function for $\alpha = 0.5$, $\alpha = 1$, $\alpha = 2$, and $\alpha = 3$. Hint: For simplicity, draw the set of points (x,y) for which $f_{X,Y}(x,y)$ equals the constant $c^2(\alpha)e^{-1}$.
  (ii) For which value of α is the joint density function invariant under rotation? What is the corresponding distribution?

(b) In general we can show that if X and Y are iid random variables and $f_{X,Y}(x,y)$ is circularly symmetric, then X and Y are Gaussian. Use the following steps to prove this.
  (i) Show that if X and Y are iid and $f_{X,Y}(x,y)$ is circularly symmetric then $f_X(x)f_Y(y) = \psi(r)$ where ψ is a univariate function and $r = \sqrt{x^2 + y^2}$.
  (ii) Take the partial derivative with respect to x and y to show that
  $$\frac{f'_X(x)}{x\, f_X(x)} = \frac{\psi'(r)}{r\, \psi(r)} = \frac{f'_Y(y)}{y\, f_Y(y)}.$$
  (iii) Argue that the only way for the above equalities to hold is that they be equal to a constant value, i.e. $\frac{f'_X(x)}{x\, f_X(x)} = \frac{\psi'(r)}{r\, \psi(r)} = \frac{f'_Y(y)}{y\, f_Y(y)} = -\frac{1}{\sigma^2}$.
  (iv) Integrate the above equations and show that X and Y should be Gaussian random variables.

Miscellaneous exercises

EXERCISE 7.11 (Real-valued versus complex-valued constellation) Consider 2-PAM and 4-QAM. The source produces iid and uniformly distributed source bits taking value in $\{\pm 1\}$, and the constellations are $\{\pm 1\}$ and $\{1 + j, -1 + j, -1 - j, 1 - j\}$, respectively. For 2-PAM, the mapping between the source bits and the channel symbols is the obvious one, i.e. bit b_i is mapped into symbol $s_i = \sqrt{\mathcal{E}_s}b_i$. For the 4-QAM constellation, pairs of bits are mapped into a symbol according to
$$b_{2i}, b_{2i+1} \to s_i = \sqrt{\mathcal{E}_s}(b_{2i} + jb_{2i+1}).$$

The symbols are mapped into a signal via symbol-by-symbol on a pulse train, where the pulse is real-valued, normalized, and orthogonal to its shifts by multiples of T. The channel adds white Gaussian noise of power spectral density $\frac{N_0}{2}$. The receiver implements an ML decoder. For the two systems, determine (if possible) and compare the following.

(a) The bit-error rate P_b.
(b) The energy per symbol $\mathcal{E}_s$.

(c) The variance σ^2 of the noise seen by the decoder. Note: when the symbols are real-valued, the decoder disregards the imaginary part of Y. In this case, what matters is the variance of the real part of the noise.
(d) The symbol-to-noise power ratio $\frac{\mathcal{E}_s}{\sigma^2}$. Write them also as a function of the power P and N_0.
(e) The bandwidth.
(f) The expression for the signals at the output of the waveform former as a function of the bit sequence produced by the source.
(g) The bit rate R.

Summarize, by comparing the two systems from a user's point of view.

EXERCISE 7.12 (Smoothness of bandlimited signals) *We show that a continuous signal of small bandwidth cannot vary much over a small interval. (This fact is used in Exercise 7.13.) Let $w(t)$ be a finite-energy continuous-time passband signal and let $w_E(t)$ be its baseband-equivalent signal. We assume that $w_E(t)$ is bandlimited to $[-B, B]$ for some positive B.*

(a) Show that the baseband-equivalent of $w(t-\tau)$ can be modeled as $w_E(t-\tau)e^{j\phi}$ for some ϕ.
(b) Let $h_{\mathcal{F}}(f)$ be the frequency response of the ideal lowpass-filter, i.e. $h_{\mathcal{F}}(f) = 1$ for $|f| \leq B$ and 0 otherwise. Show that

$$w_E(t+\tau) - w_E(t) = \int w_E(\xi)[h(t+\tau-\xi) - h(t-\xi)]d\xi. \qquad (7.58)$$

(c) Use the Cauchy–Schwarz inequality to prove that

$$|w_E(t+\tau) - w_E(t)|^2 \leq 2\mathcal{E}_w[\mathcal{E}_h - R_h(\tau)], \qquad (7.59)$$

where

$$R_h(\tau) = \int h(\xi+\tau)h(\xi)d\xi$$

is the self-similarity function of $h(t)$, $\mathcal{E}_h = R_h(0)$ is the energy of $h(t)$, and $\mathcal{E}_w = R_w(0)$ is the energy of $w(t)$.
(d) Show that $R_h(\tau) = (h \star h)(\tau)$.
(e) Show that $R_h(\tau) = h(\tau)$.
(f) Put things together to derive the upper bound

$$|w_E(t+\tau) - w_E(t)| \leq \sqrt{2\mathcal{E}_w[\mathcal{E}_h - h(\tau)]} = \sqrt{4B\mathcal{E}_w\left(1 - \operatorname{sinc}(2B\tau)\right)}. \qquad (7.60)$$

Verify that the bound is tight for $\tau = 0$.
(g) Using part (a) and part (f), conclude that if τ is small compared to $1/B$, the baseband-equivalent of $w(t-\tau)$ can be modeled as $w_E(t)e^{j\phi}$.

EXERCISE 7.13 (Antenna array) *Assume that a transmitter uses an L-element antenna array as shown in Figure 7.18 for $L = 5$.*

The receiving antenna is located in the direction pointed by the arrows, far enough that we can approximate the wavefront as being a straight line as shown in the figure. Let $\beta_i w_E(t)$ be the baseband-equivalent signal transmitted at antenna

7.10. Exercises

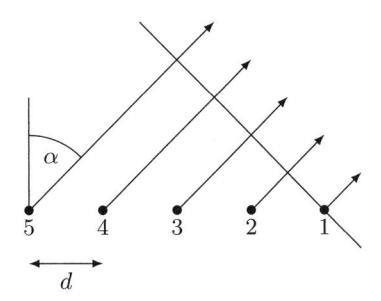

Figure 7.18.

element i, $i = 1, 2, \ldots, L$, with respect to some carrier frequency f_c. We assume that each antenna element irradiates isotropically. (More realistically, you can picture each dot as a dipole seen from above and we are interested in the radiation pattern in the plane perpendicular to the dipoles.)

(a) Argue that the received baseband-equivalent signal (at the matched filter input) can be written as

$$r_E(t) = \sum_{i=1}^{L} w_E(t - \tau_i)\beta_i \alpha_i,$$

where $\alpha_i = e^{-j2\pi f_c \tau_i}$ with $\tau_i = T + i\tau$ for some T and τ. Express τ as a function of the geometry (d and α) shown in Figure 7.18.

(b) We assume that $w_E(t)$ is a continuous bandlimited signal, which implies that for a sufficiently small τ_i, $w_E(t - \tau_i)$ is essentially $w_E(t)$ (see Exercise 7.12). We assume that all τ_i are small enough to satisfy this condition, but that $|f_c \tau_i|$ is not negligible with respect to 1, where f_c is the carrier frequency. Under these assumptions, we model the received baseband-equivalent signal as

$$r_E(t) = \sum_{i=1}^{L} w_E(t)\beta_i \alpha_i$$

plus noise. Choose the vector $\beta = (\beta_1, \beta_2, \ldots, \beta_L)^T$ that maximizes the energy $\int |r_E(t)|^2 dt$, subject to the constraint $\|\beta\| = 1$. Hint: Use the Cauchy–Schwarz inequality (Appendix 2.12).

(c) Let $\mathcal{E}$ be the energy of $w_E(t)$. Determine the energy of $r_E(t)$ as a function of L when β is selected as in part (b).

(d) In the above problem the received energy grows monotonically with L although $\|\beta\| = 1$ implies that the transmitted energy is constant. Does this violate energy conservation or some other fundamental law of physics?

EXERCISE 7.14 (Bandpass pulses) Let $p(t)$ be the pulse that has Fourier transform $p_\mathcal{F}(f)$ as in Figure 7.19.

(a) What is the expression for $p(t)$?
(b) Determine the constant c so that $\psi(t) = cp(t)$ has unit energy.

(c) Assume that $f_0 - \frac{B}{2} = B$ and consider the infinite set of functions $\{\psi(t - lT)\}_{l \in \mathbb{Z}}$. Do they form an orthonormal set for $T = \frac{1}{2B}$? (Explain.)

(d) Determine all possible values of $f_0 - \frac{B}{2}$ so that $\{\psi(t - lT)\}_{l \in \mathbb{Z}}$ forms an orthonormal set for $T = \frac{1}{2B}$.

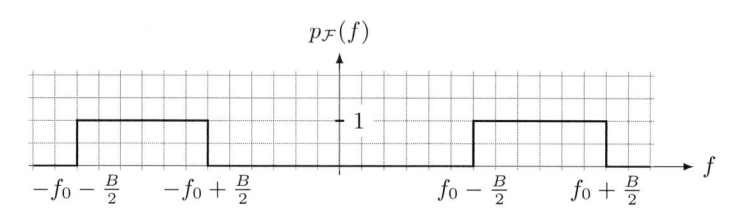

Figure 7.19.

EXERCISE 7.15 (Bandpass sampling) *The Fourier transform of a real-valued signal $w(t)$ satisfies the conjugacy constraint $w_\mathcal{F}(f) = w_\mathcal{F}^*(-f)$. Hence if $w_\mathcal{F}(f)$ is non-zero in some interval (f_L, f_H), $0 \le f_L < f_H$, then it is non-zero also in the interval $(-f_H, -f_L)$. This fact adds a complication to the extension of the sampling theorem to real-valued bandpass signals. Let $\mathcal{D}^+ = (f_L, f_H)$, $\mathcal{D}^- = (-f_H, -f_L)$, and let $\mathcal{D} = \mathcal{D}^- \cup \mathcal{D}^+$ be the passband frequency range of interest. Define $\mathcal{W}$ to be the set of $\mathcal{L}_2$ signals $w(t)$ that are continuous and for which $w_\mathcal{F}(f) = 0$, for $f \notin \mathcal{D}$.*

(a) *Assume $T > 0$ such that*

$$\left\{\frac{n}{2T}\right\}_{n \in \mathbb{Z}} \cap \mathcal{D} = \emptyset. \tag{7.61}$$

The above means that $\mathcal{D}^+ \subset [\frac{l}{2T}, \frac{l+1}{2T}]$ for some integer l. Define

$$h_\mathcal{F}(f) = \mathbb{1}\left\{\frac{l}{2T} \le |f| \le \frac{l+1}{2T}\right\} \quad \text{and} \quad \tilde{w}_\mathcal{F}(f) = \sum_{k \in \mathbb{Z}} w_\mathcal{F}\left(f - \frac{k}{T}\right),$$

where the latter is the periodic extension of $w_\mathcal{F}(f)$. Prove that for all $f \in \mathbb{R}$,

$$w_\mathcal{F}(f) = \tilde{w}_\mathcal{F}(f) h_\mathcal{F}(f).$$

Hint: Write $w_\mathcal{F}(f) = w_\mathcal{F}^-(f) + w_\mathcal{F}^+(f)$ where $w_\mathcal{F}^-(f) = 0$ for $f \ge 0$ and $w_\mathcal{F}^+(f) = 0$ for $f < 0$ and consider the support of $w_\mathcal{F}^+(f - \frac{k}{2T})$ and that of $w_\mathcal{F}^-(f - \frac{k}{2T})$, k integer.

(b) *Prove that when (7.61) holds, we can write*

$$w(t) = \sum_{k \in \mathbb{Z}} Tw(kT) h(t - kT),$$

where

$$h(t) = \frac{1}{T} \operatorname{sinc}\left(\frac{t}{2T}\right) \cos(2\pi f_c t)$$

7.10. Exercises

is the inverse Fourier transform of $h_{\mathcal{F}}(f)$ and $f_c = \frac{l}{2T} + \frac{1}{4T}$ is the center frequency of the interval $\left[\frac{l}{2T}, \frac{l+1}{2T}\right]$. Hint: Neglect convergence issues, use the Fourier series to write

$$\tilde{w}_{\mathcal{F}}(f) = \text{l.i.m.} \sum_k A_k e^{j2\pi fTk}$$

and use the result of part (a).

(c) Argue that if (7.61) is not true, then we can find two distinct signals $a(t)$ and $b(t)$ in $\mathcal{W}$ such that $a(nT) = b(nT)$ for all integers n. Hence (7.61) is necessary and sufficient to guarantee reconstruction from samples taken every T seconds. Hint: Show that we can choose $a(t)$ and $b(t)$ in $\mathcal{W}$ such that $\tilde{a}_{\mathcal{F}}(f) = \tilde{b}_{\mathcal{F}}(f)$, where tilde denotes periodic extension as in the definition of $\tilde{w}_{\mathcal{F}}(f)$.

(d) As an alternative characterization, show that (7.61) is true if and only if

$$\lfloor 2Tf_L \rfloor + 1 = \lceil 2Tf_H \rceil.$$

(e) Show that the largest T, denoted T_{max} that satisfies (7.61) is

$$T_{max} = \frac{\left\lfloor \frac{f_H}{f_H - f_L} \right\rfloor}{2f_H}.$$

Hence $1/T_{max}$ is the smallest sampling rate that permits the reconstruction of the bandpass signal from its samples.

(f) As an alternative characterization, show that $h(t)$ can be written as

$$h(t) = \left(\frac{l+1}{T}\right) \operatorname{sinc}\left(\frac{(l+1)t}{T}\right) - \frac{l}{T} \operatorname{sinc}\left(\frac{lt}{T}\right).$$

Hint: Rather than manipulating the right side of

$$h(t) = \frac{1}{T} \operatorname{sinc}\left(\frac{t}{2T}\right) \cos(2\pi f_c t),$$

start with a suitable description of $h_{\mathcal{F}}(f)$.

(g) As an application to (7.61), let $w(t)$ be a continuous finite-energy signal at the output of a filter of impulse response $h_{\mathcal{F}}(f)$ as in Figure 7.20. For which of the following sampling frequencies f_s is it possible to reconstruct $w(t)$ from its samples taken every $T = 1/f_s$ seconds: $f_s = 12, 14, 16, 18, 24$ MHz?

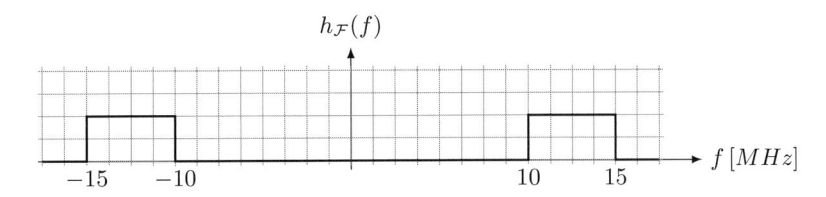

Figure 7.20.

Bibliography

[1] J. M. Wozencraft and I. M. Jacobs, *Principles of Communication Engineering*. New York: Wiley, 1965.

[2] R. G. Gallager, *Principles of Digital Communication*. New York: Cambridge University Press, 2008.

[3] A. Lapidoth, *A Foundation in Digital Communication*. New York: Cambridge University Press, 2009.

[4] J. F. Kurose and K. W. Ross, *Computer Networking*. New York: Addison Wesley, 2010.

[5] J. Gleick, *The Information: A History, a Theory, a Flood*. London: Fourth Estate, 2011.

[6] M. Vetterli, J. Kovačević, and V. Goyal, *Foundations of Signal Processing*. New York: Cambridge University Press, 2014.

[7] S. Ross, *A First Course in Probability*. New York: Macmillan & Co., 1994.

[8] W. Rudin, *Real and Complex Analysis*. New York: McGraw-Hill, 1966.

[9] T. M. Apostol, *Mathematical Analysis*. Reading, MA: Addison–Wesley, 2nd edn, 1974.

[10] S. Axler, *Linear Algebra Done Right*. New York: Springer-Verlag, 2nd edn, 1997.

[11] K. Hoffman and R. Kunze, *Linear Algebra*. Englewood Cliffs: Prentice-Hall, 2nd edn, 1971.

[12] R. A. Horn and C. R. Johnson, *Matrix Analysis*. Cambridge: Cambridge University Press, 1999.

[13] J. G. Proakis and M. Salehi, *Communication Systems Engineering*. Englewood Cliffs: Prentice-Hall, 1994.

[14] J. R. Barry, E. A. Lee, and D. G. Messerschmitt, *Digital Communication*. New York: Springer, 3rd edn, 2004.

[15] U. Madhow, *Fundamentals of Digital Communication*. New York: Cambridge University Press, 2008.

[16] S. G. Wilson, *Digital Modulation and Coding*. Englewood Cliffs: Prentice-Hall, 1996.

[17] D. Tse and P. Viswanath, *Fundamentals of Wireless Communications*. New York: Cambridge University Press, 2005.

[18] A. Goldsmith, *Wireless Communication*. New York: Cambridge University Press, 2005.

Bibliography

[19] T. M. Cover and J. A. Thomas, *Elements of Information Theory*. New York: Wiley, 2nd edn, 2006.

[20] R. G. Gallager, *Information Theory and Reliable Communication*. New York: Wiley, 1968.

[21] D. MacKay, *Information Theory, Inference, and Learning Algorithms*. New York: Cambridge University Press, 2003.

[22] S. Lin and D. J. Costello, *Error Control Coding*. Englewood Cliffs: Prentice-Hall, 2nd edn, 2004.

[23] T. Richardson and R. Urbanke, *Modern Coding Theory*. New York: Cambridge University Press, 2008.

[24] C. Shannon, "A mathematical theory of communication," *Bell System Tech. J.*, vol. **27**, pp. 379–423 and 623–656, 1948.

[25] H. Nyquist, "Thermal agitation of electric charge in conductors," *Physical Review*, vol. **32**, pp. 110–113, July 1928.

[26] A. W. Love, "Comment: On the equivalent circuit of a receiving antenna," *IEEE Antenna's and Propagation Magazine*, vol. **44**, pp. 124–125, October 2001.

[27] D. Slepian, "On bandwidth," *Proceedings of the IEEE*, vol. **64**, pp. 292–300, March 1976.

Index

a posteriori probability, *see* posterior
absorption, 235
additive white Gaussian noise (AWGN) channel, *see* channel
affine plane, 37, 38, 48, 70
amateur radio, 234, 241
antenna
    effective area, 120
    gain, 119
antenna array, 80, 81, 128, 280
autocorrelation, 117, 190, 191
autocovariance, 164, 190
automatic gain control (AGC), 112

band-edge symmetry, 169
bandwidth, 142, 149
    single-sided vs. double-sided, 144
basis, 71
Bhattacharyya bound, *see* error probability
bit-by-bit on a pulse train, 137, 145
bit-error probability (convolutional code), 219
block-orthogonal signaling, 139, 145
Bolzmann's constant, 118

carrier frequency, 232, 237–240, 244, 246, 259, 261
Cauchy–Schwarz inequality, 67
channel, 24
    baseband-equivalent, 11, 252
    binary symmetric (BSC), 51
    complex-valued AWGN, 264
    continuous-time AWGN, 25, 95, 96, 102, 103, 105, 119, 148
    discrete memoryless, 49, 89
    discrete-time AWGN, 25, 32, 39, 45, 46, 53, 97, 102, 180, 205, 208
    discrete-time vs. continuous-time, 144, 180
    impulse response, 112
    lowpass, 160
    memoryless, 24
    passband AWGN, 243, 252
    time-varying, 113
    wireless, 111
    wireline, 111
channel attenuation, 111, 256
channel capacity, 8, 14, 221
codes
    convolutional, 146, 205–208, 226–229
    low-density parity-check (LDPC), 9
    turbo, 9, 222
codeword, 205
collinear, 67
colored Gaussian noise, 113
communication, digital vs. analog, 13
complex vector space, 66
conjugate
    (anti)symmetry, 235
    transpose, 36
converter, up/down, 12, 232, 246, 264, 278
convolutional codes, *see* codes
correlated symbol sequence, 166
correlative encoding, 166, 224
correlator, 108
covariance matrix, 61, 269
cumulative distribution function, 190

decision
    function, 29, 31
    regions, 29, 31
decoder, 12, 97, 205
decoding, 26
    function, *see* decision function
    regions, *see* decision regions
delay, 256
delay locked loop, 259
detection, *see* hypothesis testing
detour, 212, 213
detour flow graph, 214
differential encoding, 202, 260

Index

diffraction, 112, 234
dimensionality, 135, 142
Dirichlet function, 181
dither, 163
down-converter, *see* converter
duration, 142

electromagnetic waves, 234
encoder, 12, 97, 102, 205
encoding
  circuit, 207
  map, 206
entropy, 19
error probability, 26, 29, 31, 35
  Bhattacharyya bound, 50, 87–90, 219
  union Bhattacharyya bound, 48, 53
  union bound, 44, 86, 91, 141, 228
events, 190
eye diagram, 173

fading, 81, 114
finite-energy functions, 161
first layer, 23
Fisher–Neyman factorization, 43, 83, 101, 130
Fourier
  series, 167, 187
  transform, 16, 184, 203
frame, 3
free-space path loss, 120
frequency hopping (FH), 261
frequency-shift keying, *see* FSK
frequency-shift property, 236
FSK, 104, 139

generating function, 214
Gram–Schmidt procedure, 72, 123, 124

Hermitian
  adjoint, 53
  matrix, 54, 61
  skew, 54
hyperplane, 69, 70
hypothesis testing, 26, 32, 102
  binary, 28
  m-ary, 30
  exercises, 74–77, 80–82, 91, 93, 125, 126, 129, 130

independent white Gaussian noises, 254
indicator function, 16, 30, 43
indistinguishable signals, 142
information theory, xi, 9, 18, 19, 103, 142, 147, 221
inner product space, 15, 65–67, 73, 96, 97, 100, 107, 132, 135, 162, 183, 244

inter-symbol interference (ISI), 173, 224
interference, 114
Internet, 1, 2, 4
Internet message access protocol (IMAP), 2
Internet protocol (IP), 5
irrelevance, 41, 84
isometry, 132

Jacobian, 59
jitter, 173

$\mathcal{L}_2$
  equivalence, 161, 183
  functions, 161, 180
Laplacian noise, 75, 77, 90
law of total expectation, 30
law of total probability, 29
layer
  application, 2
  data link, 3
  network, 3
  physical, 4
  presentation, 3
  session, 3
  transport, 3
layering, 1
LDPC code, *see* codes
Lebesgue
  integrable functions, 182
  integral, 180
  measurable sets, 182
  measure, 182
light-emitting diode (LED), 26
likelihood
  function, 27, 257
  ratio, 29, 34, 43
l. i. m., 168, 188
log likelihood ratio, 34
low-noise amplifier, 111

MAP
  rule, 27, 51, 77
  test, 29, 107
Mariner-10, 123
Markov chain, 41
matched filter, 108, 127, 128, 201
maximum a posteriori, *see* MAP
maximum likelihood, *see* ML
Mercury, 123
messages, 12
metric, 209
military communication, 261
minimum distance, 40, 136, 145, 146
minimum-distance rule, 39

ML parameter estimation, 175, 257
ML rule, 27, 39, 208, 261
    for complex-valued AWGN channels, 252
mobile communication, 113, 222
modem, 4
modulation (analog)
    single sideband (SSB), 241
    amplitude (AM), 13, 241, 278
    DSB-SC, 240
    frequency (FM), 13
multiaccess, 93
multiple antennas, *see* antenna array

n-tuple former, 12, 97, 102, 159
NASA, 123
nearest neighbors, 145
noise, 111
    correlated, 93
    data dependent, 93
    man-made, 99
    Poisson, 75
    shot, 99
    thermal (Johnson), 99, 118
    white Gaussian, 98
noise temperature, 119
non-coherent detection, 260, 263
nonlinearities, 114
norm, 67, 96
Nyquist
    bandwidth, 169
    criterion, 167, 168, 170, 179, 198–200, 281
    pulse, 198

observable, 26, 101
OFDM, 113
open system interconnection (OSI), 1, 103, 250
optical fiber, 26
orthonormal expansion, 71

packet, 3
PAM, 39, 48, 79, 105, 135, 159, 179, 208, 221, 279
parallelogram equality, 67
parameter estimation, 175, 256
Parseval's relationship, 99, 144, 161, 165, 167, 185, 238, 254
partial response signaling, 166
passband, 12, 232, 235, 276, 277
phase drift, 260
phase locked loop, 260
phase-shift keying, *see* PSK
phase synchronization, 256
    decision-directed, 260
picket fence miracle, 193, 196, 200

Plancherel's theorem, 185
Planck's constant, 119
Poisson distribution, 26
positive (semi)definite, 55
post office protocol (POP), 2
posterior, 27, 51
power
    dissipated, 121
    received, 120, 121
power spectral density (PSD), 119, 163, 191, 197, 223, 224
power spectrum, *see* power spectral density
PPM, 104, 139
prior, 51
probability measure, 190
probability of error, *see* error probability
probability space, 190
projection, 69, 72
propagation delay, 112
pseudo-covariance matrix, 269
PSK, 46, 48, 80, 106, 136, 160, 179, 248, 263
pulse, 15
pulse amplitude modulation, *see* PAM
pulse position modulation, *see* PPM
Pythagoras' theorem, 69

Q function, 31, 78, 228
QAM, 40, 48, 79, 86, 106, 126, 160, 179, 249, 263, 279
QAM (analog), 241
quantization, 20

radio spectrum, 234
raised-cosine function, 170
random process, *see* stochastic process
random variables and vectors, 61, 190
    circularly symmetric, 270, 271, 273, 274, 279
    complex-valued, 252, 267, 269
    proper, 269, 271
Rayleigh probability density, 60
real vector space, 66
receiver, 24
receiver design
    for continuous-time AWGN channels, 95
    for discrete-time AWGN channels, 32
    for discrete-time observations, 23
    for passband signals, 243
reflection, 112, 234
refraction, 234
Riemann integral, 180
root-raised-cosine (im)pulse (response), 170, 192
root-raised-cosine function, 170
rotation (of the matched filter output), 260

Index

sample space, 190
sampling theorem, 161
satellite communication, 119, 222
Schur, 54
second layer, 95
self-similarity function, 164, 172, 178, 197
Shannon, xv, 8
σ-algebra, 190
signal
   analytic-equivalent, 235, 236, 241
   baseband-equivalent, 12, 235
   energy, 46, 95, 96, 106, 111, 119, 121, 133–135, 137, 147, 166, 183, 236, 241, 264
   power, 95
signal design trade-offs, 132
simple mail transfer protocol (SMTP), 2
simulation, 115, 231
sinc
   Fourier transform, 185
   peculiarity, 201
singular value decomposition (SVD), 56, 62
sky wave, 234
Slepian, 143
software-defined radio, 162
source, 23
   binary symmetric, 19
   continuous-time, 20
   discrete, 18
   discrete memoryless, 19
   discrete-time, 20
spread spectrum (SS), 105, 261
square-root raised-cosine function, *see* root-raised-cosine function
standard inner product, 66
state diagram, 207
stationary, 191
stochastic process, 190
sufficient statistic, 41, 83, 85, 102, 130

symbol, 159
   PAM, 159, 160, 179, 208, 221, 279
   PSK, 160, 179, 248, 263
   QAM, 160, 179, 249, 263, 279
symbol-by-symbol on a pulse train, 159
synchronization, 112, 175
   Bayesian approach, 176
   DLL approach, 176
   for passband signals, 256
   LSE approach, 176
   ML approach, 175

third layer, 232
time offset, 256
training
   signal, 256, 258
   symbols, 177
transmission control protocol (TCP), 5
transmitter, 24
trellis, 209
triangle inequality, 67

uncorrelated, 64, 165
union Bhattacharyya bound, *see* error probability
union bound, *see* error probability
unitary, 53
up-converter, *see* converter
user datagram protocol (UDP), 5

vector space, 65
Viterbi algorithm (VA), 209, 222, 224
voltage-controlled oscillator (VCO), 177
Voronoi region, 39

water-filling, 113
waveform, 15
waveform former, 97, 102, 159
whitening filter, 113
wide-sense stationary (WSS), 163, 191